VANQUISHING CANCER

Winning the Fight of Your Life

STEVEN P. SHEPARD

Vanquishing Cancer: Winning the Fight of Your Life

Wolverine Rocket Publishing

To contact the author, visit www.stevenpshepard.com or email stevenpshepard@gmail.com.

ISBN: 978-1-7348201-1-9

About the Author

Steven Shepard was born in Lansing, MI, growing up in nearby Haslett for the majority of his childhood (with stops in Atlanta, GA and St. Louis, MO). He played multiple sports, learning from his dad, including football, tennis, basketball, baseball and soccer. At an early age he fell in love with Star Wars and later in high school was inspired by the book *Black Holes and Time Warps: Einstein's Outrageous Legacy*, by Kip Thorne. This led him to pursue a career in aerospace engineering, where he graduated magna cum laude with his Bachelor and Master of Science in engineering (aerospace engineering) degrees from the University of Michigan, before embarking on a career at Lockheed Martin Space (LMS) in Sunnyvale, CA. Steve specialized in gas dynamics and fluid dynamics during his studies, with an additional focus on electric propulsion while at Michigan; this included supporting NASA's Cassini mission to Saturn.

Steve spent the first part of his career designing satellites for multiple customers in LMS's Special Programs directorate, as well as helping design the propulsion system for NASA's Orion mission to return humans to the Moon, Mars and beyond. His expertise centered on propulsion and thermal spacecraft analysis and design, where he contributed to the successful launch of three satellites.

Presently Steve is a research manager at LMS where he leads a group of talented scientists and engineers in the Hypersonics and Advanced Materials directorate at LMS's Advanced Technology Center in Palo Alto, CA. Of particular interest are high-energy laser-material interactions, nanomaterials and advanced high-temperature materials and coatings for hypersonic applications.

Steve competes in multiple United States Tennis Association (USTA) tennis leagues and tournaments, trains Muay Thai, has picked up skiing, enjoys biking and hiking the many wonderful locations sprinkled throughout Northern California, is an avid traveler, voracious reader and is a member of multiple animal welfare and environmental advocate agencies. He lives in San Jose, CA with his girlfriend Moy and four cats (Callie, Spartacus, Maximus and Daenerys).

Pictured here with Moy in Sausalito, CA.

ACKNOWLEDGMENTS

When I undertook the challenge of writing this book, I knew it would be daunting. However, even going in with these expectations, the amount of time and effort spent far exceeded what I had imagined. Part of this is due to my meticulous nature and thirst for knowledge and another part owed to working full-time while pursuing another advanced degree, but a good deal is attributable to learning all the processes that go into writing a book.

I wrote a first draft introduction for this book around June of 2014, although I had plans to write ever since my cancer treatment was over in April of 2010 . It then marinated (my fancy way of saying I procrastinated) for about two years, where I fully committed to writing in the summer of 2016. Roughly two years later, I had my manuscript draft ready for editing. This editing process took around six months, followed by a couple of months working on a book cover (note that these durations included working primarily on weekends while juggling work and school).

Then came the challenge of citing sources for all my figures and contacting owners for rights to publish. After getting some quotes back and deliberating more on the flow of my book, I decided to hire an illustrator, who I worked with for eight months or so into September of 2019. Finally came formatting (which is quite involved, especially when producing multiple formats for softcover, hardcover, color, black and white, and electronic versions), creating an index and general cleanup tasks. All told this was around a four-year process! I suppose it's fitting that this book journey concludes on my birthday, 11 years to the day when my remission began.

Early on I traded whether to hire an agent and pursue a traditional publishing house or self-publish. After doing my research, I was surprised at how successful many self-published books have been (e.g. *Fifty Shades of Grey* and *The Martian*). Another option was to do a hybrid approach, where you pay money up-front for a publishing house that specializes in package deals to work with you on pseudo self-publishing. There were certain pros to self-publishing that I liked such as having more control over the overall writing, formatting and creative process for my book, as well as having a larger percentage of the royalties. For example, self-publishing authors receive 70% of Kindle sales versus around 10% using traditional publishing (depending on the publishing house). This difference is smaller when comparing printed books and dependent on variables such as printing costs and if you sell through an author website, Amazon or a brick and mortar store, yet self-publishing remains on top.

Now I did not write this book to make money, although monetary gains are always a plus. In fact, I decided ahead of time that I could lose money on this book, and while I'm hopeful that will not be the case, you must understand the challenges and costs associated with self-publishing ahead of time. More important to me is the accomplishment that comes from completing such a monumental task; even more important than that and the ultimate reason I wrote this book is the hope that it will have an impact on guiding others who have either been recently diagnosed with cancer or have a loved one diagnosed, arming them with knowledge to fight and beat their diagnosis.

Early on I had to decide about the scope of my book. One of the primary decisions was whether to focus primarily on lymphoma (I was diagnosed with Hodgkin's lymphoma) or broaden it to include all cancers. I choose the latter with the intent to reach as many individuals as possible. Cancer impacts so many humans that I really could not think of a more far-reaching self-help or educational topic than this.

Back to the editing and acknowledgments. Choosing to self-publish created quite a challenge, as you are responsible for all aspects of your book. This can be fun, but also very demanding! There were times I was often overwhelmed with how my book kept growing in scope and I wanted to research so many topics that it felt never-ending. I lost track of how many hours I spent all-told, but while the writing itself was a monumental effort, so too was hiring and working with all the different individuals who helped turn this book into a finished product.

I worked with three editors (after reviewing over thirty applications via Upwork), all of whom had different styles. In particular I would like to thank James Rosen for his enthusiasm for the book and edits, which helped improve my writing. He is a talented author and journalist in his own right and I am grateful for his feedback. Anne Belott ended up taking the book home, doing final edits as well as the index. She also is a lymphoma cancer survivor and has a nice editing style that I gel well with, including a good balance of suggestions and editing. Thank you, Anne, for all your hard work and helping me see this through to the finish line!

My book cover design was a fun process. I used 99Designs, which allows you to either have a collaborative process where multiple artists submit ideas and you iterate on a final choice, or you can hire a designer outright. After some back and forth, I came across a designer whose work I fell in love with, so I went with the sole-source option. Miladinka Milic (Mila, found under Meella in 99Designs, or Mila Book Covers: www.milagraphicartist.com) did an outstanding job and was very patient throughout the process. I have a bit of an artistic side so really got into the details and we ended up with a cover that I am thrilled with. Thank you, Mila!

Working on the figures was an incredible amount of work, where I ended up with over 750 pages of research that went into producing what you see in this book. I again used Upwork to solicit bids, where after reviewing dozens of applicants, one stood above the rest for his talent, organization and communication skills that shone through during the proposal process. Gamzat Ramazanov was a delight to work with and I am very proud of the images we produced, which spanned both anatomy and medical equipment disciplines. Thank you Gamzat for all the long hours and being patient with me throughout the creative process!

Thank you to June Casey and the American Cancer Society, as well as Raymond Bricault and Mike Mooney of AccuBoost for granting permissions and providing additional material for my book. Your generosity and collaborations are greatly appreciated!

Doug Williams (also found via Upwork) provided final formatting to really turn a Word document into a polished book that could be tailored for multiple formats. I chose Doug based on his excellent track record and portfolio; he was a great pleasure to work with. Thank you, Doug!

While I read multiple books on how to publish, there was one in particular that I found myself referring to often and I would like to recognize, to aid aspiring authors. *Book Launch Formula: How to Write, Publish, and Market Your First Non-Fiction Book (Around Your Full Time Schedule),* by Justin Ledford, proved invaluable. Many of the websites I used and my overall strategy for editing, designing, formatting and publishing my book had roots in this book. Thank you, Justin, for providing this resource!

I received support from numerous friends who encouraged me along the way. While there were many, Elisban Rodriguez, Gary Cunningham and Danny Sato were particularly supportive and are great friends; thank you guys!

My family was there for me throughout and I often leaned on them when my book was overwhelming. Thank you to my dad Bill and sister Allison, I love you! Marc, my brother, consistently encouraged me along the way and helped lift me up numerous times (as well as during my cancer treatment). He additionally provided valuable feedback to my book cover and other strategic decisions regarding my book. Thank you, bro; I'm lucky to have such a loving and amazing big brother!

I called my mom, Elizabeth, often, both for advice and encouragement. She never disappointed and I am grateful for all her time, energy and love that went into supporting me. We did it, Mom!

Thank you to my loving girlfriend of many years, Moy Truong. She set up an appointment early on with an author to discuss my book and in her own loving way, rode me until I finished! Thank you, Moy; time to celebrate!

I would be remiss to not thank my cats. Yes, I am a crazy cat guy thanking my cats! But they provide endless love and are an important part of my life. Maverick passed away last year of cancer; that crushed my soul, he was the most loving companion and we fought until the end. His surviving sister Callie, my buddy Spartacus and the two newest additions to the family, Maximus and Daenerys, are such cuties!

To Mom and Dad, with love.
The best parents a son could have.

Tattoo by Marc Shepard.

Contents

List of Figures

List of Tables

Preface

There are many cancer books out there. However, having gone through cancer myself, I found that literature was scattered and there was no consolidated location where I could go to answer my questions. Now I do not profess this book will have all the answers for everyone, but the intent is to present an easy-to-read reference for lay cancer patients and their support group as well as professionals who wish to relate better with patients and overall improve the patient-doctor relationship.

Cancer is a scare nobody is ready for. The fact that we are mortal beings who can be inflicted with a disease out of our control is a scary feeling. Yet you can fight back! Arming yourself with information can empower you to make smart decisions and, with the help of doctors, beat cancer. I found learning about cancer incredibly emotionally uplifting, and this was the biggest reason I was able to stay upbeat throughout the process.

I do not have a professional background in the medical field, yet with two degrees in aerospace engineering, I have an analytical mind and a thirst for knowledge. Anyone diagnosed with cancer can learn about their specific disease and recommended ways to treat it. I often steered my team of doctors in a certain direction based on the information I read, which led to a faster recovery. While I provide some technical information and data for those interested, my main focus is to produce an accessible text that appeals to a wide audience and enables patients to make smart decisions regarding their treatment.

Congratulations on exploring this book and being willing to attack cancer! You have already taken the first step toward successful treatment and a full recovery. I encourage you to read voraciously and become your own champion! You are the single most important person in your recovery. Along with resources and a strong support group of friends, family and doctors, you will overcome this challenge!

Kind Regards,
Steven P. Shepard
San Jose, CA
April 23, 2020

INTRODUCTION

THE WALLS COME CRUMBLING DOWN

"STEVE, THIS IS Dr. Jack. You need to come in here; it's important." That was how it started for me. My doctor never called me back the same day after an office visit.

Something must be wrong.

I drove over to Kaiser and went up to the fourth floor and checked in. After going through the traditional procedures, Dr. Jack came in and showed me an X-ray.

"See that there? That's your heart. See that next to it? That's not supposed to be there."

"You mean that blob that's larger than my heart?"

"Yes. That is abnormal."

What the fuck, went through my mind. "What is it?"

"Well, we need to do some tests, but I believe you have lymphoma."

"I've heard of lymphoma but honestly don't know exactly what it is."

"It is cancer of the lymph nodes. My dad had this and passed away due to it."

Wow. Not the news I was hoping for. My mind was going a bit numb by then as Dr. Jack politely explained the next steps.

"First we'll want to do a biopsy to confirm that this is malignant and indeed lymphoma. Then we'll want to find out what type of lymphoma you have so your oncologist can determine a treatment plan."

I felt like Walt from *Breaking Bad*; everything around me stopped and I heard mumbling. Thankfully I didn't choose the course Walt did. Rather, I went out to the car and called my brother. I paused for a bit while he waited patiently on the other line.

"What's wrong?"

I was holding back tears, trying to get it out. "I have cancer. The doctor thinks it's lymphoma. I have a tumor the size of a football inside my chest."

My brother was stunned, but to his credit he was reassuring and loving. We talked about what the next steps would be and what little information I knew. I followed this up with a call to my mom and had a similar conversation. However, I did not want to broadcast to the world what was going on, but rather I did want to learn more about what I had.

I searched Amazon for books on lymphoma and one popped up that had the best reviews: *Living with Lymphoma: A Patient's Guide*, by Elizabeth Adler.[1] Not wanting to wait, I called local bookstores to see if they had any copies. After striking out a few times, I found one that did and immediately drove over to pick up a copy. My brother and mom had already booked flights out and I told my dad, sister and girlfriend, Moy, by that point.

I spent the night reading and gathering as much information as I could. The more I read, the better I felt as I learned more about my cancer. At 28, I was at the mean age for Hodgkin's lymphoma. This was the second "youngest" cancer behind leukemia, which are both cancers of the blood. I went over the symptoms I had experienced, including incessant coughing and night sweats, which were both discussed. It made me feel somewhat better that others had gone through similar experiences and there was a good chance for a recovery. I never at any point thought about dying. I was way too young and just starting out in my life for that. In later chapters I will discuss the psychology behind treatment and how a positive outlook can improve your recovery.

The next day I told my boss that I had been diagnosed and would be out indefinitely. I also requested my diagnosis be kept confidential because I wanted time to prepare for my treatment before hearing from a lot of people providing support. That was the best approach for me. I was so focused on attacking my cancer and getting the right treatment that I needed as few distractions as possible. Later, say a month post-diagnosis and steadily into chemotherapy, I welcomed and greatly appreciated the support once I decided to tell more people what was going on.

1 Elizabeth Adler, *Living with Lymphoma* (Baltimore: John Hopkins University Press, 2015).

For me it was all about timing. This is my recommendation for those experiencing a surprise diagnosis: discuss with close family while you learn about your cancer and how to treat it, opening up to friends and colleagues for emotional support once you have set a path and initiated your treatment.

So began my journey to beat cancer. In the following chapters I am going to break down what to expect throughout, including decisions that you will need to make, treatment impacts and how to recover. This is a very trying time in your life, but you will get through it and come out a stronger and more enlightened individual!

Chapter 1

Symptoms

I found that diagnosing cancer is not as easy as one might think. Although I had symptoms early, I did not for once think they were attributed to cancer. While some of the symptoms I experienced are specific to Hodgkin's lymphoma, many are prevalent in other types of cancers as well.

Fatigue

Working out at the gym, I first noticed my jugular vein was sticking out abnormally when doing squats. I also felt a bit light-headed during this motion, so I knew something was not quite right with my body. This is very important for life or for any type of medical condition—as my dad always tells me: nobody knows your body better than you. That would lead to my diagnosis, albeit not until a couple of months after I first started experiencing that symptom. I had no idea at the time that a giant tumor was pushing on my heart and making it more difficult to pump blood to my head, which was causing the jugular to stick out.

I ended up going to see my doctor to have my neck checked out. Being an active athlete, between tennis and weightlifting, I was not someone you would expect to have cancer. He examined my breathing and heart rate along with lymph nodes in my neck and the usual checks, concluding I was healthy and had nothing to worry about. This is the first lesson I learned—doctors are human and do their best to help you; however, it is up to you to follow through when it comes to your health.

Night Sweats

Living with Lymphoma does a nice job of discussing night sweats. A few months leading up to my diagnosis I would wake up drenched in sweat, my bed completely soaked. While this had never happened before and I thought it was odd, I assumed I was too warm at night and did not think much of it—certainly not enough to get checked out. However, this can be a clear sign of many cancers such as lymphoma, leukemia, carcinoid tumors (present in cancers including gastrointestinal (GI) tract cancers: notably stomach, intestinal, lung, pancreatic, testicular and ovarian cancers) and adrenal tumors (adrenocortical cancer, malignant adrenal pheochromocytomas and malignant paragangliomas). Now, my night sweats did come after the tumor had grown quite a bit, starting three to four months prior to my diagnosis and one to two months before my jugular incident. Had I been diagnosed at this time with a smaller tumor, it is possible I could have received less radiation treatment while still on the same chemotherapy plan.

Night sweats are classified as Ann Arbor B symptoms, which are systemic symptoms that are impacted throughout the body and include night sweats, fevers and weight loss. Ann Arbor is a specific classification system created for lymphoma (both Hodgkin's and non-Hodgkin's) and will be discussed in more detail later on in the book.

Cytokines are proteins involved in signaling between cells and are believed to be the cause of night sweat symptoms. Lymphocytes are one of five white blood cell (WBC) types[2] and are found in both your blood and lymph, unlike the other four that are found only in your blood. Lymph is a clear fluid that flows throughout the body's lymphatic system. Along with other cells in the body, lymphocytes can produce cytokines. Cancerous lymphocytes can produce increased levels of cytokines or result in other immune cells increasing their levels to fight the cancerous cells, both of which will create night sweats. Repeated, drenching night sweats can be a clear sign of cancer and must be taken seriously.

Weight Loss

Lymphoma, and cancer in general, can lead to weight loss due to a loss of appetite. I did not notice a large weight loss before my diagnosis but

2 The five white blood cell types are neutrophils (defend against bacterial and fungal infections), eosinophils (fight parasitic infections), basophils (deal with allergic and antigen reactions), lymphocytes (regulate the body's immune system) and monocytes (guard against repeated infections).

was also very physically active and ate large portions of protein, which perhaps offset this impact. Appetite loss can be attributed to various factors, including increased levels of cytokines from a tumor or immune system cells or feeling full if your tumor is near your gut.

You are considered to have B-symptoms attributable to lymphoma if you lose more than 10% of your body weight. The timeframe for this loss is somewhat vague. *Living with Lymphoma* says over a six-month period while other sources have no timeframe at all. Since individuals differ in their metabolism, diet, physical activity and eating patterns, the 10% is merely a guideline. For instance, I often fluctuate up to five pounds over a couple of days depending on diet and exercise since I am very active.

The important point is to understand your normal eating and exercise patterns—along with metabolism and your normal weight fluctuation—to determine if sudden weight loss may be due to a cancer such as lymphoma. Further sections will discuss weight loss during chemotherapy, which has completely different causes and resulted in me losing a large percentage of my body weight.

Fever

Lymphoma and other cancers can cause a fever. Most commonly it will be a slight fever hovering around 100°F that persists for multiple days, weeks or months (if left undiagnosed). I do not recall experiencing a fever, so this is not a requisite for lymphoma or other types of cancers and since fevers can be caused by many factors, other symptoms such as night sweats and fatigue are more telling. There is also a more extreme fever condition among Hodgkin's lymphoma patients called Pel-Ebstein fever, where fevers cyclically increase—up to 106°F—and decrease, typically over a period of seven to ten days. This is a rare and specific condition; if you experience this type of fever there is a strong likelihood that you have Hodgkin's lymphoma.

Lymph Node Pain

One of the most common signs of cancer is pain in the lymph nodes, which are located throughout your body. Doctors typically check for enlarged lymph nodes in your neck during a routine physical, although armpits and the groin region are likewise locations where copious amounts of cancer cells may congregate. Lymphoma starts in a lymph node, but other cancers also migrate through the lymphatic system after their genesis in or near a body organ. Any soreness or swelling in lymphatic regions may be attributed to cancer, yet it should be noted that many sicknesses cause this as well and is why doctors check lymph nodes during routine exams. Common colds, the flu and other ailments all can cause enlarged lymph nodes. These will be in the neck region, however, making soreness in the groin or armpits more telling for potential lymphoma.

I do not recall feeling any lymph node pain prior to my diagnosis and my doctor did check lymph nodes in my neck when I visited, so I presume they were not enlarged. While my cancer eventually spread to my neck and groin lymph nodes, at the time of my initial doctor visit, external swelling would not suggest the massive tumor inside my chest. Furthermore, I have had pain before in my armpits prior to lymphoma, though this was rare and lasted only for a day. Therefore, like some of the other symptoms mentioned here, lymph node soreness is meant to be a potential sign for cancer but needs to be properly diagnosed by your physician.

Body Itching

While rare, chemicals secreted by lymphoma cells can cause itching all over your body, clinically called *pruritus*. Again, there are a multitude of conditions that could cause body itching; some are obvious topical ones like poison ivy or poison oak and some are less obvious ones such as the liver or kidney not processing waste properly.

Specifically for Hodgkin's lymphoma, itching can occur in up to 25% of individuals. This itching[3] generally manifests itself in your lower legs and is referred to as *Hodgkin itch*. Itching can be a sign of a more advanced stage of lymphoma where an abnormal amount of histamines is released to combat the cancer. Additional cancers such as the blood cancer polycythemia vera and pancreatic cancer may also cause itching.

Coughing

Coughing was the major symptom I experienced that told me something was wrong. What started with my jugular sticking out of my neck during

3 Itching is medically referred to as pruritus and itching caused by cancer is deemed paraneoplastic pruritus.

physical activity led to incessant coughing. There was no sore throat or stuffed up nose that may typically accompany a cough, so this was odd. I recall going to my doctor for a second follow-up visit and spending a long time in the waiting room, impatiently debating whether I should leave and go back to work. Thank God I did not!

This symptom may have saved my life since that second visit led to the chest X-ray that revealed the large tumor wrapped around my heart and collapsing one of my lungs. How I managed to work out with this I do not know, but my radiation oncologist told me that had this gone undiagnosed for another month or two, the tumor would have likely punctured my heart and killed me (I pointedly asked if he had ever seen this happen and he said he had). That bit of news floored me, so this symptom above all I regard as the most important to get checked out if you are experiencing excessive coughing.

The coughing symptom is unique from others in that it is constant. I coughed several times a minute and it was unclear at the time what brought it on—unless of course I knew there was a tumor in my chest. That symptom was specific to the location of my tumor since lymphoma tumors can form all throughout the body, yet mine happened to form where my heart and lung were—not the best place for a tumor! The coughing was also uncontrollable and distinct from the kind of cough to clear your throat due to phlegm or bronchial congestion (I have taken QVAR, an asthma medication for chest congestion, so I understand the difference in symptoms). If you are experiencing similar symptoms, please get it checked out.

Lump

Lumps can be an easily detectable potential sign of cancer and can be checked for rather easily. Most commonly, lumps show up in the breasts, testicles or lymph nodes. You should immediately see a doctor if you discover lumps in these locations because it is generally a sign of a tumor—either benign (non-cancerous) or malignant (cancerous)—underneath the skin.

Skin Growth and Discoloration

Any suspicious growth on your skin or change in skin appearance, such as an enlarged or discolored mole, a new wart or skin tag that appears abnormal, or a freckle that changes size and appearance can be a potential sign of skin cancer and should be examined by a dermatologist immediately. While there are plenty of non-cancerous reasons for these growths, if it is skin cancer, receiving an early diagnosis is key to your chances of eradicating the cancer before it spreads. Breast cancer can also show skin redness and thickening of the skin as signs of cancer.

Trouble Swallowing or Indigestion

Although there are many reasons for issues with swallowing and indigestion that are non-cancerous, these symptoms can be indicative of cancers of the esophagus (tube between your throat and stomach that food and liquid travels through), stomach or pharynx (i.e. your throat, which connects the mouth to your esophagus). If you notice swallowing changes or stomach indigestion that is otherwise unexplained, ensure your doctor performs additional checks for signs of cancer.

Abnormal Bladder or Bowel Functionality

More or less frequent urination, blood in your urine, or pain while urinating can all be signs of bladder or prostate cancer and should be examined by your doctor immediately. Consistent changes to your defecation patterns such as constipation, diarrhea or changes in the size of your stool are possible indications of colon cancer and should also be examined immediately.

Wounds or Sores That Have Difficulty Healing

Sores and wounds that persist without healing can be signs of cancer. These can occur anywhere on the skin, which could point to skin cancer. Sores in the mouth can also persist, potentially representing oral cancer. Sores on the penis or vagina can also be signs of cancer, although there are many other causes of such ailments. Seek your doctor if you have any of these symptoms.

White Patches on the Tongue or Inside Your Mouth

White patches on the inside of your mouth—generally on the gums, the bottom of your mouth, underneath your tongue, on your inner cheeks or on your tongue—are likely leukoplakia, which is irritation linked to tobacco use (smoked, chewed or dipped). While many patches are benign, some are early signs of cancer. See your doctor for a diagnosis if you have this symptom. Additionally, hairy leukoplakia is a specific form of leukoplakia that primarily impacts individuals with weakened immune systems caused by human immunodeficiency virus (HIV)/acquired immunodeficiency syndrome (AIDS). This form of leukoplakia does not lead to mouth cancer but is an important sign of a weakened immune system.

CHAPTER 2

STAGING

WHEN FIRST DIAGNOSED with cancer, your oncologist will discuss the staging, i.e. how advanced your cancer is. Multiple tests will be performed to help determine your cancer stage. Understanding the staging and severity of your cancer is critical to choosing a treatment plan and fighting your disease.

Most cancers are staged using the American Joint Committee on Cancer's tumor-node-metastasis (TNM) system. Cancers not included in this system that have their own unique staging systems include childhood cancers, blood cancers, brain cancers and spinal cord cancers. Note there are three main types of blood cancers: leukemia, lymphoma and myeloma.

TNM STAGING

TNM staging is broken down into three main categories and then assigned an overall number.

TUMOR (T)

What is the size of the tumor, its growth and how far has it spread into nearby tissue? Numbers are assigned from 0 to 4 (0 being no cancer present) with a higher number indicating a more developed cancer. Specific suffixes are as follows:

- ***TX:*** Tumor cannot be measured.
- ***Tis:*** Carcinoma in-situ (CIS) cells are observed, which are a group of abnormal cells on superficial rather than deep tissues. These are sometimes referred to as pre-cancer cells, as they may lead to cancer.
- ***T0:*** No tumor is present.

- ***T1–T4:*** Descriptions for these numbers are specific to the type of cancer since one size for, say, bladder cancer may be a different stage than for liver cancer. The higher the number, the larger the tumor and the more it has spread outside of the primary organ and into nearby tissue. As an example, see the diagram for bladder cancer staging displayed in Figure 1. Note the tumor growth and penetration through the bladder lining, tissue, muscle and fat as the number increases.

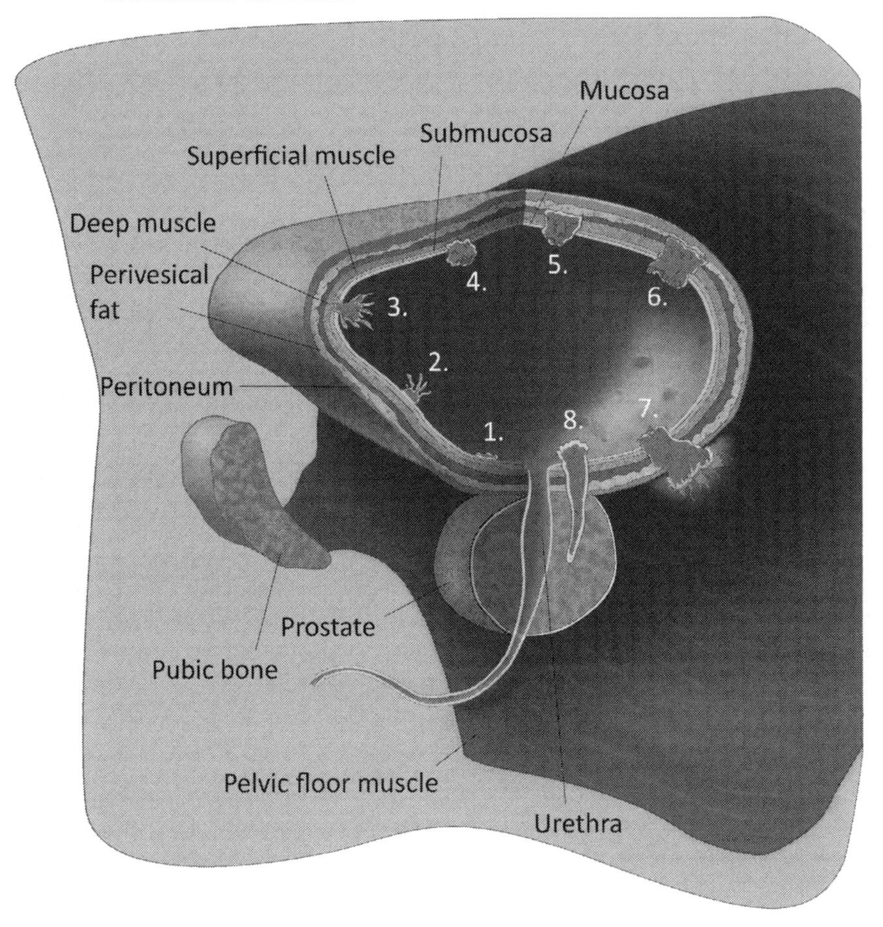

TNM Classification: 1. Tis; 2. Ta; 3. T1; 4. T2; 5. T3a; 6. T3b; 7. T4b; 8. T4a

Figure 1: Bladder cancer staging.

Specific T-staging for bladder cancer is as follows:

- ***TX and T0:*** Same as above.

- ***Tis:*** Non-invasive flat CIS, which means abnormal cells that can grow into a tumor across the surface of the bladder are present.
- ***Ta:*** Non-invasive papillary carcinoma, which means abnormal cells that can grow into a papillary tumor projecting into the bladder lumen (cavity where urine collects) are present.
- ***T1:*** Tumor has penetrated the mucosa and possibly submucosa layers of tissue lining the inner bladder cavity.
- ***T2a:*** Tumor has grown into the inner layer of muscle tissue.
- ***T2b:*** Tumor has grown into the outer layer of muscle tissue.
- ***T3a:*** Tumor has grown into the outer fatty tissue of the bladder, but the penetration can only be seen under a microscope.
- ***T3b:*** Tumor has grown farther into the outer fatty tissue of the bladder such that the penetration can be seen via an image (or your surgeon is able to feel the tumor).
- ***T4a:*** Tumor has punctured the outer fat layer and is spreading into nearby organs such as the prostate or seminal vesicles in men (the latter secrete protein, mucus and fructose alkaline liquid semen contributors that feed and help sperm survive in the acidic environment of the vagina) and the vagina or uterus in women.
- ***T4b:*** Tumor has punctured the outer fat layer and has spread to the pelvic or abdominal wall.[4]

Lymph Nodes (N)

Has the cancer spread to your lymph nodes and if so, how many? A range of 0 to 3 is used (0 being cancer does not exist in any lymph nodes) with a higher number corresponding to a higher amount of spreading. A specific suffix breakdown includes:

- ***NX:*** Spread to lymph nodes cannot be ascertained.
- ***N0:*** No cancer exists in lymph nodes.
- ***N1–N3:*** Describes the degree of cancer spreading to lymph nodes. In addition to location, the quantity and sizes are taken into account. Specific breakdowns are particular to the type of cancer being evaluated.

4 "How Cancers Grow," *Cancer Research UK,* https://www.cancerresearchuk.org/about-cancer/what-is-cancer/how-cancers-grow.

Following the bladder cancer example, here are specific breakdowns for N-staging:

- ***NX:*** Spread to lymph nodes cannot be ascertained.
- ***N0:*** No cancer exists in lymph nodes.
- ***N1:*** Cancer has spread to a single lymph node in the true pelvis, the cavity formed by a region of the pelvic bone, shown in Figure 2.
- ***N2:*** Cancer has spread to two or more lymph nodes in the true pelvis.
- ***N3:*** Cancer has spread to lymph nodes lining the common iliac artery and shown in Figure 3.

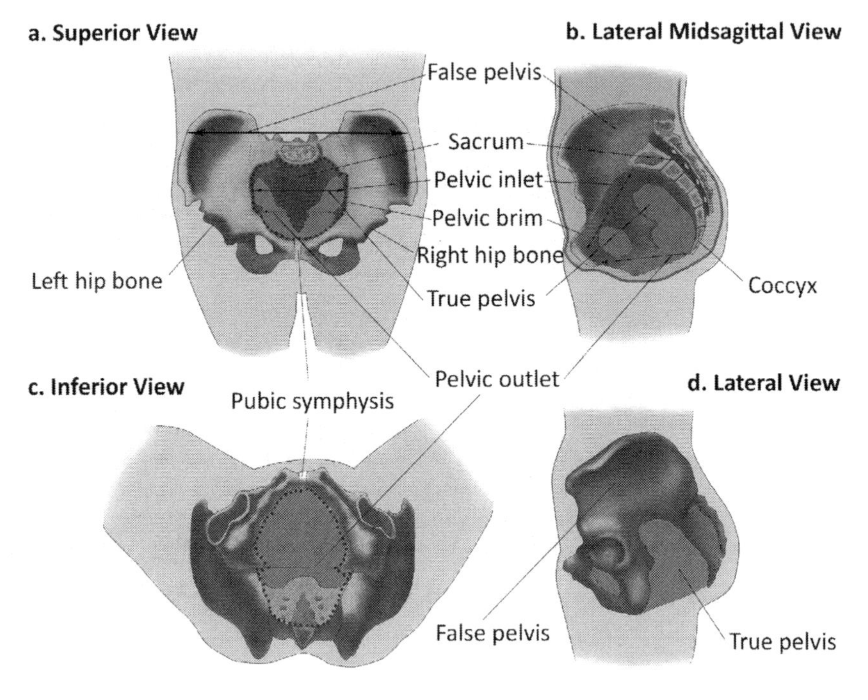

Figure 2: Diagram showing region of true pelvis (male).

As another example, liver cancer has a simplified N-staging as follows:

- ***NX:*** Spread to lymph nodes cannot be ascertained.
- ***N0:*** No cancer exists in lymph nodes.
- ***N1:*** Cancer has spread to lymph nodes.

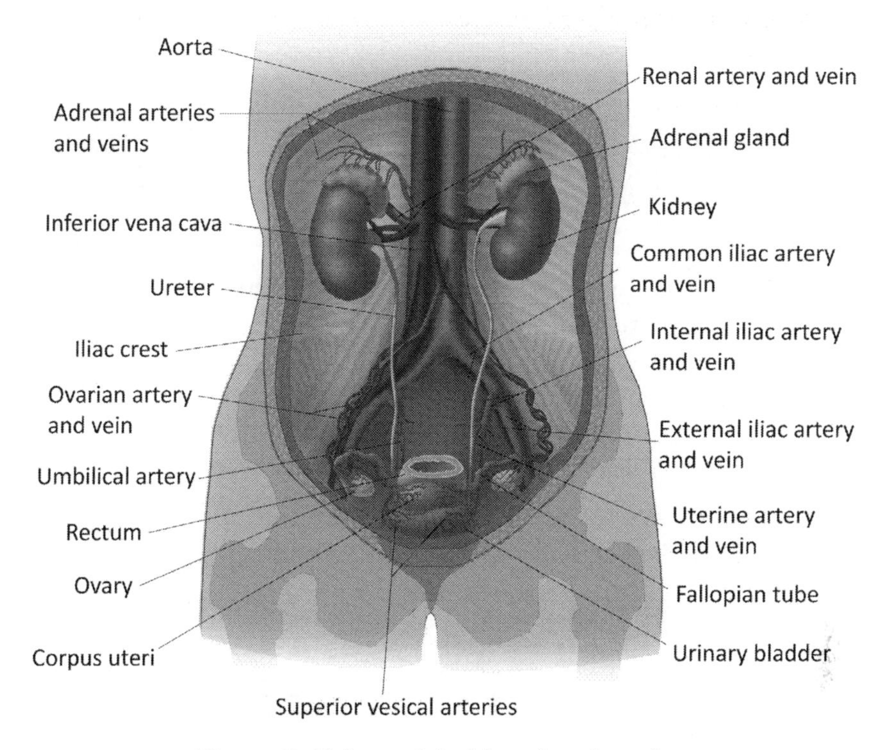

Figure 3: Urinary bladder showing the location of the iliac artery (female).

State of Metastasis (M)

Has the cancer spread to distant parts of the body (i.e. metastasized) and if so, by how much?

- ***M0:*** The cancer has not spread to other organs, bones or distant lymph nodes or tissues in the body and remains localized to the initial cancerous organ or tissue.
- ***M1:*** The cancer has spread to distant parts of the body and is invading other organs, bones or tissues and lymph nodes far away from the initial cancerous region.

Overall Cancer Stage

Once the three cancer staging categories above have been determined, a composite score from I to IV is used, with I being the least advanced and IV the most advanced. Since these scores are influenced by the three categories above, which are unique to each cancer, the composite score is also unique. Additional subcategories such as A or B may be assigned to the stage and the description is likewise unique to each type of cancer.

An example of the scoring for bladder cancer is as follows:

- ***Stage I (T1, N0, M0):*** The cancer has penetrated the inner two layers of the bladder wall lining, the mucosa and submucosa layers, but not the muscle layers (superficial and deep muscles). There are no signs of the cancer spreading to nearby lymph nodes or organs throughout the body.
- ***Stage II (T2a or T2b, N0, M0):*** The tumor has penetrated the inner or inner and outer layers of muscle (superficial and deep muscles); however, there remains no spreading to lymph nodes or distant organs in the body.
- ***Stage III (T3a, T3b or T4A, N0, M0):*** The tumor has penetrated the fatty layer in the bladder wall (perivesical fat layer) or has punctured the fat layer and is spreading into a nearby organ such as the prostate or uterus. There remains no spreading of the cancer to lymph nodes or distant organs in the body.
- ***Stage IV:*** One of the following scenarios is present:
    - ***T4b, N0, M0:*** The tumor has punctured the outer perivesical fat layer and is spreading into the abdominal or pelvic wall. No cancer is present in lymph nodes or distant organs.
    - ***Tis, Ta or T1–T4, N1–N3, M0:*** Any stage of tumor is present and the cancer has spread to lymph nodes but has not metastasized into distant organs.
    - ***Tis, Ta or T1–T4, N0–N3, M1:*** Any stage of tumor is present and the cancer may or may not have spread to lymph nodes but has metastasized and spread into distant organs.

For the liver we have:

- ***Stage I (T1, N0, M0):*** A single tumor is present and the cancer has not spread to lymph nodes or distant organs.

- ***Stage II (T2, N0, M0):*** Either a single tumor has grown new blood vessels, or more than one tumor exists with neither tumor being more than 5 cm across. The cancer has not spread to lymph nodes or distant organs.
- ***Stage IIIA (T3a, N0, M0):*** Multiple tumors are present with at least one exceeding 5 cm across. Cancer has not spread to lymph nodes or distant organs.
- ***Stage IIIB (T3b, N0, MO):*** At least one of the present tumors has grown into the hepatic portal vein or hepatic veins (large veins carrying blood to the liver, see Figure 4). No spreading has occurred to lymph nodes or distant organs.

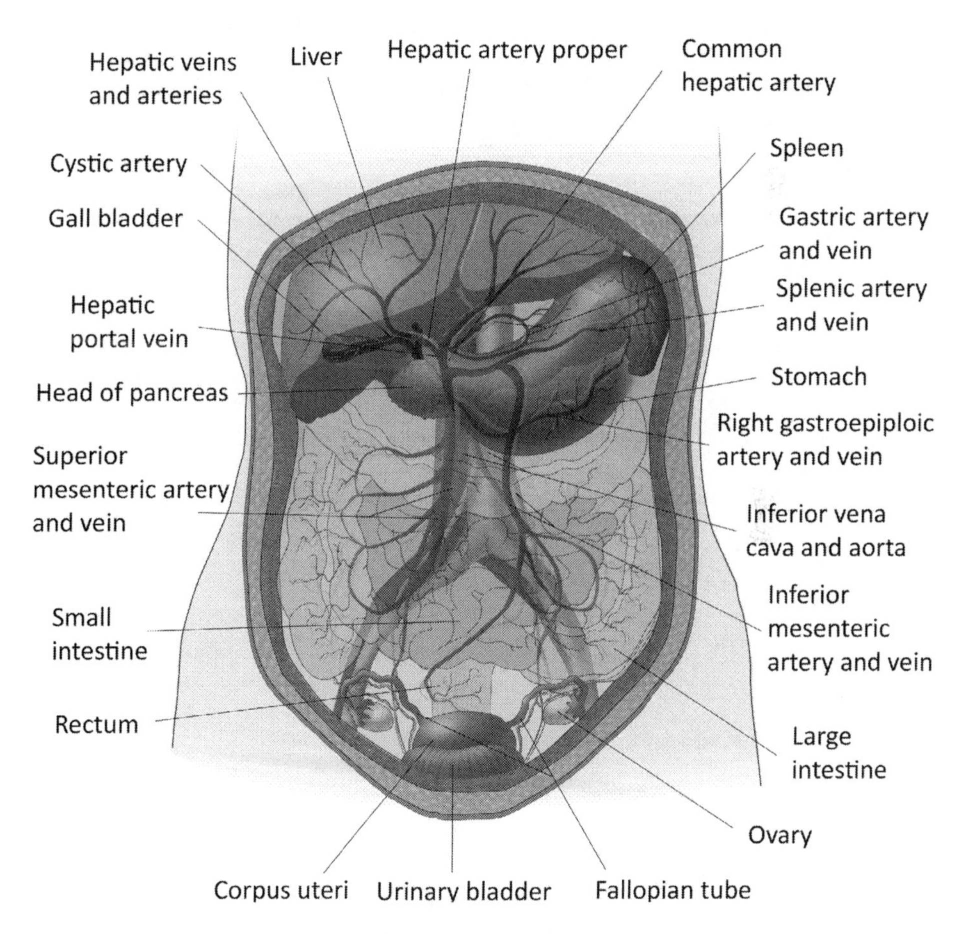

Figure 4: Schematic showing location of hepatic blood vessels (female).

- ***Stage IIIC (T4, N0, M0):*** The tumor has grown into a nearby organ (excluding the gallbladder) or into the visceral peritoneum (thin layer of tissue covering the liver). Cancer has not spread to lymph nodes or distant organs.
- ***Stage IVA (Tis, Ta or T1–T4, N1, M0):*** Any stage of tumor is present. Cancer has spread to lymph nodes but has not metastasized into distant organs.
- ***Stage IVB (Tis, Ta or T1–T4, N0 or N1, M1):*** Any stage of tumor is present. Cancer may or may not have spread to lymph nodes but has metastasized and spread into distant organs.[5]

Brain, spinal cord and blood cancers utilize different staging systems than the TNM system, which will be discussed in the following sections.

Lugano Classification (Ann Arbor Staging)

The Lugano classification system is derived from the Ann Arbor classification system and covers Hodgkin's and non-Hodgkin's lymphoma. Starting with the Rye classification system from 1965, the Ann Arbor classification system was developed in 1971 and is named after the city of Ann Arbor (location of my alma matter, the University of Michigan!), where the Committee on Hodgkin's Disease Staging Classification met to discuss lymphoma staging with experts from the USA, UK, France and Germany.[6] Ann Arbor staging remains the basis for the Lugano classification system and is often the more referenced classification system. However, updates such as accounting for the bulkiness of the tumor have been made (via the Cotswolds modification), making the Lugano classification the most up-to-date system.

Similar to the TNM system, the Lugano classification system is broken down into four stages as shown in Figure 5:

- ***Stage I:*** The cancer is located in one localized region of the body involving one lymph node and the surrounding tissue.

5 "Liver Disorders," *Gastroenterology Consultants CFL, PA,* https://www.orlandogidoc.com/liver-disorders.php.

6 Additional updates were made to this system in 1988 (Cotswolds modification, held in the English Cotswolds), 1999 (NCI criteria), 2007 (IWG revised guidelines), 2011 (Workshop at 11-ICML), 2013 (2nd Workshop at 12-ICML) and finally 2014 (Lugano classification system).

- ***Stage IE:*** Stage I with an additional extranodal lesion without nodal involvement, meaning a lesion has formed on a nearby organ but is not from the lymph node tumor protruding into the organ.
- ***Stage II:*** The cancer has spread to a second lymph node localized region with both regions on one side of the diaphragm, or there remains one region but the cancerous lymph node is impacting a nearby organ. All cancerous regions of the body are either above or below the diaphragm.
- ***Stage IIE:*** Stage II but with limited, contiguous extranodal involvement (i.e. one of the regional tumors has grown and is impacting a nearby organ).
- ***Stage III:*** The cancer has spread to localized lymph node regions both above and below the diaphragm or is in a lymph node region above the diaphragm and in the spleen below the diaphragm.
- ***Stage IV:*** More severe spreading of the cancer has occurred, including disseminated involvement of at least one major extranodal organ or bone marrow (i.e. the cancer has spread throughout an extranodal organ—commonly the lungs, liver or spleen—or penetrated into the bone marrow).

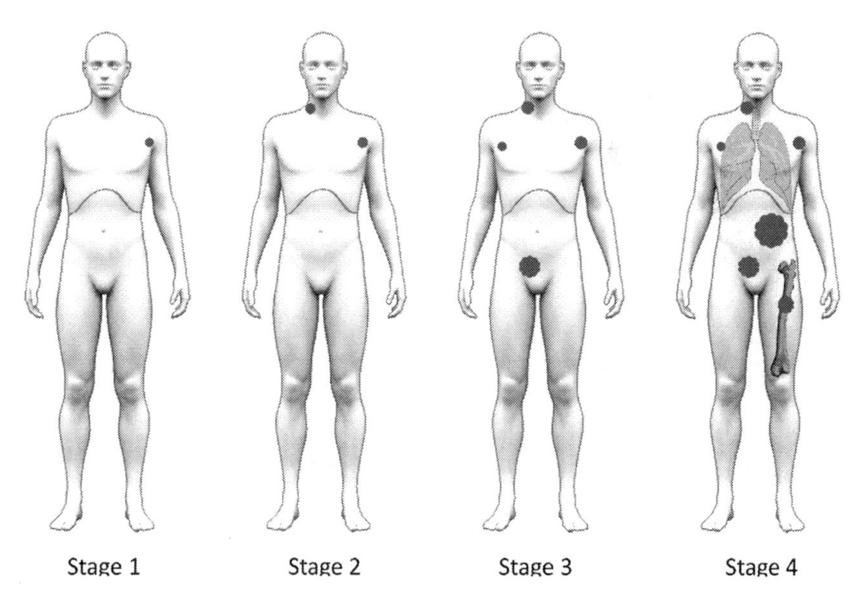

Figure 5: Lugano classification system cancerous lymph node locations.

ADDITIONAL MODIFIERS

Additional modifiers may also be added to the Lugano classification system.

A or B: An A or B is used to designate whether or not B symptoms are present, which were discussed in depth at the beginning of this book. Specific B symptoms include:

- Drenching night sweats.
- Fever (100.4°F is listed as a minimum on the American Cancer Society webpage, although *Living with Lymphoma* states low grade fevers of less than 100°F may persist for weeks or months).
- Unexpected loss of body weight (generally 10% or more).

If you experience any of the above symptoms, a B will be added to your staging (e.g. stage IIIB) and if not, the letter A is added to denote no B symptoms. B symptoms indicate the cancer is more advanced and will require aggressive treatment.

E or S: These modifiers are used to specify how much the cancer has spread to organs. The E modifier was included above in stages I and II since it is quite a common modifier. When the lymphoma has impacted an extranodal organ (i.e. an organ outside of the lymphatic system), tissue or bone marrow, the letter E is assigned. When the cancer has specifically spread to the spleen, a letter S is appended.

X: When the tumor is larger than 10 cm, it is considered bulky. Bulky tumors require additional considerations due to their size, such as the chemotherapy type and additional radiation treatment. Recently the X modifier was dropped and the tumor's longest dimension is instead recorded.

My prognosis was Hodgkin's lymphoma stage IVB since I had cancerous lymph nodes on both sides of my diaphragm in my neck and groin as well as a chest tumor larger than 10 cm in length along with B symptoms. Additionally, the tumor collapsed my left lung and displaced my heart down in my chest cavity as it grew.

CENTRAL NERVOUS SYSTEM STAGING

Central nervous system (CNS) cancer staging differs from other cancer staging in that it places increased significance on the cancer grade (the biological behavior of the cells under a microscope) rather than just the size and location of cancer cells, the latter of which is primarily used in the majority of cancer staging. While there are various staging systems out there, the most prevalent relies on a grading system developed by the World Health Organization (WHO) and published in a 2016 report.[7] This breaks cancer down into various grades.

Grade 1: These tumors grow slowly and are less likely to spread to nearby tissues. When viewed under a microscope, cancer cells behave rather normally compared to other grades of CNS cancer cells. Surgery may be able to remove the cancer completely without further treatment. These cancers are benign, although they can become malignant over long periods of time if left untreated (e.g. on rare occasions craniopharyngioma has become malignant over multiple years). CNS cancers falling under this grade are provided in Table 1.

Grade 2: Cancers in this grade are slow growing but appear slightly more abnormal than those in Grade 1. Also, these cancers are more likely than Grade 1 cancers to spread into nearby tissue. Grade 2 cancers are considered benign, although they can develop into malignant cancer cells over time. Table 1 provides cancer types that fall under this grade.

Grade 3: These tumors have a history of malignancy and therefore are considered cancerous. Cancer cells divide rapidly and spread to nearby tissue. Recurrence is common, often coming back as a grade IV cancer. Tumor cells are highly abnormal when examined under a microscope. Examples of Grade 3 CNS cancers are provided in Table 1.

Grade 4: These cancer cells are the most abnormal of the CNS grades and divide rapidly and aggressively throughout the brain. Tumor cells form new blood vessels to increase the rate at which they divide and contain dead cells at their center. Forms of this cancer are provided in Table 1, with glioblastoma being the most common form.

Many cancer types also metastasize to the brain and are sometimes not diagnosed until discovering the cancerous brain cells. Lung cancer is the most common type of cancer to metastasize to the brain, followed by breast, colon, kidney, thyroid and uterine cancers, as well as melanoma.

---

7 David N. Louis, Arie Perry, Guido Reifenberger et al., "The 2016 World Health Organization Classification of Tumors of the Central Nervous System: A Summary," *Acta Neuropathol* 131, no. 6 (June 2016): 803-820, https://www.ncbi.nlm.nih.gov/pubmed/27157931.

One unfortunate result of the increased efficacy of cancer treatments is a prolonged chance for cancers to metastasize to the brain. When cancer has originated in other parts of the body, the TNM system will generally be used for staging, except for lymphoma (where the Lugano classification system is used) or leukemia (which uses various staging based on the cancer subtype).

Table 1: WHO staging for CNS tumors.[8]

Diffuse Astrocytic and Oligodendroglial Tumors		Rosette-Forming Glioneuronal Tumor	I
Diffuse Astrocytoma, Idh-Mutant	II	Central neurocyto-ma	II
Anaplastic Astrocytoma, Idh-Mutant	III	Extraventricular neurocytoma	II
Glioblastoma, Idg-Wild-type	IV	Cerebellar liponeu-rocytoma	II
Glioblastoma, Idh-Mutant	IV	Tumors of the pineal region	
Diffuse Midline Glioma, H2k17m-Mutant	IV	Pineocytoma	I
Oligodendroglio-ma, Idh-Mutant And 1p/19q-Codeleted	II	Pineal parenchymal tumor of intermedi-ate differentiation	II or III
Anaplastic Oligodendro-glioma, Idh-Mutant and 1p/19q-Codeleted	III	Pineoblastoma	IV
Other Astrocytic Tumors		Papillary tumor of the pineal region	II or III
Pilocytic Astrocytoma	I	Embryonal tumors	
Subependymal Giant Cell Astrocytoma	I	Medulloblastoma (all subtypes)	IV
Pleomorphic Xanthoastro-cytoma	II	Embryonal tumor with multilayered rosettes, C19MC-al-tered	IV
Anaplastic Pleomorphic Xanthoastrocytoma	III	Medulloepithelioma	IV
Ependymal Tumors		CNS embryonal tumor, NOS	IV
Subependymoma	I	Atypical teratoid/ rhabdoid tumor	IV

8 Louis, "The 2016 World Health Organization Classification of Tumors of the Central Nervous System," 803-820.

Diffuse Astrocytic and Oligodendroglial Tumors		Rosette-Forming Glioneuronal Tumor	I
Myxopapillary Ependymoma	I	CNS embryonal tumor with rhabdoid features	IV
Ependymoma	II	Tumors of the cranial and paraspinal nerves	
Ependymoma, Rela Fusion-Positive	II or III	Schwannoma	I
Anaplastic Ependymoma	III	Neurofibroma	I
Other Gliomas		Perineurioma	I
Angiocentric Glioma	I	Malignant peripheral nerve sheath tumor (MPNST)	II, III or IV
Chordoid Glioma Of Third Ventricle	II	Meningiomas	
Choroid Plexus Tumors		Meningioma	I
Choroid Plexus Papilloma	I	Atypical meningioma	II
Atypical Choroid Plexus Papilloma	II	Anaplastic (malignant) meningioma	III
Choroid Plexus Carcinoma	III	Mesenchymal, non-meningothelial tumors	
Neuronal And Mixed Neuronal-Glial Tumors		Solitary fibrous tumor / haemangiopericytoma	I, II or III
Dysembryoplastic Neuroepithelial Tumor	I	Hemangioblastoma	I
Gangliocytoma	I	Tumors of the sellar region	
Ganglioglioma	I	Craniopharyngioma	I
Anaplastic Ganglioglioma	III	Granular cell tumor	I
Dysplastic Gangliocytoma Of Cerebellum (Lhermitte-Duclos)	I	Pituicytoma	I
Desmoplastic Infantile Astrocytoma and Ganglioglioma	I	Spindle cell oncocytoma	I
Papillary Glioneuronal Tumor	I		

LEUKEMIA STAGING

Leukemia has its own unique staging since this type of cancer does not generally form tumors but rather is prevalent throughout the bone marrow. Therefore, rather than staging based on tumor size and how much the cancer has spread, leukemia is broken down into subtypes with tailored treatments and varying prognoses.

ACUTE LYMPHOCYTIC LEUKEMIA (ALL)

ALL staging can be subdivided into two categories based on how many of B-cell and T-cell lymphocytes are present and how mature the leukemia cells are. Basing the staging on these two factors is deemed classification by the immunophenotype of leukemia.

B-CELL ALL

- Early pre-B ALL (also called pro-B ALL) ~10% of cases.
- Common ALL ~50% of cases.
- Pre-B ALL ~10% of cases.
- Mature B-cell ALL (Burkitt leukemia) ~4% of cases.

T-CELL ALL

- Pre-T ALL ~5 to 10% of cases.
- Mature T-cell ALL ~15 to 20% of cases.

Each type and subtype has a different prognosis, with T-cell ALL having overall better treatment success rates than B-cell ALL.

ACUTE MYELOGENOUS LEUKEMIA (AML)

Succeeding a system called the French-American-British (FAB) classification system that looked at what type of cells the leukemia cells came from and their maturity, the WHO created a new classification system. Leveraging the FAB, the WHO classification system for AML also considers the patient's prognosis. The categories are divided and subdivided as follows:

AML with Certain Genetic Abnormalities

- AML (megakaryoblastic) with a translocation between chromosomes 1 and 22.

- AML with a translocation or inversion in chromosome 3.
- AML with a translocation between chromosomes 6 and 9.
- AML with a translocation between chromosomes 8 and 21.
- AML with a translocation between chromosomes 9 and 11.
- Acute promyelocytic leukemia (APL) (M3) with a translocation between chromosomes 15 and 17.
- AML with a translocation or inversion in chromosome 16.

AML with Myelodysplasia-Related Changes

AML Related to Previous Chemotherapy or Radiation

AML Not Otherwise Specified

This includes cases of AML that do not fall into one of the above groups and is like the FAB classification.

- AML with minimal differentiation (M0).
- AML without maturation (M1).
- AML with maturation (M2).
- Acute myelomonocytic leukemia (M4).
- Acute monocytic leukemia (M5).
- Acute erythroid leukemia (M6).
- Acute megakaryoblastic leukemia (M7).
- Acute basophilic leukemia.
- Acute panmyelosis with fibrosis.

Myeloid Sarcoma (Granulocytic Sarcoma or Chloroma)

Myeloid Proliferations Related to Down Syndrome

Undifferentiated and Biphenotypic Acute Leukemias

This refers to leukemias that have both lymphocytic and myeloid features and is sometimes called ALL with myeloid markers, AML with lymphoid markers or mixed phenotype acute leukemias.

CHRONIC LYMPHOCYTIC LEUKEMIA (CLL)

There are two common forms of staging used for CLL: the Rai system and the Binet system. The former is more prevalent in the United States while the latter is used primarily in Europe.

RAI STAGING SYSTEM

Developed in 1968, this system relies on a prognosis of lymphocytosis, i.e. when a high number of lymphocytes exist in the blood and bone marrow that are not due to a known cause such as an infection. At the time, the cutoff for lymphocytosis was more than 15,000 lymphocytes/mm^3 of blood and 40% or greater of the bone marrow consisting of lymphocytes. This has since been revised so that for a patient to be diagnosed with CLL, he or she must have 5,000 monoclonal lymphocytes/mm^3. Here monoclonal refers to all lymphocytes coming from the same original cell.

- ***Rai Stage 0:*** Lymphocytosis and no enlargement of the lymph nodes, spleen or liver and with near normal red blood cell (RBC) and platelet counts.
- ***Rai Stage I:*** Lymphocytosis plus enlarged lymph nodes. The spleen and liver are not enlarged and the RBC and platelet counts are near normal.
- ***Rai Stage II:*** Lymphocytosis plus an enlarged spleen (and possibly an enlarged liver), with or without enlarged lymph nodes. The RBC and platelet counts are near normal.
- ***Rai Stage III:*** Lymphocytosis plus anemia (too few RBCs), with or without enlarged lymph nodes, spleen, or liver. Platelet counts are near normal.
- ***Rai Stage IV:*** Lymphocytosis plus thrombocytopenia (too few blood platelets), with or without anemia, enlarged lymph nodes, spleen, or liver.

The four stages above are divided into three categories of risk as follows:

- Stage 0 is low-risk.
- Stages I and II are intermediate-risk.
- Stages III and IV are high-risk.

BINET STAGING SYSTEM

The Binet staging system looks at the number of impacted lymphoid tissue groups such as in or around the groin, liver, neck, spleen and underarm areas, as well as whether a patient has anemia (low RBC count) or thrombocytopenia (low platelet count).

- ***Binet Stage A:*** Fewer than three areas of lymphoid tissue are enlarged, with no anemia or thrombocytopenia.
- ***Binet Stage B:*** Three or more areas of lymphoid tissue are enlarged, with no anemia or thrombocytopenia.
- ***Binet Stage C:*** Anemia, thrombocytopenia or both are present.

CHRONIC MYELOID LEUKEMIA (CML)

Staging of CML is done in phases, with three different categories of advancing symptoms as outlined below.

Chronic Phase: Early phase of CML where fatigue is more prominent along with additional symptoms such as a fever, reduced appetite and weight loss. Immature non-lymphocyte WBCs—stem cells in the bone marrow called myeloblasts or "blasts" for short—are less than 10% of the patient's total WBC count in this phase.

Accelerated Phase: If symptoms do not subside during the chronic phase, the CML moves to an accelerated phase where the fatigue worsens. Any of the following will place a patient in this phase:

- Blasts are between 10 or 20% of the patient's total WBC count when bone marrow or blood samples are taken.
- Basophils make up 20% or greater of the patient's WBC count.
- Abnormally high WBC counts do not respond to treatment.
- High or low platelet counts (not caused by treatment).
- New chromosome changes are observed in leukemia cells.

Blast Phase: More than 20% of a patient's WBCs are blasts that are present in blood or bone marrow samples. Blast cells generally have spread beyond the bone marrow to nearby tissues and organs. Symptoms worsen, including a fever, poor appetite and weight loss. CML acts like aggressive acute leukemia in this phase.

CHAPTER 3

TREATMENT LOGISTICS

CHOOSING A FACILITY AND DOCTORS

CHOOSING A TOP-NOTCH facility and group of doctors is very important since you are your main champion! Nobody has as much interest in getting better than you do, so you must take ownership of your recovery.

I live in the Bay Area and have insurance through my work with Kaiser. Once a year in November, I am allowed to update my benefits, which turned out to be about a month after my initial diagnosis. That time sticks out as I recall having Thanksgiving dinner with my dad and brother at a restaurant—the first time ever doing that—but we were a bit too preoccupied to cook.

The first step I took toward recovery was to meet with Kaiser's head of oncology, Dr. Chang, because I needed to receive a definitive diagnosis immediately. Looking at his biography, I saw he trained at Stanford with excellent credentials and had strong medical experience fighting various forms of cancer. Also, his being the head of the department made me feel that I would be getting the best treatment possible at Kaiser.

Stanford and the University of California, San Francisco (UCSF) were the other two locations I considered, as both are world-renowned facilities within an hour drive of me. My brother, dad and I made an appointment with Stanford to discuss treatment options, facilities, equipment, doctors and costs. These were provided in the form of informational packets as we discussed with a non-physician representative. I would have preferred to speak with a doctor in the department, yet that is how they set up their informational sessions.

Unfortunately, we did not have a tour and instead spent a lot of time going over costs and insurance. We were not given an official quote; however, for chemotherapy alone, the estimate was at least $35K. The radiation following chemotherapy would also cost a significant amount, on top of expensive prescription costs. Since we did not travel up to San Francisco to meet with a representative or doctor from UCSF in person, I relied on what I could find via the Internet, as I wanted to begin treatment as soon as possible.

After culling this information, my options were either to use Kaiser for a month and then switch my insurance over so that I could go to Stanford/UCSF or stick with Kaiser throughout. I could also pay for Stanford/UCSF out of pocket for the first month, but that made little sense.

UCSF was an hour away for multiple visits a week and since Stanford was rated just as highly while being closer, I ruled out UCSF; however, choosing between Kaiser and Stanford took some deliberation. Starting a treatment at one facility and then transferring over to another sounded complicated from both an insurance and treatment perspective. It would entail more paperwork on top of the multitude of paperwork I already had as well as having all my results transferred between facilities. Additionally, I would have the burden of starting over with a whole new team of doctors.

What sealed the deal for me was when I visited Kaiser's Radiation Oncology Department, which was a new facility with its own building. The staff was friendly and most importantly, I developed an immediate rapport with who would become my future radiation oncologist, Dr. Anderson. He was very personable, patiently answered all my questions about the process and gave the impression that he would give my treatment the attention it deserved.

Furthermore, the facility I was shown was brand new and featured radiation equipment that was like what I would have used at Stanford. Talking to other patients, I learned that people were coming from hours away to be treated at the facility—which was 15 minutes from where I lived—and that it was considered one of the best in the Bay Area.

Like Dr. Chang, Dr. Anderson also had an impressive résumé. That coupled with the high-quality facility, convenient location, ability to have one team throughout the process and lower cost (my new insurance would have higher co-pays and premiums if I switched to be treated at Stanford) made Kaiser the best choice.

The point of sharing this experience is that it is important to do your homework and find out all your treatment options in terms of facilities and doctors. There are many factors, including the competence and experience of the doctors you will be working with, insurance costs, quality of the facilities and medical equipment, bedside manner of the doctors and overall convenience, as you will be making multiple visits a week for the next six to twelve months of your life. Arm yourself with knowledge to maximize your chance of success!

Insurance Forms

Understanding your insurance is vital to your treatment. Depending on your employer, you may or may not have options when it comes to choosing your insurance. At the time of my diagnosis, we were able to use either a health maintenance organization (HMO) via Kaiser where I could go to Kaiser only, or a preferred provider organization (PPO) where I could choose facilities outside of Kaiser but could not be treated at Kaiser.

Generally, PPOs are preferred so that you have more options; however, as mentioned previously, my situation was such that an HMO was the better choice since I wanted to go to Kaiser. There are too many variables with insurance plans and most people do not plan to get cancer. If you do have a choice, make certain you understand all the co-pays, deductibles and more importantly, what insurance is required for the facility you want to be treated at. If your insurance is already set by your employer, make sure you understand your options for facilities and how much you will be expected to pay.

Short-Term Disability

The abundance of paperwork patients or caretakers need to complete throughout the cancer process can be quite daunting. Policies vary by state and employer, so it is important to research the topic to fully understand your work benefit and disability options.

My company, Lockheed Martin, picked up approximately half of my salary while the State of California covered the remainder through my short-term disability benefit. The state's disability insurance is funded through deductions from employee's earnings. I paid state taxes the entire five years I worked in California prior to my diagnosis so it was nice to see that the state would cover a portion of my salary while I was unable to work.

Working either full-time or part-time during treatment is a trade that I contemplated. This is a personal decision that each patient must make. While my personality leads me to work hard and through pain (based on my sports background), you also must think long-term and what is best for your health.

I made the decision to take off work to focus completely on my treatment and ensure that I was able to recover fully. Additionally, I chose an aggressive chemotherapy and radiation plan, which factored into my decision to stay home. Work presents obvious stressors and in my opinion, all your energy and mental strength should be used toward treating yourself, being your own doctor, researching treatments, dealing with side effects, eating healthy, exercising and being your best friend to yourself. You will be most effective in all these facets if you are able to devote yourself 100% to your treatment.

Now, the severity of cancers and treatment plans certainly differs, so this is wide-reaching and generalized advice. There will be treatments for early detection and various cancer types that make it easier to continue working. Also the type of work is a factor, which is why ultimately this decision is unique to each individual and prognosis. Do listen to your family, though. We are often our toughest critics and they tend to be easier on us. Weigh all the pros and cons. Money should not be a factor if you are able to take advantage of employer and state benefit/disability programs effectively.

I was fortunate that California is currently among only a handful of states offering up to a year of short-term disability from the state directly, paying 55% of the employee's salary. New Jersey provides up to six months at two-thirds an employee's salary, up to a maximum value. Rhode Island allows up to thirty weeks of disability with the amount based on a complex formula that only certain employees are eligible for. Hawaii offers half a year of benefits at around 58% of the employee's salary. Finally, New York does not pay individuals directly but rather mandates that employers offer short-term disability, paying 50% of the employee's wage for up to six months.

If you live in one of the above states, as of 2019, your income will be covered via a combined employer and state contribution between 50 and 100% depending on the employer and state. If you do not, find out how much your employer pays and for how long. Typical percentage and duration include 60% for up to half a year. One factor that should be considered (but you likely will not be told about) is that choosing your treatment plan should also consider how much disability you will receive and for how long. I chose an aggressive treatment so I could be cured quickly, as I was confident my body could handle the increased concentration of toxicity; however, this plan also ensured that my treatment ended within the six months I would be receiving disability.

Chapter 4

Tests

There are a multitude of tests you will undergo during your cancer experience, starting with diagnosis through post-treatment checkups. This chapter outlines the various tests you may endure, their purpose, how they work and what to expect.

Being of a scientific background, I could not help but provide some of the engineering details on how they work. Furthermore, understanding how tests work can not only arm patients with their purpose, but also reveal levels of machine accuracy and the amount of potential harm your body will undergo.

X-Rays

X-rays are electromagnetic waves that have a shorter wavelength than visible light. They are generated by an electrical current that passes through an X-ray tube, producing a beam of radiation that is directed at the part of the body being tested. When the X-rays travel through the body, they are absorbed in different amounts by different tissues, depending on the density of the tissues they pass through, allowing an image to be created of your body's interior.

X-rays are among the simplest and least harmful radiation tests performed. Unfortunately, they also provide the least amount of information. However, for an initial diagnosis X-rays often can identify large tumors that can alert doctors to possible cancer, which was the case in my situation when a chest X-ray identified a football-sized tumor next to my heart. Smaller tumors and trace cancer cells will not be identified by an X-ray; that will require a combined PET/CT scan.

Traditionally, X-rays have been produced using film detectors, where the X-rays interact chemically with a film to produce an image. This process uses photochemistry, whereby electromagnetic radiation causes a chemical change in a particle (the same process occurs with your skin's pigments when you get a tan). The film itself has many constituents, the most important being a layer of plastic on the back away from the incoming X-ray or visible light for cameras, with multiple layers of silver halide crystals suspended in gelatin on the front. The former chemically change to create the image while the latter position these crystals.

More recently, however, film is being replaced by digital images that can be produced faster and with less radiation exposure to the patient. Scintillators, also referred to as scintillation crystals, produce visible light when impacted by radiation and are used in conjunction with amorphous silicon, which produces electrons when hit with visible light via the photoelectric effect. Additional details will be discussed regarding these techniques in subsequent sections.

Computed Tomography (CT) Scan

A CT scan (also known as a CAT scan) is a three-dimensional X-ray that utilizes a computer to create cross-sectional images, or slices, of the bones, blood vessels and soft tissue inside your body. This enables doctors to make diagnoses without having to perform invasive surgery. Because a CT scan is essentially a series of X-rays rather than a single X-ray, this test does subject the patient to greater amounts of radiation, but the exposure remains low enough to remain safe.

CT contrast agents, often called dyes, are used to highlight specific areas so that organs, blood vessels and tissue are more visible. By increasing the visibility of all organ surfaces or tissue being studied, these dyes help doctors determine the presence and extent of a disease like cancer. Note that dyes are generally injected through a vein in the arm. Iodine-based contrasting agents are primarily used based on X-ray absorbing qualities while also being safe for patients. Barium sulfate is another contrasting agent that is used when specifically investigating the digestive system.

X-rays are produced by a cathode/anode emitter located at one end of the hardware ring surrounding the patient. The cathode is a large tungsten filament inside a vacuum that is heated via current passing through. Once the filament reaches a high temperature, electrons begin sputtering off the filament. A large voltage differential is set across the cathode and anode, which causes electrons to fly toward the tungsten anode (incidentally similar to an ion engine in space). Electrons knock off a lower electron orbiting a tungsten atom, causing an electron in a higher orbit to fall inward to take the ejected electron's position. Electrons moving orbital rings inward reduce energy, causing photons to be released (conservation of energy) in the form of high energy X-rays (Figure 6). This process is called *characteristic X-ray production*. A second process, called *Bremsstrahlung X-ray production*, occurs when the incoming electrons come near but do not interact with the atom. This causes a deceleration of the electrons, thereby reducing energy and again an X-ray is given off.

Lead is placed around the emitter in all directions save one to have a focused X-ray beam emitted (lead has a higher atomic number and absorbs X-ray energy without letting the photons pass through). Outer skin cells are made primarily of smaller atoms of carbon (C), oxygen (O) and hydrogen (H), which allow X-rays to pass through without being absorbed. Calcium in bones, however, has a larger atomic number and is therefore absorbed. Organs have similar C, O and H atoms as skin cells and therefore require a contrast to block X-rays and be seen. X-rays not colliding with larger atoms pass through to a detector that captures the incoming radiation via the photoelectric effect (like a regular camera but X-rays instead of visible light produce metal electron movement).

CT detectors (Figure 7) are photodetectors of various types, most commonly either a photomultiplier tube (PMT) or a photodiode. Both take incoming visible light produced by the scintillating crystals and convert it into electrons that can then be converted from an analog to a digital signal and read out by computers. Photodiodes generate electrons via the photoelectric effect—knocking electrons from their valance band to their conduction band—using a semiconductor such as silicon.

PMTs use the photoelectric effect as well but additionally rely on secondary emissions, whereby electrons or ions in a gas or vacuum generate more particles via interactions with a metal surface. The net effect of a PMT is therefore first moving electrons from their valance to conduction band using the photoelectric effect and then to additional free electrons from secondary radiation.

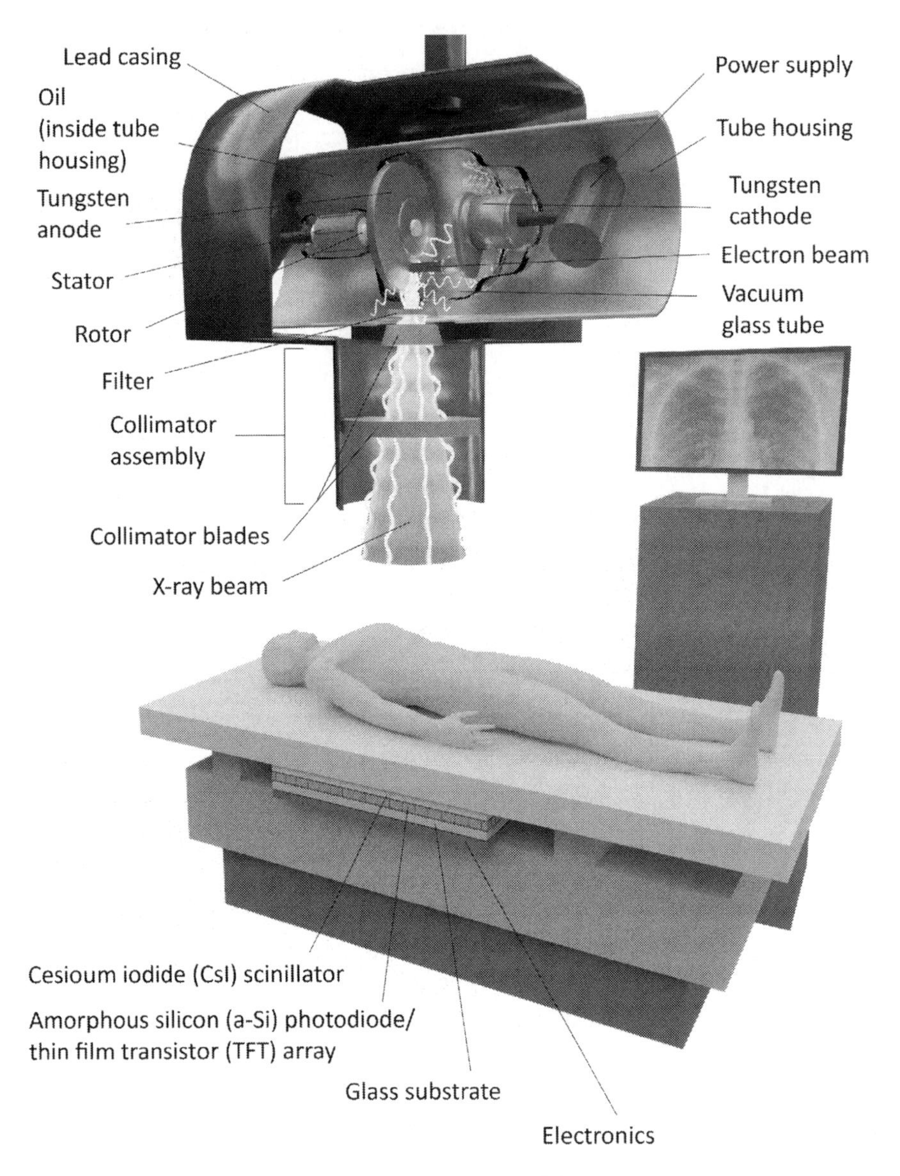

Figure 6: X-ray machine.

X-ray signals are reconstructed using computer algorithms that determine the relative opacity of the material inside the body. The body is divided up into numerous voxels—the 3-D equivalent of a pixel—and the intensity of X-rays through four different angles on each voxel is measured. The ability of a voxel to absorb radiation is provided a number one through eight, with eight being the most opaque. Using the information delivered by the CT scan, the algorithm can reconstruct the intensity of all these voxels within the body to provide doctors with 3-D images inside our body.

Since most cells within the body absorb little X-ray radiation (except for bones), a contrasting agent can be injected into the circulatory system (generally via veins in the arm). This allows tissues and organs within the body to show up on the image. Iodine-based contrasting agents are generally used based on their X-ray absorbing qualities, while also being safe for patients. Barium sulfate is another contrasting agent that is used when specifically investigating the digestive system. A picture of a CT scanner is shown in Figure 8.

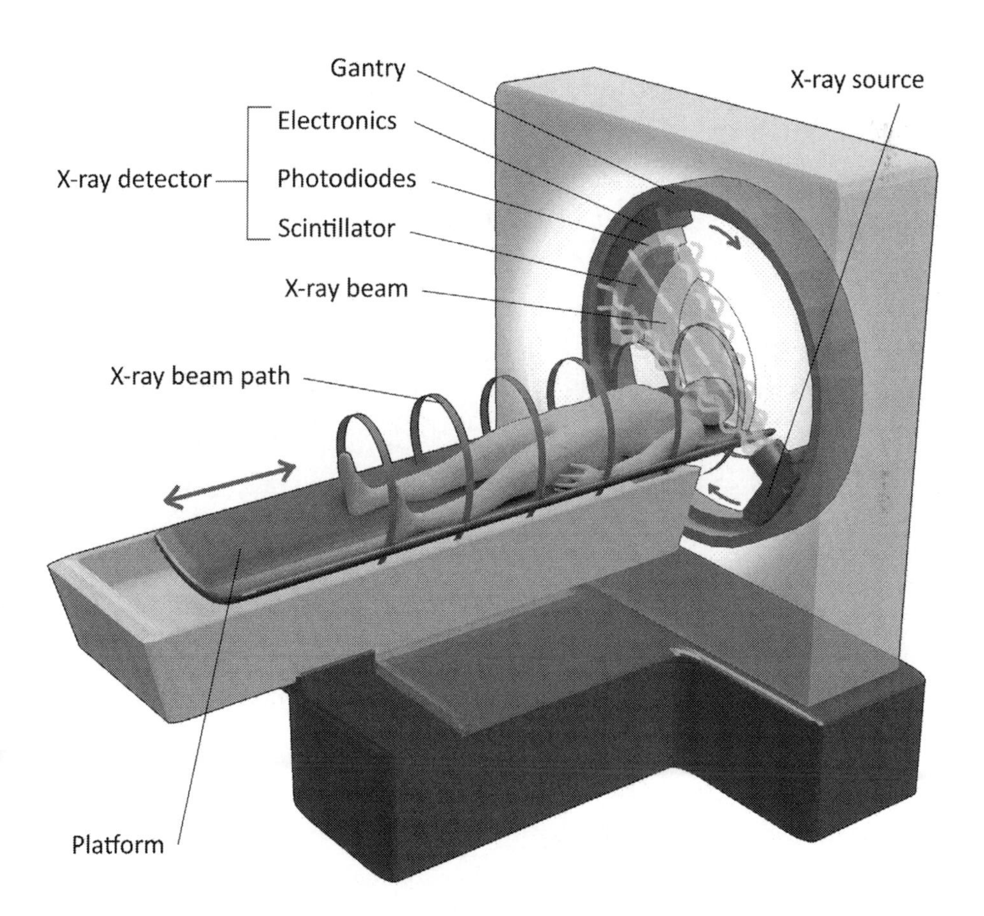

Figure 7: CT scanner schematic.

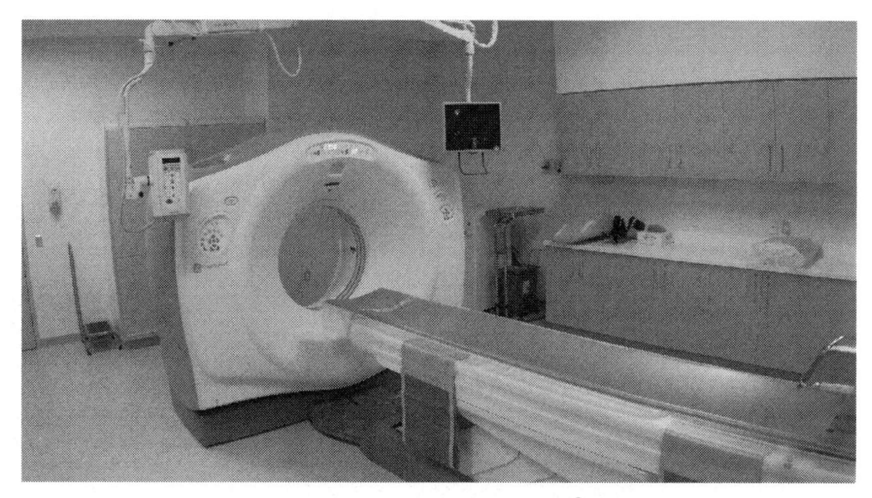

Figure 8: CT scanner.[9]

POSITRON EMISSION TOMOGRAPHY (PET) SCAN

A PET scan is an imaging test that allows doctors to check for disease in your body. The scan uses radioactive tracers in a special dye that are injected into a vein in your arm and are then absorbed by your organs and tissues.

PET scans are used in conjunction with CT scans to identify the location of cancer cells throughout the body. While a CT scan provides a three-dimensional view of your body and internal organs, a PET scan is required to show where active cancer cells are located and can differentiate between a benign and active tumor.

This is possible because a tracer will collect in areas of higher cellular activity associated with cancer. These areas of disease will show up as bright spots on the PET scan, which can also measure blood flow and oxygen use.

PET scans work by injecting fludeoxyglucose, which has the same molecular structure as glucose (a simple sugar) apart from radioactive fluorine (fluorine-18) replacing one of the OH (hydroxyl) pairs. Since cancer cells use higher amounts of glucose compared to normal healthy cells, glucose gets absorbed at much higher rates by cancer

9 Daveynin, "GE LightSpeed CT scanner at open house, Monroeville, PA, UPMC East," *Wikimedia Commons*, https://commons.wikimedia.org/wiki/File:UPMCEast_CTscan.jpg.

cells than by healthy cells. While both healthy and cancer cells use the anaerobic glycolysis metabolic process to produce adenosine triphosphate (ATP, chemical formula $C_{10}H_{16}N_5O_{13}P_3$: the currency of life that provides cell energy), healthy cells also rely heavily on an additional aerobic metabolic process in the mitochondria to convert pyruvate into ATP.

Pyruvate is a byproduct of the glycolysis process and its conversion into ATP is more efficient than the direct conversion of glucose into ATP. Therefore cancer cells need to ingest additional glucose in order to compensate for their lack of ATP production via the mitochondria, which does not function properly in cancer cells. As a result of this increased glucose ingestion, cancer cells light up like a Christmas tree when isolated via their glucose absorption!

In order to properly capture these high rates of ingestion, the added radioactive fluorine isotope emits positrons (analogous to an electron in size and mass but with a positive charge) as part of natural radioactive decay. With a half-life of roughly 2 hours (109.8 min), these emitted positrons collide with electrons in nearby tissues, annihilating one another and generating two gamma rays that shoot out in opposite directions (due to the conservation of momentum). These high-energy gamma rays are detected by a surrounding ring of scintillating crystals, generally either germanium oxide, gadolinium oxyorthosilicate or lutetium oxyorthosilicate crystals. Scintillating crystals are excited by the incoming gamma ray photons and generate multiple lower-energy photons that can be seen by the visible eye (Figure 9).

Visible photons are converted to electrons using a photomultiplier tube (PMT) via the photoelectric effect (which Albert Einstein received the Nobel Prize in physics for); see Figure 10. Thousands of scintillation crystals and hundreds of PMTs surround the patient in a circular ring in order to detect the gamma ray signals and convert them into electrons for computer processing and image reconstruction; cancer cells within the body are pinpointed based on multiple gamma-ray ejections.

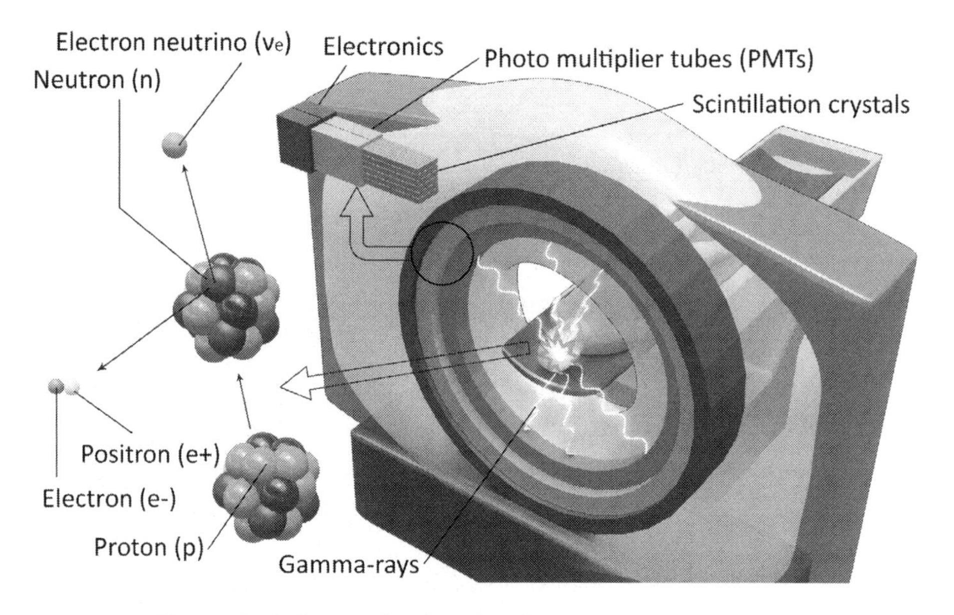

Figure 9: Schematic showing how PET scan gamma rays are used to pinpoint cancer cell locations.

Note that often only one gamma ray makes its way out of the body and these lone signals must be discarded; without the accompanying opposite direction gamma ray, the source could come from an infinite amount of directions. A photograph of a PET scanner is displayed in Figure 11.

In terms of what to expect for a PET scan, this procedure is more rigorous than a CT scan. You will lie on a similar movable table where you are fed through a circular section while your body is imaged. However, unlike the CT scan where a tracer is given while you are on the table right before the procedure, for the PET scan you go into a room where a nurse injects you with a radioactive fluorine isotope. You are required to rest between 30 and 90 minutes while the solution spreads and absorbs throughout your body. The actual PET scan itself lasts between 30 and 45 minutes.

Figures 12 and 13 show selected cross-sections from my PET scan two days before chemotherapy began and after my chemotherapy, three days before my first radiation treatment. The first image shows the size of my tumor being larger than my heart and glowing, along with my brain and heart, based on active cells consuming the sugar-laced tracer I was given.

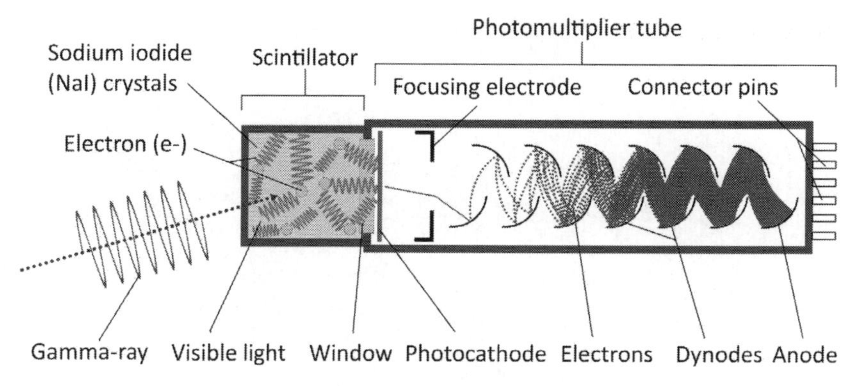

Figure 10: Photomultiplier tube with coupled scintillator schematic.

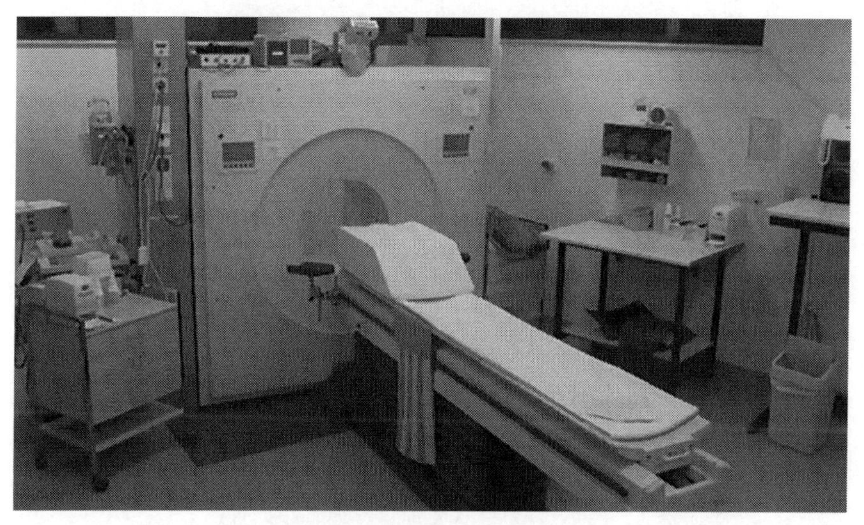

Figure 11: PET scanner.[10]

The subsequent image taken roughly three months later shows the tumor's metabolic activity essentially halted, as there is no glowing. While you can still see a substantial mass above my heart, which is primarily dead cancer cells and scar tissue, the tumor's size is drastically smaller.

What was alarming to me when I compared these two pictures was my heart was significantly lower in the first picture based on the tumor displacing anything in its path. These before and after pictures both capture the extent of my tumor and the power of a PET scan to show both geometry and metabolic activity in one image.

10 Jens Maus, "PET Scanner," *Wikimedia Commons*, https://commons.wikimedia.org/wiki/File:ECAT-Exact-HR--PET-Scanner.jpg.

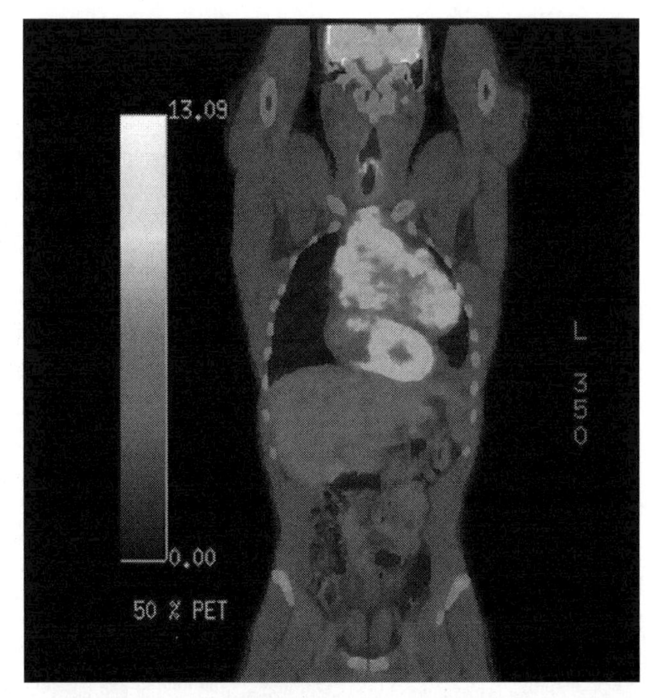

Figure 12: My pre-cancer treatment PET scan.

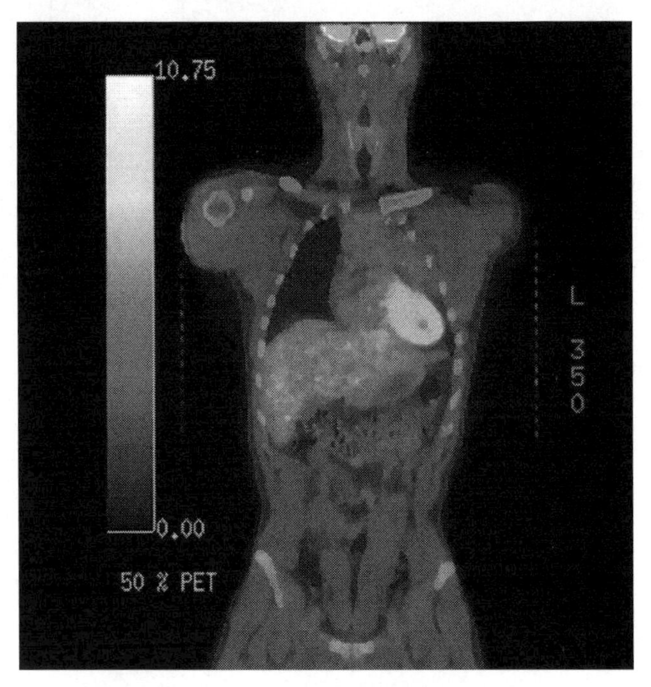

Figure 13: My post-chemotherapy and pre-radiation treatment PET scan.

In order to capture these images, I requested the raw files from the images that were taken. When your radiation oncologist goes through your images, he or she will likely scroll through real-time to show you various slices in an attempt to find the best ones to show your cancer or cancer treatment progress. I scrolled through and found similar cross-section depths to compare the two images side-by-side.

Once your cancer treatment is completed, you will require checkups. Most likely this will entail a PET scan, although your oncologist will weigh the benefits of the scan with the negative radiation dosage. I came in yearly for appointments for the first five years or so, with the first couple being PET scans and then moving to lower dosage (but less conclusive) CT scans. After five years I stopped going in for regular checkups.

COMBINED PET/CT SCAN

Combining the CT and PET scan into one device is more efficient, so hybrid machines are being manufactured. Check with your nuclear medicine department to see if they have this capability. Traditionally, software is used to perform image fusion where the separate CT and PET scans are superimposed on top of one another, allowing doctors and patients to view the information in one set of images. Having a device that does both scans not only reduces the amount of time a patient must spend testing, it also reduces the processing time and labor associated with superimposing images. PET scans and magnetic resonance imaging (MRI) scans are less likely to be superimposed and at the time of this writing, there are not machines available that do both an MRI and a PET scan.

MAGNETIC RESONANCE IMAGING (MRI)

An MRI scan is also used to get detailed images of the organs and tissues in the body, albeit with a differing focus. MRIs are useful for imaging soft tissue, which includes every tissue in the human body other than bone, considered a hard tissue. There are four categories of animal tissue: muscle, connective, epithelium and nerve. Muscle and nerve tissue are rather straightforward, while connective tissue includes bones, ligaments (bone to bone), tendons (bone to muscle), adipose tissue (fat), fascia areolar tissue (binds skin to muscle) and synovial membranes (found between bones and containing synovial fluid as a lubricant). Epithelial tissue includes the outer skin, tissue covering cavities such as the mouth and rectum, as well as tissue surrounding internal organs.

MRIs are particularly useful for imaging brain and spinal cord tumors due to the soft tissue involved. A contrast may be injected into your veins to speed up the response time for the hydrogen protons inside your body. This imaging technique is also useful for determining if a cancer has metastasized and to plan additional treatments, such as surgery or radiation. MRIs are often used for breast cancer after diagnosis to measure the size of the tumor and look for additional tumors.

Many feats of science occurred to reach the eventual MRI used today to image inside the human body. This included a 1952 Nobel Prize in physics for Felix Bloch and Edward Purcell, along with the first MRI taken on a human in 1977 by Raymond Damadian.

The MRI utilizes external magnetic fields to align hydrogen protons inside the body. Hydrogen exists throughout the body, as approximately 70% of the body is made of water and hydrogen also exists in tissues and fats. With one proton and no neutron, hydrogen is significant since it creates a magnetic field naturally in the body with the direction of the field being random. Under normal circumstances, humans are not in the presence of a significant magnetic field—the Earth's field is 60,000 times smaller than an MRI machine's—so the hydrogen proton magnetic vectors cancel out. However, when exposed to a magnetic field, the hydrogen proton magnetic fields align with the external magnetic field.

In a nearly fifty-fifty split, these protons align with the magnetic fields in the direction of the external magnetic field (i.e. parallel to) or opposite the external magnetic field (i.e. anti-parallel to). The parallel facing protons have less energy than the anti-parallel facing protons, but the number of protons in the parallel direction is slightly higher for a net magnetic field in the direction of the external magnetic field.

The main MRI magnetic field is produced by an electromagnetic coil magnet. A secondary magnetic field is superimposed onto the primary field via gradient coils, which allows the direction of the hydrogen magnetic fields to be further manipulated for more precise imaging in 3-D. Radio frequency (RF) coils then put out radio waves that cause the hydrogen protons to precess with a frequency matching the source, creating resonance. The energy absorbed from the radio wave is re-emitted when the RF signal is turned off and the hydrogen proton magnetic fields realign to their position without the signal. This directionality is complex, but the net magnetic vector created by the protons is going between vertical (aligned with the patient height) to transverse (perpendicular to his or her height) and back again. RF coils have both a transmitter and a receiver, so they also receive back the radio waves generated by the hydrogen protons.

As water and hydrogen are found in different concentrations for various tissues, fats, bones, etc., an MRI is able to decipher the different matter using what is termed the relaxation time for the proton's magnetic vector to return to its pre-RF-induced state. The differences in concentrations result in longer relaxation times, which are tracked in both the longitudinal magnetic vector (T1-weighted) and transverse magnetic vector (T2-weighted) directions. Using these different rates allows the MRI to map out your body's constituents.

RF signals are sent to an analog-to-digital converter and Fourier transform equations are necessary to get the final image you see on a computer. Various grey tints are assigned to different tissues, bones, fats, etc. that allow the image to clearly show the makeup. Additionally, tumors will have different water content and can therefore be identified as well; depending on the type of tumor and location, oncologists may be able to determine if the tumor is cancerous. Schematics of an MRI are shown in Figures 14 and 15

The experience of an MRI is the most unpleasant of the imaging devices. While the table you lie on and the circular opening are like a CT or PET scanner, the MRI machine has a longer depth, which results in more claustrophobia than a CT or PET scan. Perhaps more importantly, an MRI machine is loud! You are provided earplugs, but you still hear what sounds like shrieking (similar to a loud thermal vacuum machine) caused by the oscillating electromagnetic coils. These coil deflections cause vibrations in the air, which is what you end up hearing. You also may be in an uncomfortable position that you must maintain for long periods of time with little movement (which is also the case for CT and PET scans but seems even more uncomfortable with the confined volume and loud noise). MRIs generally take between 45 minutes and an hour.

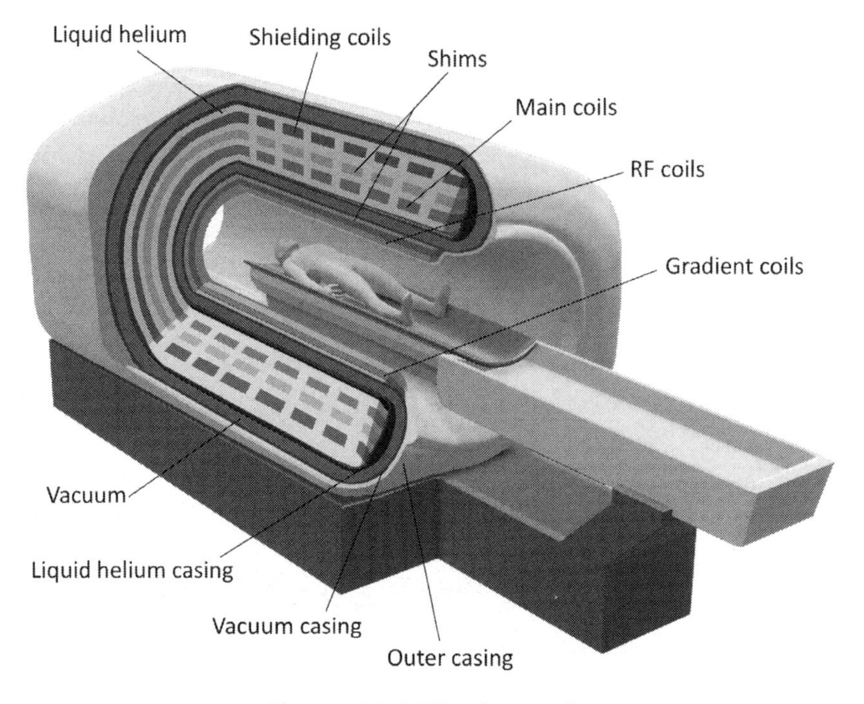

Figure 14: MRI schematic.

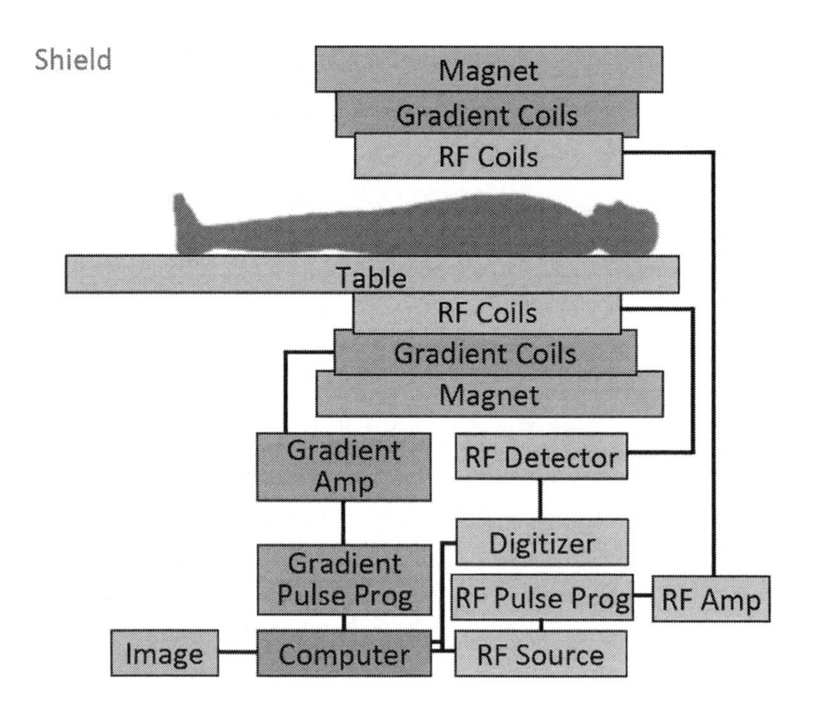

Figure 15: MRI hardware constituents.

COMBINED PET/MRI SCAN

Similar to the PET/CT scan, there is an obvious efficiency (and potentially more accurate) advantage to combining PET and MRI technologies. Machines have been developed that implement this concept; however, at the time of this writing the technology is still being researched prior to widespread adoption by the medical community. Pre-clinical and clinical studies are ongoing to understand the advantages and disadvantages of a combined system and how to best utilize the technology for cancer detection.

ULTRASOUND

Ultrasound imaging uses sound waves rather than electromagnetic radiation to map out interior regions of the body. While not as comprehensive as a CT scan or MRI, ultrasounds are useful for locating tumors without the negative consequences of radiation.

Certain soft tissue diseases can be mapped more effectively with an ultrasound than an X-ray. Fluid-filled cysts can be distinguished from solid tumors using ultrasound due to their differences in echoes. Sound waves do not travel well through dense bone, making organs surrounded by bone (such as the brain) poor candidates for imaging. Additionally, lungs are generally not imaged as the air inside them dissipates the sound waves, providing weak reflections. Doctors will often use ultrasound to direct their needles during a biopsy. I had an ultrasound for a blood clot, which is another application since sound wave reflections can help determine if blood flow is constricted.

Humans can hear frequencies in the range of 20 to 20,000 Hz. Below this range, sound is deemed infrasound while above it is referred to as ultrasound. A transducer is used to generate sound with a frequency much higher than 20 kHz (generally between 2 and 15 MHz, depending on the part of body being examined). The transducer consists of a crystal such as quartz, which can be induced to vibrate very quickly with alternating voltages, thereby producing sound waves. This effect is referred to as piezoelectricity and the transducer is referred to as a piezoelectric transducer, which sits within the ultrasound probe. In addition to producing sound waves, the crystals are also sensitive to the sound wave echoes and will generate an electric current that is measured and displayed on a monitor.

A water-based gel (which is quite cold and slimy) is applied to the surface of the probe in order to make firm contact with the patient. Air bubbles are unwanted, as the signal will be disrupted. The ultrasound process is more convenient when compared with other imaging techniques but must be used for the appropriate examination due to the limitations discussed.

FLUOROSCOPY

A fluoroscopy is essentially a CT movie. Rather than comparing still images, a doctor can look inside real-time, like an ultrasound. While a fluoroscopy can be used for various purposes and as a stand-alone imaging technique, its main role in cancer treatment is assisting a biopsy.

Depending on the type of biopsy, your surgeon may use this tool to help guide the needle into the tumor. Before the procedure, a contrasting agent or dye may be used and is generally administered via an IV, orally or via an enema. A fluoroscopy was used when my tumor was biopsied but not during my bone marrow biopsy, as the latter was easier to locate; thankfully an IV was used for administering the contrasting agent rather than an enema!

BONE SCAN

A bone scan is used to determine whether you have cancer of the bones or if your cancer has metastasized into your bones. Similar to a PET scan, a bone scan uses a nuclear medicine tracer to locate the cancer. The most common radiopharmaceutical (radioactive chemical used for medical purposes) used is technetium-99m (^{99m}Tc) methylene diphosphonate (MDP).[11] It takes between 1 and 4 hours for your body to absorb the tracer. After absorption, the scan begins and will take roughly an hour to complete.

Generally you will not have both a PET scan and a bone scan since the former encompasses the latter and is considered more accurate while not taking as long. This was the case for me, as I received a PET scan only.

BIOPSY

A biopsy involves removal of a tumor or organ sample to determine if the tissue is malignant or benign. While noninvasive images are the first line of testing in the cancer diagnosis process, biopsies are what definitively determine your prognosis and if cancerous, what type. Your medical team

11 ^{99m}Tc is a radioactive isotope of technetium.

will be more informed of your condition and treatment options once you have your biopsy results.

ENDOSCOPIC BIOPSY

For an endoscopic biopsy, a long, thin, flexible tube called a fiberoptic endoscope—with a camera at the end—is inserted into the patient either through a natural orifice or an incision. Forceps inside the tube allow the doctor to remove pieces of the tumor or nearby tissue for diagnosis. In addition to tumors, this procedure is typical for taking tissue samples from the lining of organs such as the intestines, liver and lungs. This is the most common type of biopsy. Specific types of endoscopic biopsies include a cystoscopy (collects bladder tissue), a bronchoscopy (collects lung tissue) and a colonoscopy (collects colon tissue).

NEEDLE BIOPSY

As the name implies, a needle biopsy uses a special needle that is injected through the skin to remove a sample. This procedure is generally performed on tumors that can be felt near the surface, such as a breast lump, whereas endoscopic biopsies are used for cell samples of organ cancers. There are various types of needle biopsies.

FINE NEEDLE ASPIRATION

A fine needle aspiration is the least invasive type of needle biopsy. A thin needle is inserted into the tumor and fluid is extracted in a syringe-like manner for sampling.

CORE NEEDLE BIOPSY

If a larger sample of tissue is needed, a core needle biopsy can be performed. Using a slightly larger needle with a cutting tool at the end, a column (or core) sample of tissue is extracted.

VACUUM-ASSISTED BIOPSY

As the name implies, this form of needle biopsy includes a vacuum to suck out additional tissue and fluid for sampling. This may be used to limit the number of times a needle needs to be inserted and extracted.

IMAGE-GUIDED BIOPSY

Any three of the above needle biopsy techniques can be combined with real-time imaging. The most common way to image real-time is via an ultrasound, which allows the doctor to maneuver the needle to the exact location desired to remove a sample.

An additional imaging technique utilizes a CT scan, where X-ray images are taken before the needle is inserted and used by the physician to locate the needle, followed by more images to confirm the location before extraction. A CT fluoroscopy also takes X-ray images, but does so in real-time and frequently enough to create a movie that the physician can use to locate the needle and extract the tissue sample.

In order to diagnose my cancer, a mediastinum biopsy was performed, which utilized a CT-guided core needle biopsy. The mediastinum is the cavity between the lungs that contains your heart, esophagus, trachea, cardiac and phrenic nerves, thoracic duct, thymus and local lymph nodes.

I recall being slightly lucid and in a dreamlike state during my biopsy (like when getting my wisdom teeth out), seeing a doctor go in and out of my chest as he collected samples. A tube was inserted to enable the forceps that collected the sample. The procedure itself was rather painless since I was given anesthetics to induce unconsciousness.

Skin Biopsy

Skin biopsies can be either incisional, where a portion of a tumor is removed, or excisional, having the entire tumor removed. Soft and connective tissue tumors including muscle, fat, cartilage, vascular and hematopoietic[12] tissues are all candidate tissues for a skin biopsy. Tissue cancers (along with cancellous bone[13] cancer) are termed *sarcomas*, with soft and connective tissue cancers termed *soft tissue sarcomas* and cancellous bone cancers coined *osteosarcomas*.

Superficial Shave Biopsy

This is used to remove or sample external skin growths including warts, skin tags, papillomas, seborrheic keratoses (benign skin growths that can appear with age), actinic keratoses (precancerous patches of thick, crusty growth) and protruding skin cancers (superficial basal cell carcinoma, squamous cell carcinoma and melanoma). Portions of

12 Tissue in which new blood cells are formed.

13 There are two types of bone tissues that form bones: cancellous (spongy) and cortical (compact).

suspicious growths on the external skin (epidermis) can be removed for malignancy tests using a superficial shave biopsy, which is considered an incisional biopsy. A double-edged blade is generally used for this procedure.

SAUCERIZATION BIOPSY (DEEP SHAVE)

For slightly deeper growths up to 4 mm deep and into the dermis, or if the entire growth can be removed, a saucerization biopsy may be used. This is considered an excisional biopsy since the entire growth is removed with a curved blade. The procedure is also considered a shave biopsy so can uniquely be referred to as a deep shave biopsy.

PUNCH BIOPSY

A punch biopsy penetrates deeper than a shave biopsy, puncturing the fat layer (hypodermis, also called subcutis) below the skin layers (epidermis and dermis). An instrument is used to penetrate and extract a cylindrical core sample including all three layers for diagnosis. Local anesthesia will be used to numb the site and reduce pain.

Punch biopsies are performed when a shave biopsy is deemed insufficient to diagnose a suspicious, penetrating growth. A punch biopsy is generally considered incisional. Figure 16 shows the various layers of skin and fat that are poked through during this procedure.

SURGICAL EXCISIONAL BIOPSY

A surgical excisional biopsy is a specific type of excisional biopsy that is performed in a surgical room with a scalpel. A patient will undergo general anesthesia and the surgeon will remove the entire growth and area around it. This procedure is recommended only after a shave or punch biopsy to ensure a proper diagnosis before surgery (if benign, surgery may not be necessary).

Generally used for tumors on the surface of the skin, excisional biopsies also may be done on tumors just under the skin, although needle biopsies are more common for subsurface tumors. This procedure may also be used for lumps in the breast or lymph nodes as well as organs such as the spleen if an initial needle biopsy is inconclusive (e.g. a splenectomy would be performed).

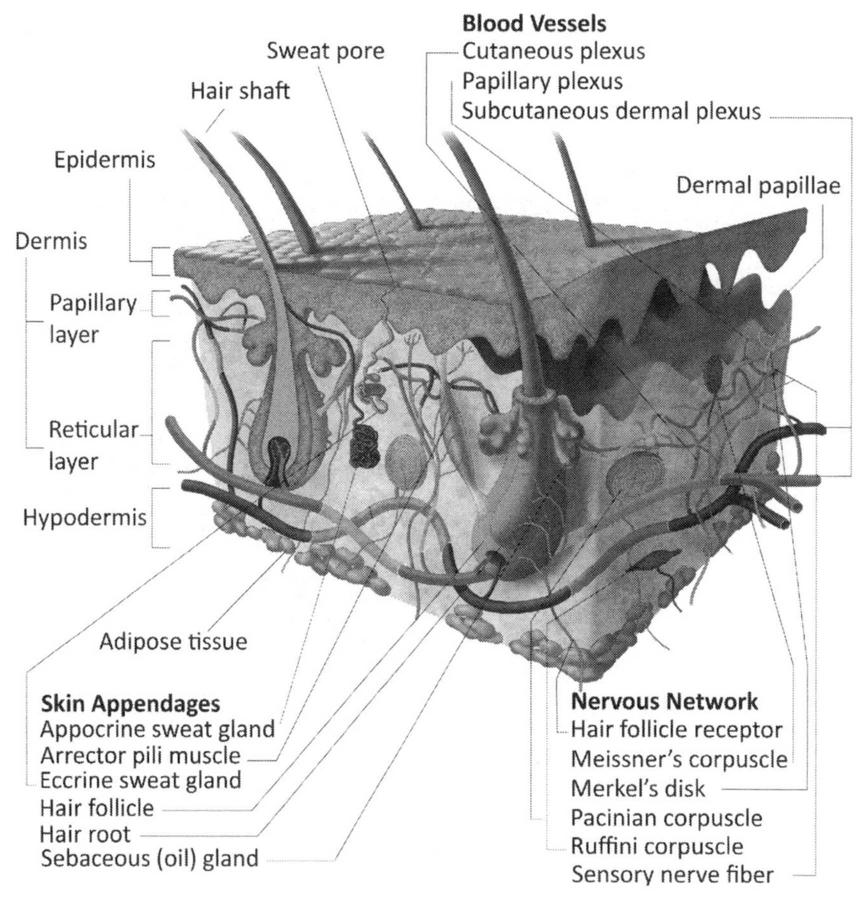

Figure 16: Skin and fat layers.

Bone Marrow Biopsy

A bone marrow biopsy is ordered to help identify staging if your oncologist is concerned that cancer has spread to the marrow or if there are abnormalities in your blood cell count, since blood cells are produced in the bone marrow. This is particularly important in blood cancers such as leukemia, lymphoma and myeloma.

The preferred location of the procedure is the rear upper pelvic bone below your lower back and above your tailbone because it is among the thickest bones in the body (after only the femur) and is not near any major blood vessels. Your oncologist will first insert a needle to numb the area, followed by a larger needle that penetrates the skin and punctures the bone, removing a core sample.

This was the most painful procedure I endured during my cancer treatment. The needle to numb the area was uncomfortable, but the extracting needle was the most painful. My girlfriend was in the room and had to leave during the procedure at the sight of blood gushing out. I recall my oncologist putting all his weight on the needle and bouncing up and down multiple times attempting to penetrate the bone and when he did, a very deep pain shot through the core of my body—not a pleasant feeling!

I had been told it would be like a hard kick in the butt, but I would say it is more like a hammer. There is something unsettling about feeling a piece of your bone get removed as well! The good news is that although the discomfort lasts for a while after the procedure, you should be feeling better by the next day and expect the soreness to subside after a few days. If the soreness lingers more than a few days, notify your oncologist. There will be an incision that needs to be treated afterward and your oncologist will likely request to examine the site the following day.

Perioperative Biopsy

A perioperative biopsy is when a sample is taken during surgery. If a lump is discovered during surgery, diagnosing the tumor in real-time will help the surgeon determine if the lump needs to be removed. Patients must give consent ahead of time for this procedure.

Chapter 5

Body Systems

In order to better understand cancer and how to treat it, you should have a general familiarization with the human body. Human anatomy can be divided into 11 body systems. While all these systems can be impacted by cancer, certain ones play a more prominent role. Furthermore, some systems are impacted more than others depending on the cancer type. A brief overview of each system will be provided and those that are more pertinent to cancer will be discussed in greater detail.

The goal is to educate the cancer patient and caretaker on the human anatomy to become more informed. While your oncologists are the specialists, the more you know, the better equipped you will be to ask questions and play an active role in your treatment—a common theme throughout this book. Knowledge is power; in this case, the power to help cure your cancer!

Cardiovascular (Circulatory System): The cardiovascular system's primary goal is to circulate blood throughout your body, delivering oxygen and nutrients to organs and cells while carrying waste products away. Components of this system include the heart, veins and arteries.

Digestive System: The digestive system processes foods, absorbing nutrients into the bloodstream. Components include the mouth, esophagus, stomach, small and large intestines (colon). Accessory organs include the gallbladder, liver, pancreas and salivary gland.

Endocrine System: The endocrine system helps regulate cells, organs and other bodily systems by producing and releasing hormones (chemical messengers) into the bloodstream. Components include the ovaries (women), pituitary gland, testes (men), thyroid gland, parathyroid gland, thymus gland, pineal gland and adrenal gland.

Integumentary System: The integumentary system protects the body from external threats such as bacteria and viruses while regulating body temperature. It is also important for sensory purposes. Components include the skin, hair, nails, sweat and oil glands.

Lymphatic System: Involved in the body's immune defense, the lymphatic system produces lymphocyte WBCs and transports them via lymph (fluid that circulates throughout the lymphatic system) to destroy harmful invaders. Components include lymph, lymphocytes, lymph nodes, lymphatic vessels, tonsils, the thymus gland and spleen.

Muscular System: Muscles are connected to your bones; in addition to allowing the body to move, they also play a role in body temperature regulation. The latter occurs via arrector pili muscles that control hair follicles, causing hair to be flat against the skin to increase airflow, cooling when the outside environment is hot. Components include skeletal muscles.

Nervous System: The nervous system controls all bodily systems via the brain and is therefore involved in all cognitive functions such as thoughts and feelings. This system also includes nerve receptors throughout the body that provide your ability to touch and feel, alerting the body to pain and thus preventing injury. Components include the brain, spinal cord, nerves and receptors.

Reproductive System: This system's main purpose is to produce offspring via the production of gametes (sperm and eggs) and incubating the fetus. Components include the testes, prostate gland, penis, ductus (vas) deferens and seminal vesicles for men. For women, they include the uterus, ovaries, vagina and fallopian tubes.

Respiratory System: The respiratory system allows the body to breathe and transport oxygen to the bloodstream for circulation to organs and cells via the cardiovascular system. Components include the nasal cavity, larynx (voice box), trachea (windpipe) and lungs.

Skeletal System: Holding your body together, the skeletal system provides structure and protects internal organs. Your bones also store calcium and produce blood cells. Components include skeletal bones.

Urinary System: The urinary system is responsible for filtering blood and secreting the resultant waste via urine. Urine production, storage and secretion involve the regulation of water, electrolytes and blood pH. Components include the kidneys, ureters, bladder and urethra.

I will delve further into the cardiovascular and lymphatic systems, as they are particularly important to understand when it comes to cancer and treatment.

CARDIOVASCULAR SYSTEM

The cardiovascular system transports blood throughout your body via a series of blood vessels. A diagram of the system with its complex network of blood vessels is displayed in Figures 17 through 21.

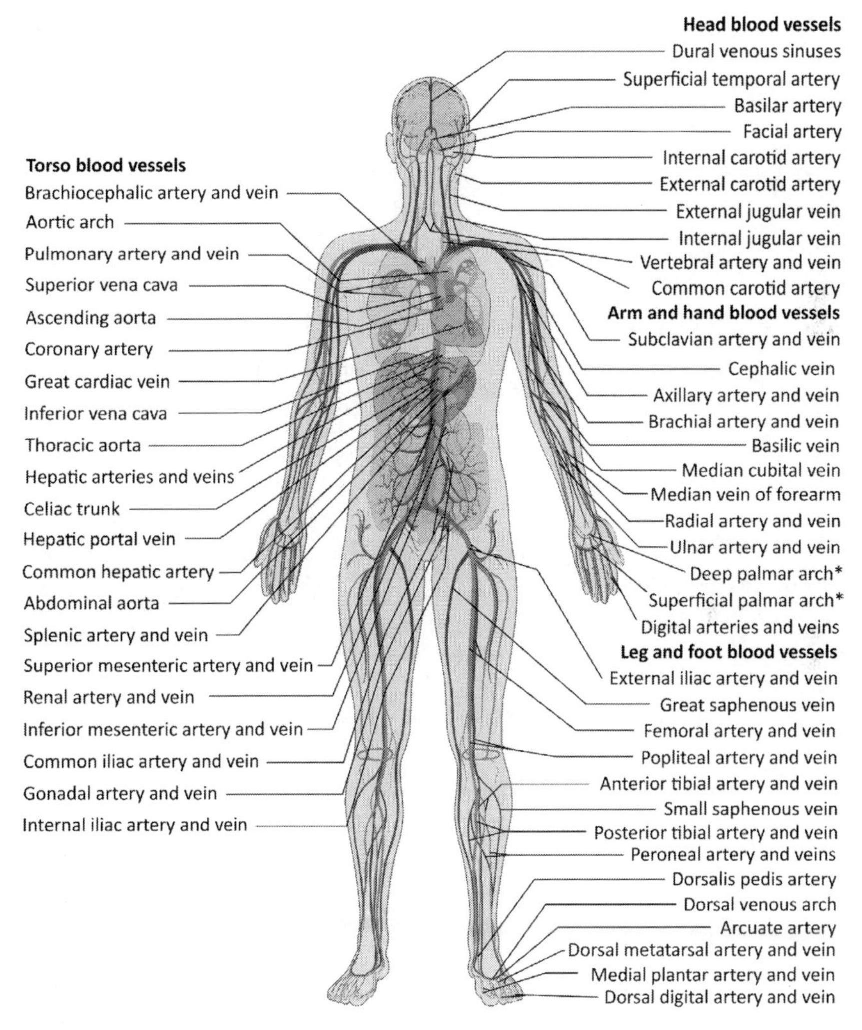

Figure 17: Cardiovascular system (full body).

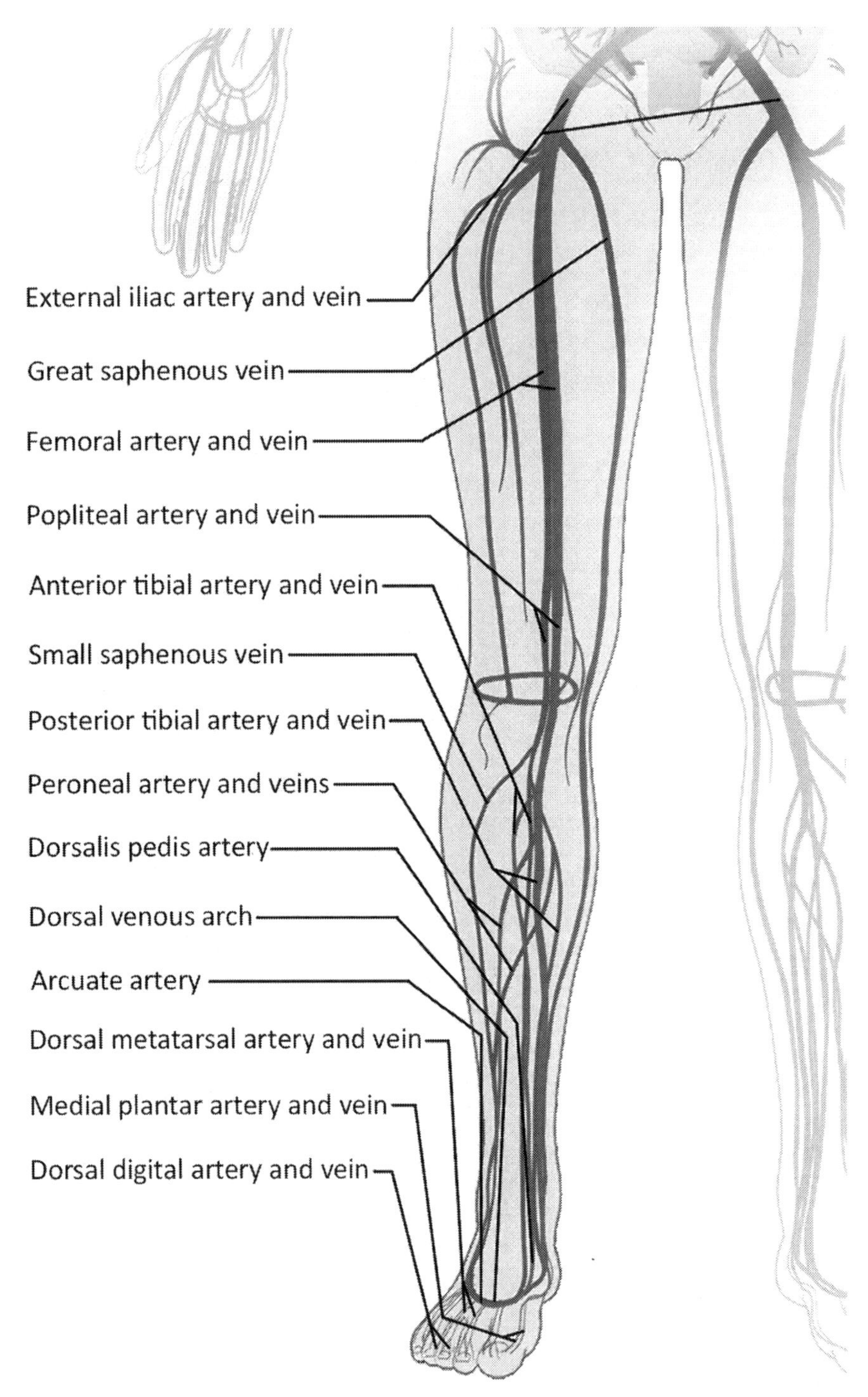

Figure 18: Cardiovascular system (lower limbs).

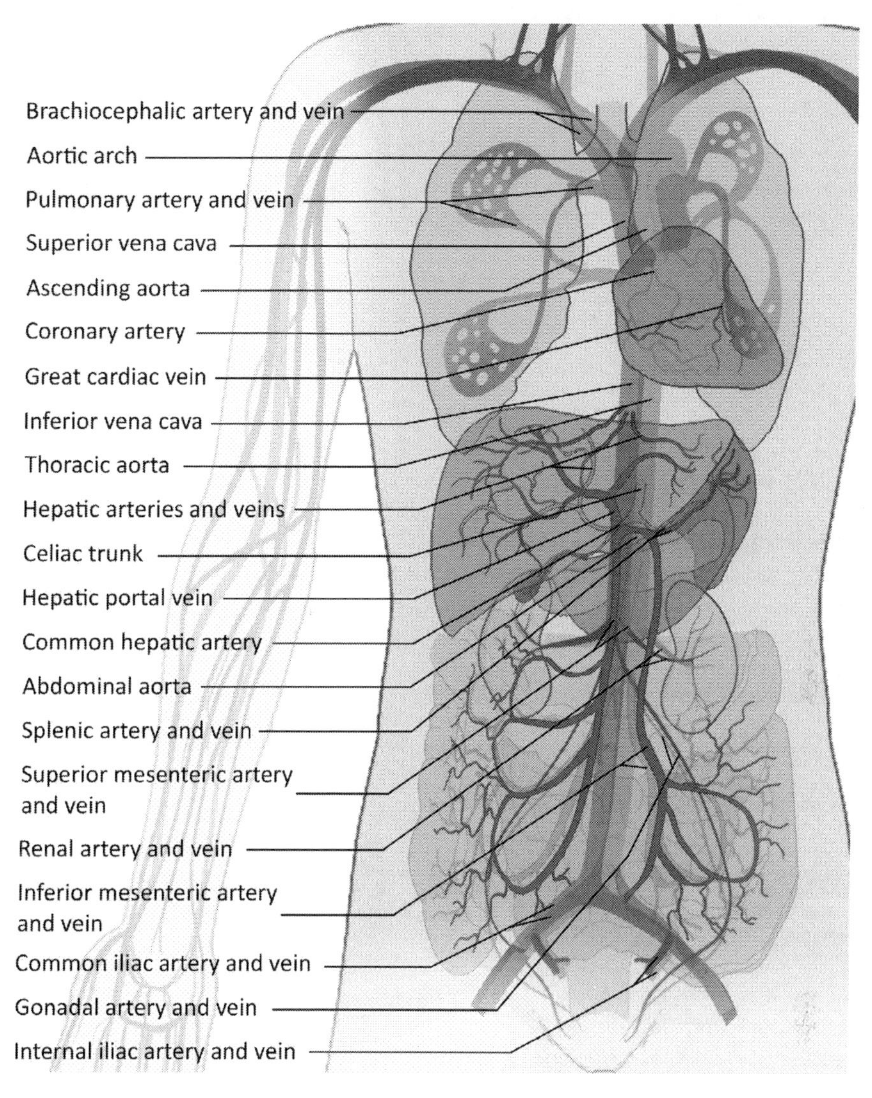

Figure 19: Cardiovascular system (torso).

Vascular Network

There are three types of blood vessels that, along with the heart and blood itself, make up the cardiovascular system:

- ***Arteries:*** Carry blood away from the heart, either oxygenated to the body or deoxygenated to the lungs.
- ***Veins:*** Carry blood toward the heart, either deoxygenated from the body or oxygenated from the lungs.

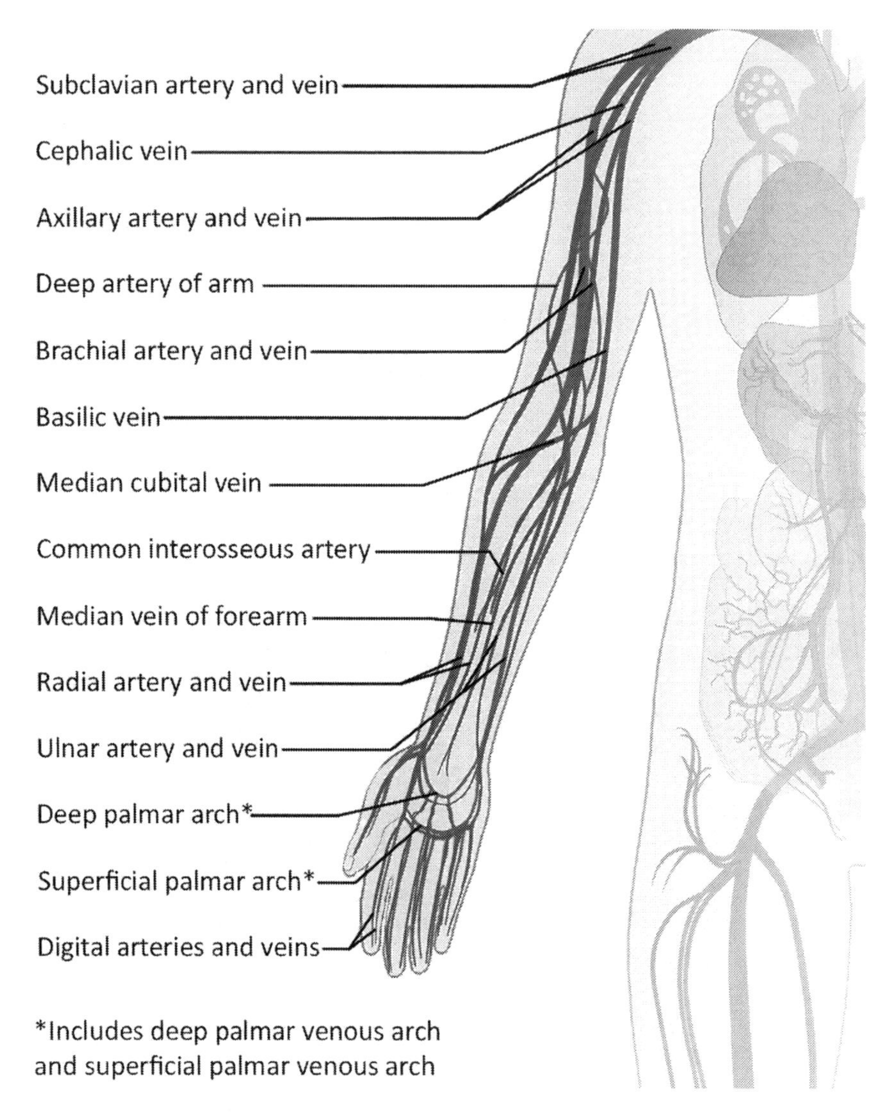

Figure 20: Cardiovascular system (upper limbs).

- ***Capillaries:*** Connect the arteries and veins while allowing the transport of water, oxygen, carbon dioxide, vitamins, minerals and waste substances between blood and tissue.

Blood flow can be grouped into two pathways, i.e. a pulmonary route between the heart and lungs and a systemic pathway between the heart and the rest of the body. Exiting blood from the heart is either heavily oxygenated or oxygen deficient, with the former representing blood travel to the body while the latter is blood flow to the lungs for oxygenation.

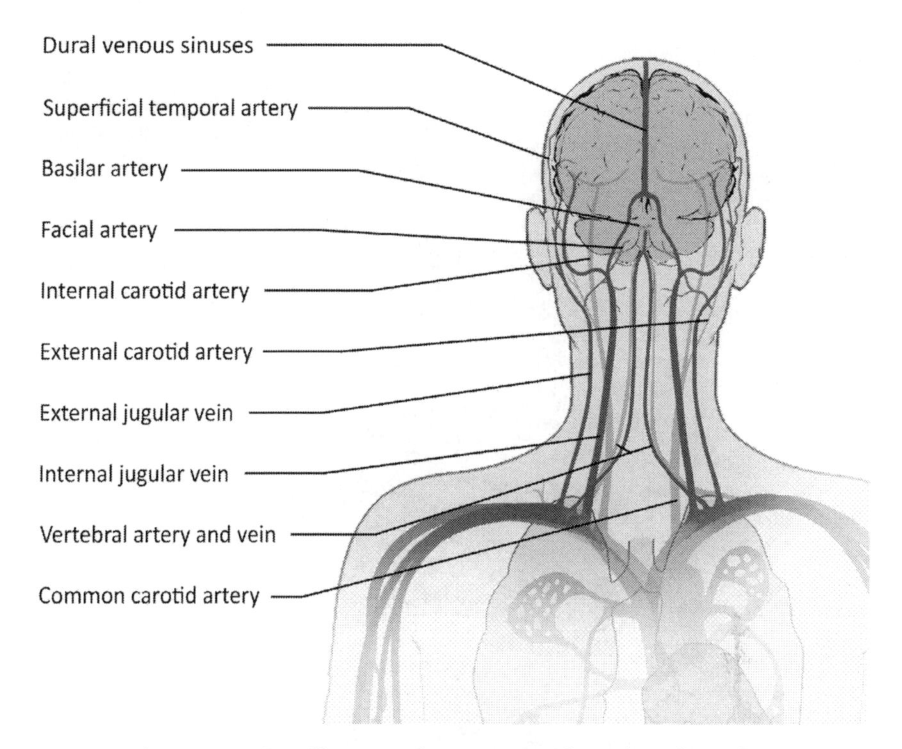

Figure 21: Cardiovascular system (head and neck).

Conversely, returning blood from the body via the systemic circuit has depleted levels of oxygen while blood returning to the heart from the lungs through the pulmonary circuit is oxygen-rich. Blood leaves the heart through elastic arteries such as the aorta (Figure 22), with elastic arteries being the largest subtype of arteries in the body (1.0 to 2.5 cm inner (lumen) diameter; 1.0 mm wall thickness). These arteries get slowly constricted the farther away from the heart you go, branching out into smaller arteries called muscular arteries (0.3 mm to 1.0 cm inner diameter; 1.0 mm thickness).

Eventually, these arteries become arterioles, which are small diameter arteries (Figure 23). Capillaries allow components of your blood (e.g. plasma and nutrients) to flow across its permeable walls and into interstitial fluid, which is fluid that surrounds cell tissue and delivers oxygen to cells. Interstitial fluid interacts with the lymphatic system that will be explained further in that section.

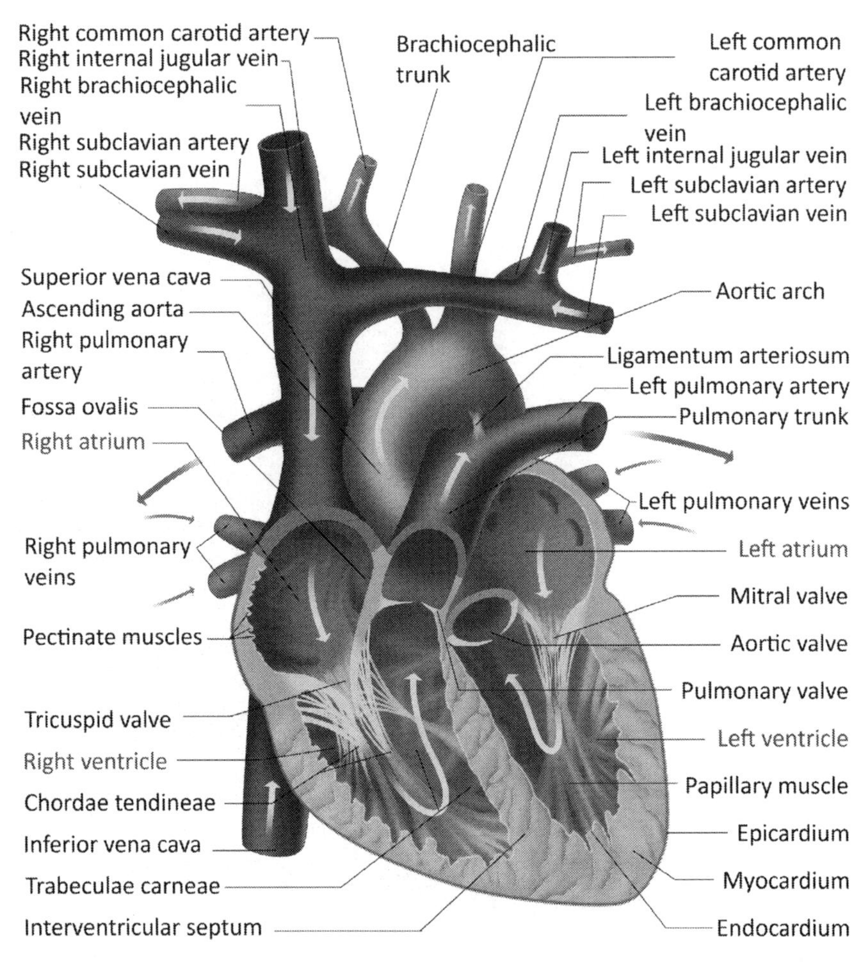

Figure 22: Schematic of the heart.

Deoxygenated blood travels through capillaries and into venules, which are analogous to arterioles for veins in that they are the smallest diameter veins in the body (8 to 100 mm inner diameter; 1.0 μm wall thickness). Traveling toward the heart, veins become larger and various branches coalesce into wide diameter veins such as the inferior vena cava and superior vena cava. Veins have average diameters of 5.0 mm with wall thicknesses of 0.5 mm. Much like plumbing in your home or any engineering problem involving a pressure gradient through tubes, there is a pressure drop as blood travels away from and back to your heart due to system friction, particularly across capillaries. To account for this, blood pressure is higher in your arteries than in your veins. This also explains why arteries have thicker walls than veins—to withstand the higher pressures.

At any point in time, approximately one-third of your blood is in your arteries while two-thirds is in your veins. The total length of all the blood vessels in an adult body is roughly 100,000 miles, which is approximately two and a half times around the Earth! Additionally, roughly 1,800 gallons of blood flows through your circulatory system every day and with an average heart rate of 75 beats per minute, a human heart will beat around 3 billion times over 76 years. That is a lot of work!

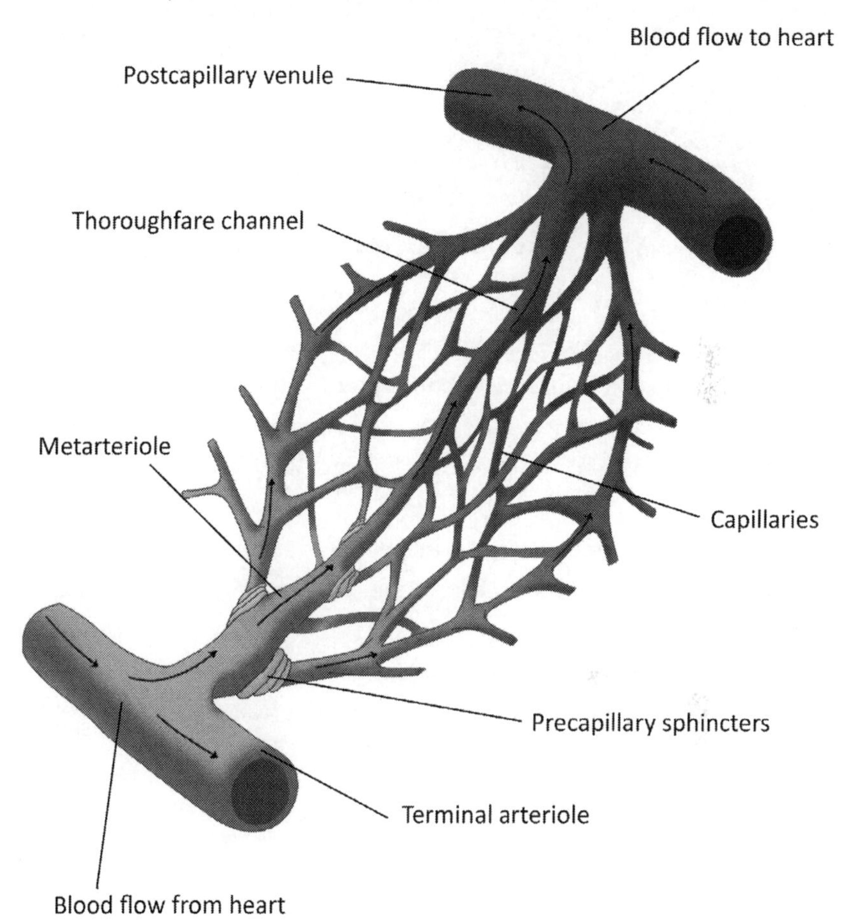

Figure 23: Artery and vein interaction via capillaries.

Blood Anatomy

Blood consists of 45% by volume blood cells (WBCs, RBCs and platelets) suspended in liquid blood plasma (55% by volume as shown in Figure 24). Aside from blood cells themselves, blood transports oxygen, carbon dioxide, glucose, proteins, vitamins, minerals and hormones throughout the body.

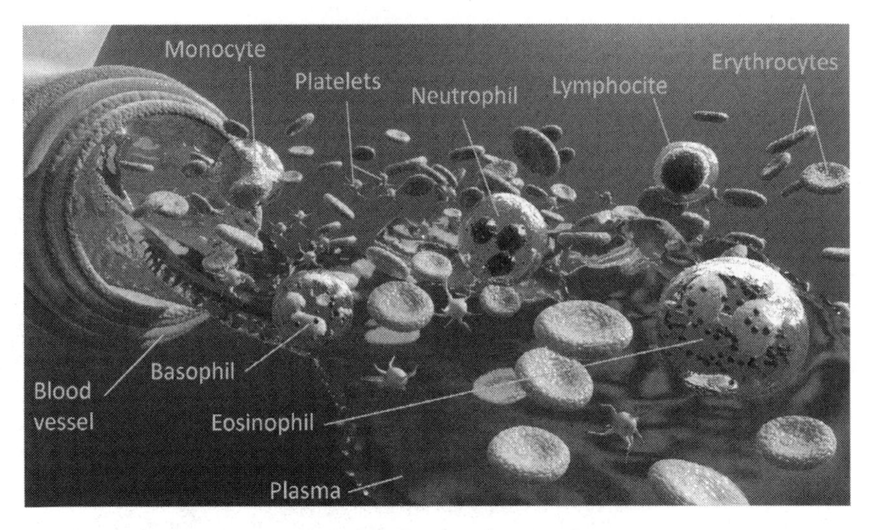

Figure 24: Blood vessel constituents.

The latter consist of three types: eicosanoids, steroids[14] and amino acid derivatives (amines, peptides and proteins). Eicosanoids are signaling molecules involved in the immune system response. Steroids consist of the sex hormones androgens, estrogens and progestogens as well as corticosteroids and anabolic steroids (natural and synthetic). Proteins, like vitamins and minerals, will be discussed in detail later in the book.

Hormones are created and transported via the endocrine system, which is a collection of glands that secrete hormones into the circulatory system to be carried away to organs and tissue. These glands include the pineal gland, pituitary gland, thyroid gland, thymus gland, adrenal gland, pancreas, ovaries and testes. Hormones impact physiological and behavioral functions including hair growth, vocal cords, sexual desire, mood, sleep and muscle/tissue growth.

Blood Cell Development (Hematopoiesis)

All blood cells are created inside the bone marrow, a process called *hematopoiesis,* from the Greek words for "to make" and "blood". Starting from a single type of stem cell (hematopoietic stem cell), blood cells mature into a sequence of cells until they end up as either mature white, red or platelet blood cells. This is done from various cell lines, of which there are seven primary.

These can first be grouped by the two stem cells that the primary

14 Hormonal steroids form a subset of steroids, which are organic compounds with four rings of carbon atoms. Cholesterol is an example of a non-hormonal steroid.

hematopoietic stem cells can turn into, namely lymphoid stem cells that all lymphocyte WBCs come from or myeloid stem cells, from which RBCs, platelets and the remaining non-lymphocyte white cells derive.

Lymphoid stem cells become either B- or T-lymphocytes, with the former producing a plasma cell and the latter an effector T-cell. Myeloid stem cells become proerythroblasts, myeloblasts, monoblasts or megakaryoblasts. These cells are termed committed cells in that once this stage is reached, their destiny is written as to what final blood cells they will be. Proerythroblasts eventually become erythrocytes (RBCs), monoblasts make their way into WBC monocytes (which produce macrophages) and megakaryoblasts become platelets.

Finally, myeloblasts have three lines that end in granular leukocyte WBCs, i.e. eosinophils, basophils and neutrophils. Note that myeloid stem cells can also produce antigen-presenting dendritic cells that are involved in the immune system; mast cells (WBCs similar to basophils that contain histamine and heparin) and osteoclasts, which are bone cells involved in the resorption of bone. Figure 25 provides an overview of the hematopoiesis process.

White Blood Cells (WBCs)

WBCs, also called leukocytes, protect your body from invading pathogens. In order to understand the different types of WBCs and their role in fighting cancer, a brief review of their structure is in order. These cells can be thought of as analogous to the human body, which has various organs surrounded by fluid with an outer skin and tissue for structure (the main difference being that human bodies are much more rigid where cells can change shapes and divide). This analogy is helpful since cells have organelles, which can be thought of as organs in a body that perform specific functions and are surrounded by fluid to allow interaction between the organs.

The major components of a WBC[15] are discussed below, along with a diagram showing their location within the cell in Figure 26.

- ***Centrosome:*** The centrosome controls the production of microtubule proteins and manipulates the cell's shape, such as during phagocytosis, the act of engulfing an external particle, or pinocytosis, when extracellular fluid is ingested.

15 Jon Mallatt, Elaine Marieb and Brady Wilhelm. *Human Anatomy*, 7th Edition (Glenview: Pearson Education, Inc., 2014), 25–37.

- ***Cytoskeletal Network:*** These are protein filaments and microtubules that provide structure for the cell.
- ***Golgi Apparatus:*** This organelle processes and stores macromolecules such as proteins produced by ribosomes.
- ***Mitochondrion:*** An organelle that specializes in ATP production and cellular respiration.
- ***Nucleolus:*** This is the largest structure in the nucleus of a cell where ribosome biogenesis occurs and signal recognition particles are created.

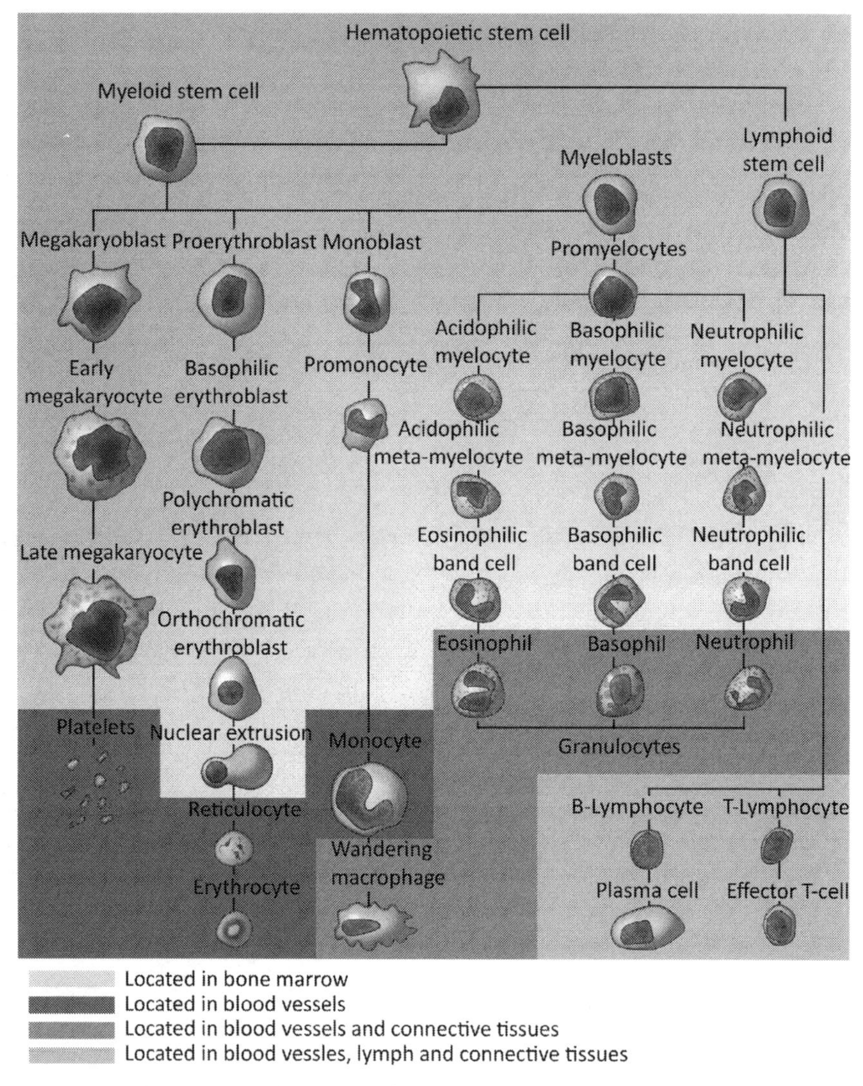

Figure 25: Blood cell formation (hematopoiesis).

- ***Nucleus:*** The nucleus is the center of the cell containing the nucleolus and deoxyribonucleic acid (DNA).
- ***Plasma Membrane:*** The outer cell structure is made of lipids and proteins that separate the inner cellular cytoplasm from the outer body fluid.
- ***Ribosome:*** Ribosomes are where protein synthesis via messenger ribonucleic acid (mRNA) and amino acids occurs.
- ***Rough Endoplasmic Reticulum:*** This membrane houses ribosomes.
- ***Smooth Endoplasmic Reticulum (SER):*** Primarily this organelle is involved in lipid (fat) and carbohydrate synthesis.
- ***Vacuole:*** Vacuoles are large membrane-bound organelles used for cell storage and transportation of larger particles. They can be formed by merging smaller vesicles and used to transport particles to lysosome cells or the plasma membrane for secretion. Vacuoles are also formed from pinching off the plasma membrane after a pathogen or nutrient cluster is surrounded for phagocytosis. Vacuoles include specific types such as phagosomes and autophagosomes, which are food vacuoles involved in phagocytosis and autophagocytosis (engulfing internal cellular components), respectively.
- ***Vesicle:*** These organelles are fluid (cytoplasm)-filled regions that are smaller than vacuoles (~100 nm) and serve various functions, including transporting and breaking down material. Here are some important vesicles inside a WBC.
    - ***Transport Vesicle:*** Transport vesicles move material between the endoplasmic rectums and Golgi apparatus and within the Golgi apparatus.
    - ***Secretory Vesicle:*** Once particles are ready to be removed from the cell, they are secreted through the plasma membrane via secretory vesicles.
    - ***Endosome:*** These vesicles are formed by pinching off sections of plasma membrane during pinocytosis, where they eventually are either secreted back out of the plasma membrane or fuse with vesicles from the Golgi apparatus to make lysosomes.

- **Lysosome:** Lysosomes utilize hydrolytic enzymes to digest large molecules that can originate within (e.g. malfunctioning organelle) or external to (e.g. bacteria) the cell. Lysosomes are the main 'garbage disposals' of the cell.
- **Peroxisome:** Peroxisomes are responsible for the oxidation of molecules such as fatty acids and amino acids via enzymes. Enzymes also convert toxic hydrogen peroxide byproducts to oxygen and water in order to prevent unwanted reactions with other cells.

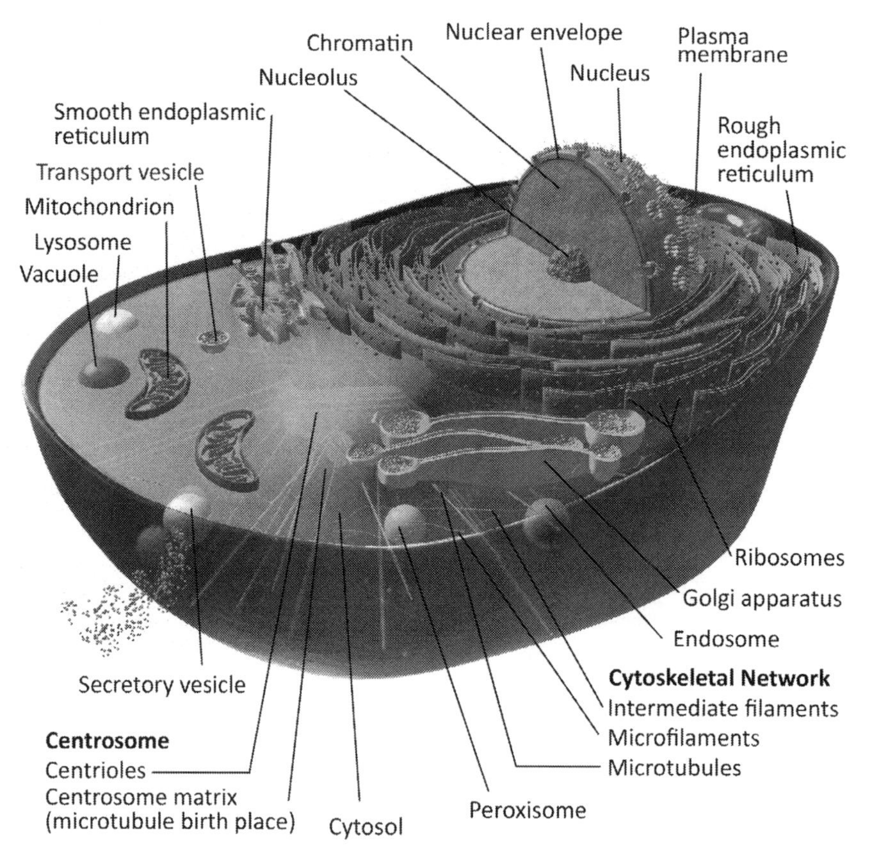

Figure 26: White blood cell constituents.

There are five types of WBCs that aid the body's immune system in fending off unwanted particles, bacteria and viruses: neutrophils, eosinophils, basophils, lymphocytes and monocytes.

Neutrophils

Neutrophils are the first responders in the immune system and comprise 50 to 70% of your WBCs. Neutrophils are phagocytic, meaning they ingest invaders (including bacterial and fungal infections). Pus formed around a wound is filled with neutrophils. These cells have an average lifespan of five days and die after ingesting a few pathogens since they are unable to renew their lysosomes.

Monocytes

Monocytes are the largest of the WBCs at roughly two to three times the size of others. Like neutrophils, monocytes are phagocytic. In addition to destroying invaders directly, these kidney-shaped cells present pathogens to T-cells inside the bloodstream for future recognition. They make up roughly 3 to 8% of the body's overall WBC count. Unlike neutrophils, monocytes can regenerate their lysosomes and therefore have longer lives.

Monocytes eventually move from the bloodstream to tissues, where they are termed macrophages, which fight microorganisms such as cancer cells and remove dead cell tissue/debris. Note that monocytes undergo a transformation and it is only when their macrophages form inside tissues that they perform phagocytosis. Macrophages also play an anti-inflammatory role through the release of cytokines (proteins used for cell signaling).

Basophils

This WBC produces chemicals in response to allergic and antigen reactions. One group of chemicals is histamines, which are organic nitrogenous (containing nitrogen) molecules that aid in the immune response. Specifically, histamines increase capillary permeability to WBCs (most notably neutrophils) and clotting proteins during an invasion. Histamines also dilate blood vessels, making blood flow and WBC transport easier[16].

Heparin is another chemical released by basophil WBCs. This chemical serves as a blood-thinning agent, promoting the flow of WBCs to the infection site. Basophils release chemical signals that recruit other WBCs to the infection site, including neutrophils and eosinophils. At 0.5% of the total WBC count, basophils are the least abundant of the five WBC types.

16 Note: antihistamines are used to reduce the efficacy of histamines, thereby preventing inflammation and a buildup of excessive WBCs due to an allergic reaction.

Mast Cells: Mast cells are similar in structure and function to basophil cells but come from a different branch in the hematopoiesis cycle and act primarily within tissues.

EOSINOPHILS

Although found in blood, eosinophils are predominantly located in the digestive, lower urinary and respiratory tracts and help fight diseases of the spleen and central nervous system. Their primary use is to destroy parasitic organisms such as tapeworms and hookworms inside the body that are too large for WBCs to phagocytize. They do this by secreting chemicals that kill the invaders. Additionally, eosinophils are the chief WBCs involved in inflammation responses to allergies such as asthma, hay fever and hives. This WBC is present inside blood and tissues with an overall WBC constituent percentage in the 2 to 4% range.

LYMPHOCYTES

These cells are more common in the lymphatic system than in the bloodstream and make up 25 to 45% of a human's WBC count. They can be divided into two main subtypes: B-cells and T-cells.

B-Cells: B-cells make antibodies, which are y-shaped proteins that contain a paratope (antibody locking mechanism) that mates up with a unique[17] epitope (antigen locking mechanism). An antigen is any molecule capable of eliciting an immune response. While usually this is a foreign or unwanted molecule produced outside of the body, it can also include molecules produced within the body.

T-Cells: T-cells act as helpers, killers and for memory retention to recognize returning invaders. There are various subcategories of T-cells:

- ***CD4+ T-Cells (Helper & Memory):*** These cells use receptors and CD4 (cluster of differentiation 4) molecules to bind antigenic peptides—short chain amino acids—thereby releasing cytokines. Receptors are protein molecules found on the outside of cells that are used to receive chemical signals that illicit a cellular response. Cytokines help in the recruitment and retention of other WBCs used to fight infections. After an attack, some of these cells stick around and are considered memory cells that recognize antigens when they return.

17 In some exceptions, a paratope can lock up with more than one epitope type.

- ***CD8+ Cytotoxic T-Cells (Killer & Memory):*** Using T-cell receptors (TCRs) along with the glycoprotein CD8 (cluster of differentiation 8), these cells are killer cells and help destroy pathogens such as cancer cells and cells infected with a virus. Various receptors specialize in recognizing specific antigens (over 100 billion). Like helper T-cells, killer T-cells (KT-cells) can turn into memory T-cells that recognize returning invaders.
- ***Gamma-Delta T-Cells (γδ T-Cells):*** While most T-cells consist of one α and one β TCR glycoprotein chain, gamma-delta cells have one γ and one δ receptor. The gamma-delta T-cells are rarer (0.5 to 10% in humans) and are concentrated in gut mucosa, skin, lungs and the uterus. These cells help with immune initiation and propagation.
- ***Regulatory (Suppressor) T-Cells:*** These cells prevent WBCs from attacking the body's healthy cells, which would lead to an autoimmune disease.

Natural Killer Cells: Unlike KT-cells, natural killer cells (NK-cells) do not require antibodies or major histocompatibility complex (MHC) molecules (which display pathogen peptide fragments to T-cells for recognition) to recognize pathogens. This allows for NK-cells to respond much quicker to pathogens than T-cells. Like KT-cells, NK-cells are cytotoxic, meaning they can be toxic to cells. A WBC destroying a pathogen is shown in Figure 27.

Red Blood Cells (RBCs)

RBCs, also referred to as erythrocytes, are primarily responsible for transferring oxygen throughout the body; similarly, they also transport carbon dioxide out of your body via the lungs when you exhale. Their cytoplasm is rich in hemoglobin, an oxygen-rich metalloprotein—a protein that contains a metal ion cofactor—that is required for a protein's biological activity. Note that an RBC's cytoplasm consists only of cytosol while WBCs have both cytosol and organelles within the cytoplasm. RBCs also do not contain a nucleus, which means they do not contain the DNA required for cell division and must therefore be replenished solely by production within bone marrow. However, this also makes them immune from invading viruses, which are unable to hijack the cell's nucleus for replication. Since RBCs do not contain organelles, this prevents them from being able to repair themselves. RBCs survive around 120 days.

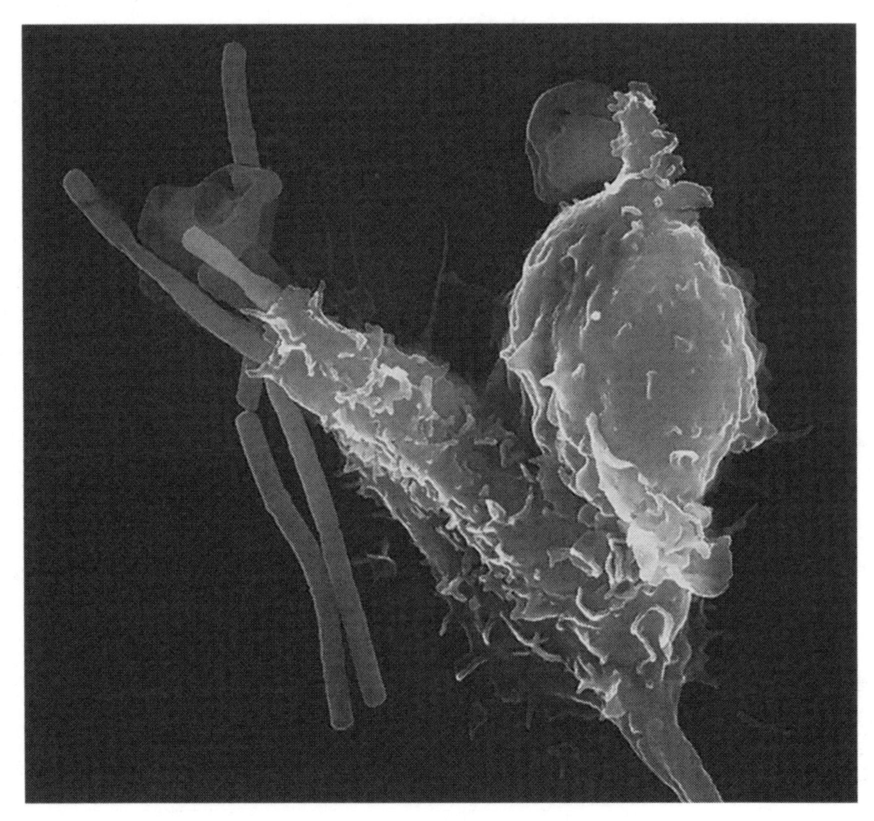

Figure 27: A neutrophil engulfing a *Bacillus anthracis* bacteria (anthrax) pathogen.[18]

Platelets

Also referred to as thrombocytes, the main purpose of platelet cells is to provide blood clotting during injuries in a process called hemostasis. This is done in a three-step process:

1. ***Adhesion:*** Platelets are recruited to the wound location and adhere to the endothelium, a tissue layer of endothelial cells that line the inside of a blood vessel, shown in Figure 28, in order to stop the

18 Volker Brinkmann, "Neutrophil Engulfing Bacillus Anthracis," *PLOS Pathogens* 1, no. 3 (Nov. 2005), https://www.ncbi.nlm.nih.gov/pmc/articles/PMC1283252.

bleeding. Endothelium tissue is a subset of epithelium tissue, which is mentioned in the MRI discussion as one of four human tissue types.

2. ***Activation:*** Chemical messages are sent out via platelet receptors to recruit additional platelets.
3. ***Aggregation:*** Platelets connect to one another by linking their receptors to close a wound if the opening is small enough to close with platelets alone.

Platelets are approximately 20% the size of RBCs and are present in between a 1:10 and 1:20 ratio of platelets to RBCs for healthy humans. Unlike RBCs, platelets contain various cellular organelle similar to WBCs but without a DNA-producing nucleus. Since no nucleus exists, platelets are like RBCs in that they cannot reproduce and rely solely on production via bone marrow.

A cartoon depicting the various blood cells types is shown in Figure 29. Figure 30 shows an image of blood cells to get a feel for what they look like and relative sizes.

Blood Plasma

Blood plasma is the fluid inside veins, arteries and capillaries where blood cells are suspended. In addition to cells, the plasma itself, which has a yellowish hue, contains proteins, CO_2, hormones, electrolytes, glucose and clotting factors. Electrolytes are molecules that, when absorbed in a polar solvent such as water, are able to conduct current. These electrolytes—found in sports drinks for rehydration purposes—include bicarbonate (HCO3-), calcium (Ca++), chloride (Cl-), magnesium (Mg++), potassium (K+), phosphate (HPO4) and sodium (Na+).

Approximately 92–95% by volume of blood plasma is water, with protein making up the other significant portion (5–8%). Some prominent families of protein include:

- ***Serum Albumin:*** This protein regulates osmotic pressure, i.e. the minimum pressure required to prevent water osmosis through tissue.
- ***Fibrinogen:*** The main role of fibrinogen is to help with blood clotting.

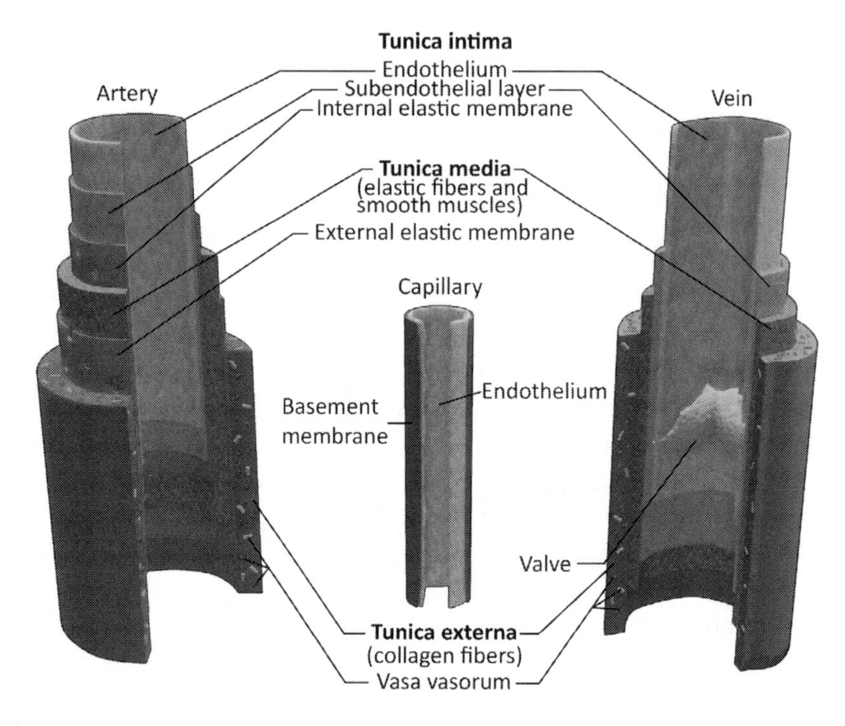

Figure 28: Cross-section of blood vessels showing the location of endothelial cells where platelets attach during blood clotting.

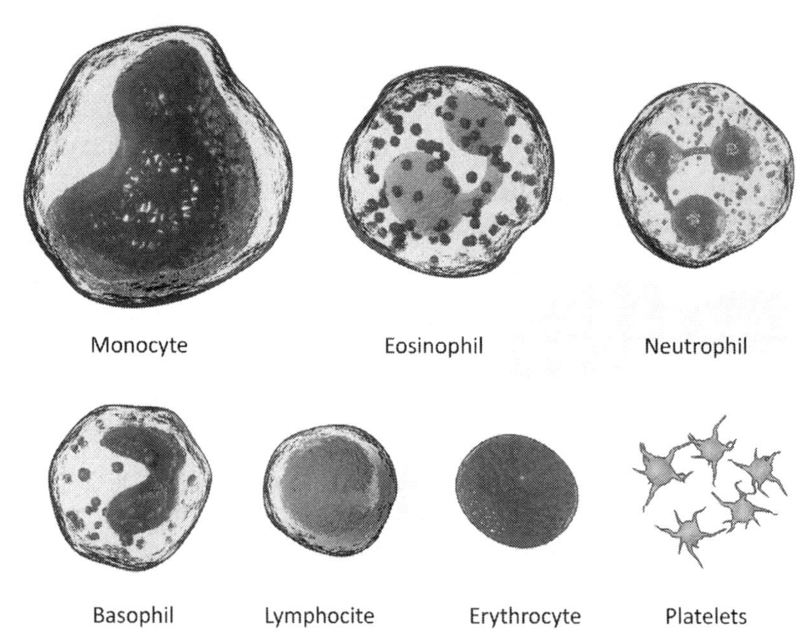

Figure 29: Comparison of white blood cells, red blood cells and platelets (sizes are to scale).

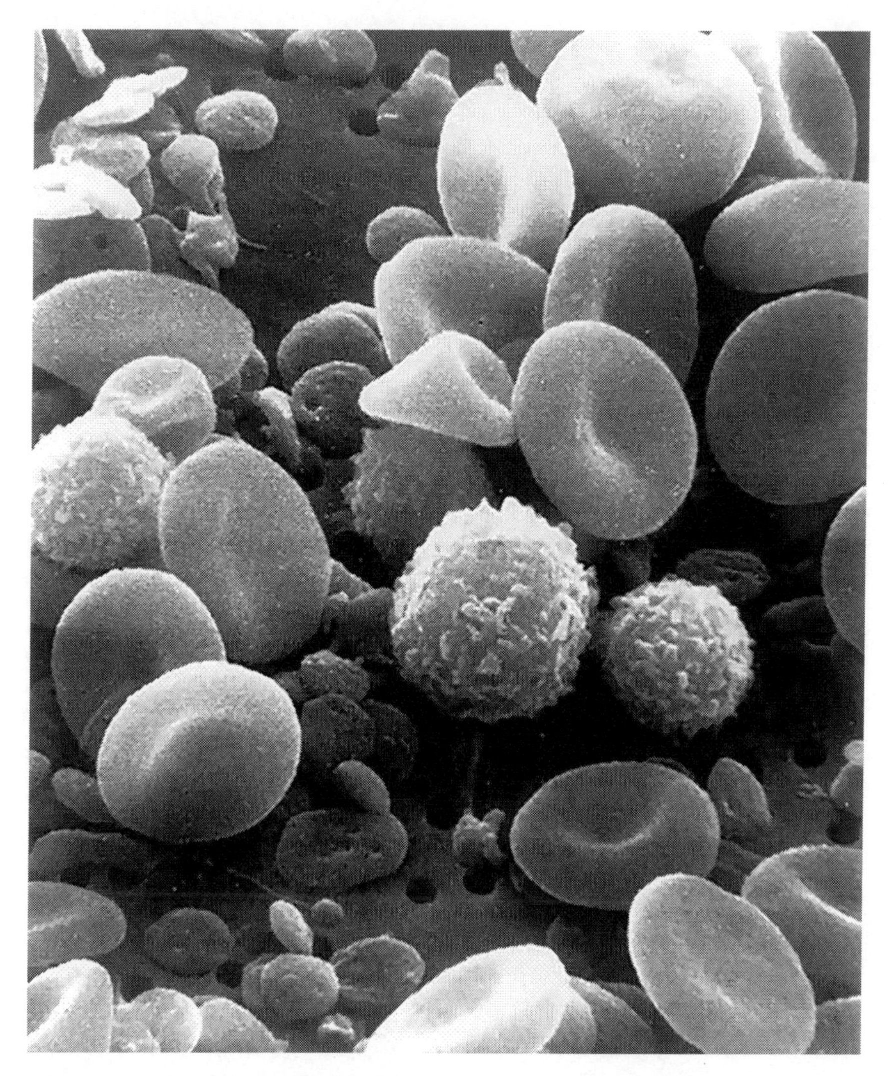

Figure 30: Scanning electron microscope image showing a cluster of white, red and platelet blood cells suspended in plasma.[19]

- ***Globulin***: This protein has various messenger and regulatory roles. Globulin can act as an enzyme by catalyzing organic reactions, a messenger via hormones (e.g. insulin), a transporter of molecules through membranes, a structural protein and an amino acid reserve. Additionally, globulin helps regulate cell motility, division, wall structure and signaling.

19 Bruce Wetzel and Harry Schaefer, "Electron Microscope Image of Blood," *National Cancer Institute*, https://www.cancer.gov.

LYMPHATIC SYSTEM

The lymphatic system is a vital portion of the immune defense network and, along with your cardiovascular system, makes up the circulatory system. Your body can be broken up into two main categories of fluids, i.e. intracellular fluid (ICF) and extracellular fluid (ECF), with the former being fluid inside cell walls and the latter being fluid outside cell walls. ECF can further be broken down into intravascular fluid (IVF) and interstitial fluid (IF), with IVF located inside blood vessel walls (i.e. blood plasma) and IF located outside blood vessel walls, in the region between blood vessels and cell tissue where connective tissue is located.

Water makes up roughly 60% of a human's mass, with two-thirds being ICF and one-third being ECF. Segregating ECF further, three-fourths of it is IF while one-fourth is IVF. Doing the math leads to a total body water breakdown of eight-twelfths ICF, three-twelfths IF and one-twelfth IVF. For some numbers, a 220 lbm male would have on average 88 lbm of solid tissue and 132 lbm of water.[20] Of the latter, 88 lbm is ICF, 33 lbm is IF and 11 lbm is IVF. Put into liquid measurements, that is 39.6 L (10.5 gal) of ICF, 14.9 L (3.9 gal) of IF and 5.0 L (1.3 gal) of IVF. Note that IVF is for the plasma only and recall that blood is approximately 55% plasma and 45% blood cells, the latter of which contribute to ICF. For a 220 lbm man, this yields a mass and blood volume of approximately 20 lbm and 8.99 L (2.4 gal), respectively.[21] Sources vary on these breakdowns, with one quoting that 20% of ECF is plasma, which for our 220 lbm male yields a lower blood volume value of 16 lbm and 7.2 L (1.9 gal). The amount of blood as a percentage of total body mass is approximately 7 to 9% (depending on the source), which yields a blood quantity range of 15.4 to 19.8 lbm (6.9 to 8.9 L, or 1.8 to 2.4 gal) for a 220 lbm male.

Blood is constantly being circulated throughout the body to transfer oxygen, which requires pressure created by your beating heart. As blood

20 A pound mass, abbreviated lbm, is distinguishable from a pound force, or lbf. Note 1 lbm = 1 lbf at Earth's surface and since the two terms are indistinguishable under this condition, the more common pound or lb abbreviation is often used. On the Moon, for example, a 180 lbm astronaut would weigh approximately 29.8 lbf due to the lower gravity, versus 180 lbf on the Earth. The SI unit of kilogram has a conversion of 1 kg = 2.205 lbm.

21 Emma Jakoi, "Homeostasis and Fluid Compartments," *Duke University*, https://www.coursera.org/lecture/physiology/homeostasis-and-fluid-compartments-1dqad.

circulates, this pressure causes some blood to escape via capillary blood vessels and enter regions of your body outside of the cardiovascular system, thereby becoming IF. IF makes its way into the lymphatic system by way of lymphatic capillaries, as displayed in Figure 31. Of this liquid, approximately 85 to 90% is directed back into the bloodstream. The other 10 to 15% stays in the lymphatic system as a fluid referred to as lymph. While the cardiovascular system completes a full circulation to and from the heart, the lymphatic system only flows in one direction toward the neck. Without the heart as a pump, this is done by a system of muscles and joint pumps coupled with lymph vessel valves that prevent backflow, as shown in Figure 32.

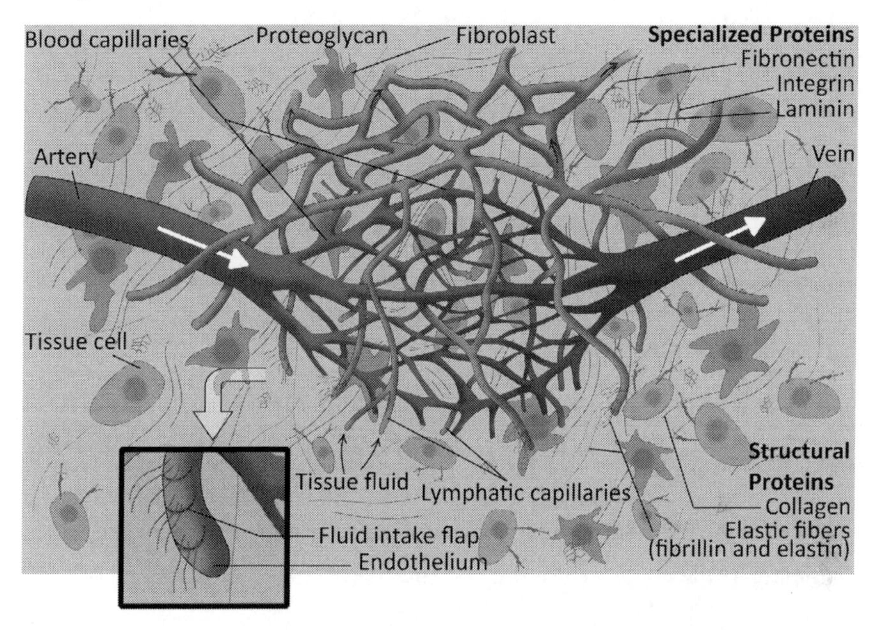

Figure 31: Blood/lymph exchange via capillaries.

Blood filtration through capillaries is driven by the Starling equation, which relates internal capillary blood pressure to IF pressure in order to govern which way fluid flows. Capillary hydrostatic pressure and IF oncotic pressure contribute to fluid seeking to exit the capillaries (i.e. a positive net filtration rate) while IF hydrostatic pressure and plasma protein oncotic pressure contribute to IF wanting to re-enter the bloodstream (i.e. a negative net filtration rate). At the arteriole end of capillaries, the net pressure is around 9 mmHg, causing fluid to pass from the bloodstream into IF surrounding cell tissue.

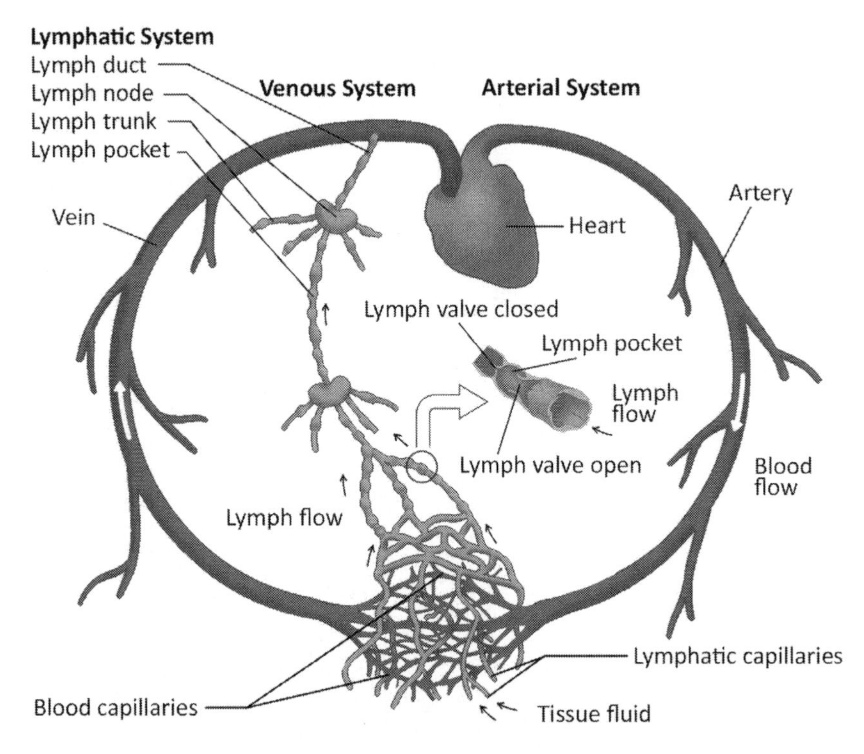

Figure 32: Lymphatic system interaction with circulatory network.

Conversely at the venule end, the net pressure is approximately -8 mmHg, causing IF to re-enter the bloodstream. Note that the difference in pressures creates excess IF fluid, which forms the approximately 10 to 15% lymph fluid mentioned above. This capillary filtration process is depicted in Figure 31, showing the interchange between blood, lymph, tissue cells, blood capillaries and lymph capillaries.

Various organs including the tonsils, thymus gland, spleen and lymph nodes—the lymph glands—are scattered throughout the lymphatic system in Figure 33, where lymph passes through on its journey to the neck. Lymph travels via lymphatic vessels, which flow to lymphatic organs housing lymphocyte WBCs (B, T and NK cells) as well as macrophages. Recall that lymphocytes identify and kill pathogens while macrophages break down large pathogens and remove their waste. Although lymphocytes are created in bone marrow, they mature in the lymphatic system's organs.

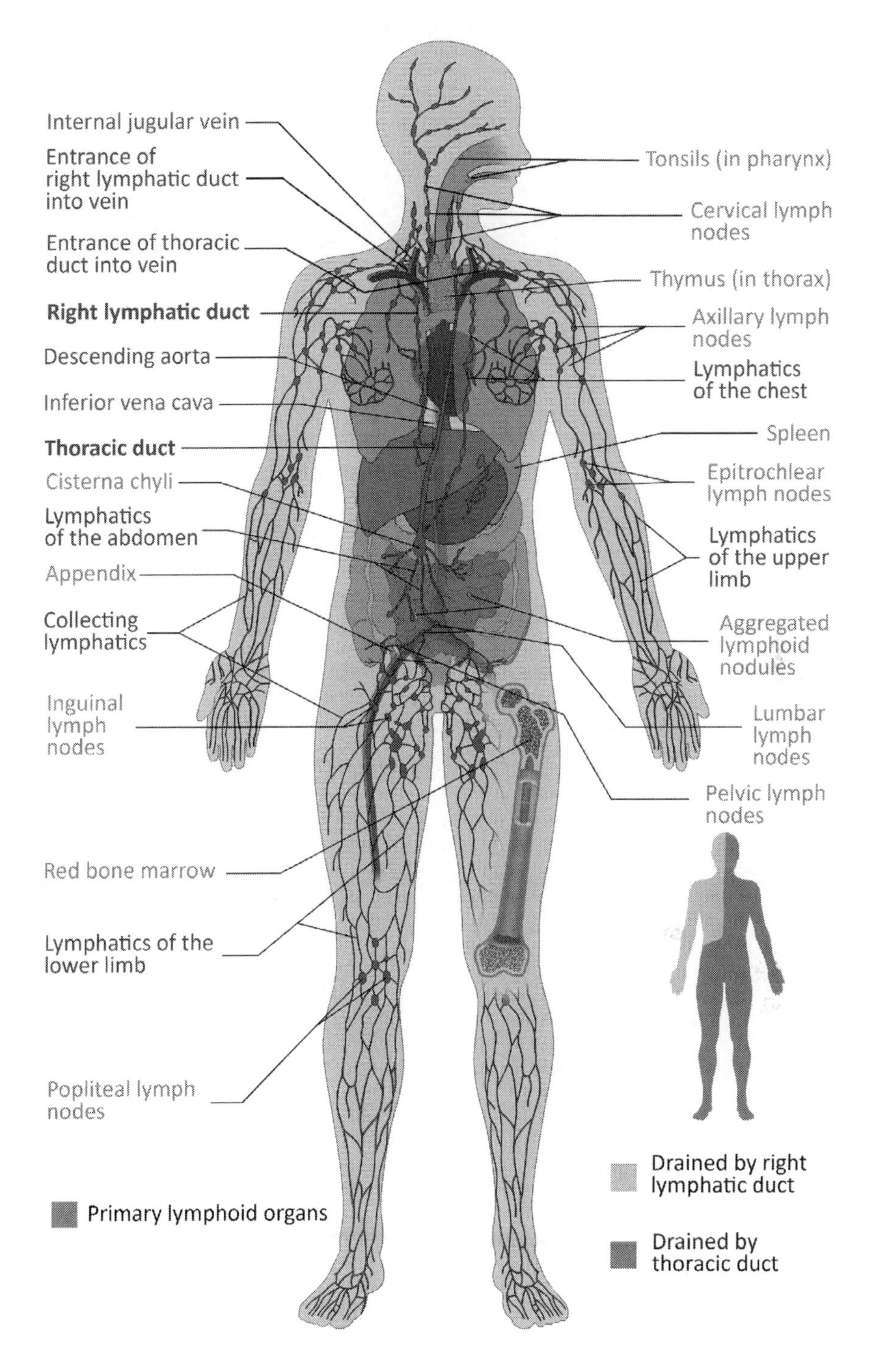

Figure 33: An overview of the human lymphatic system.

Specifically, T-cells mature in the thymus, which is why 'T' is used in the naming of these cells. Recent studies have shown that T-cells can mature in the tonsils as well. B-cells and NK-cells mature primarily in the tonsils, spleen and lymph nodes. Macrophages come from monocytes, which are also produced in bone marrow and travel to cell tissue where they become macrophages.

The lymphatic system can be divided into primary and secondary tissues.

Primary Tissue

Bone Marrow: There are two types of bone marrow: red and yellow, as shown in Figure 34. The former is responsible for RBC, platelet and most WBC production while the latter produces the remaining WBCs. At birth, your marrow is primarily red; however, yellow marrow increases with age until around a 50–50 split is achieved for adults. Red bone marrow is present primarily in flat bones such as the pelvis and skull, while yellow bone marrow is found mostly in large, hollow bones such as those of the arms and legs.

Thymus: Lymphocyte T-cell maturation occurs in the thymus, which is located right above the heart. These are the helper, memory and killer cells vital to the immune system that were discussed earlier. Once matured in the thymus, T-cells get sent throughout the lymphatic system to identify and destroy invaders.

Secondary Tissue

Spleen: In addition to maturing B and NK-cells, the spleen houses more than 50% of the body's monocytes. These are sent out to various cell tissues throughout the body, where they are converted to dendritic (antigen-presenting) cells and macrophages. The spleen also serves as a blood filter, removing dead RBCs while holding a reserve of new blood cells. Consisting of white and red pulp, the spleen removes antibody-coated bacteria found in blood and lymph. Red pulp is where the monocytes are located.

Tonsils: B-cell and NK-cells mature here and recent research has shown T-cells do as well. Located near the throat, tonsils are the first line of defense against invaders entering the oral or nasal cavities—for instance, a common cold. Tonsils swell up with blood during an immune response and are coated with M-cells, which are a class of epithelial cells that alert B and T-cells in the tonsils to the presence of pathogens.

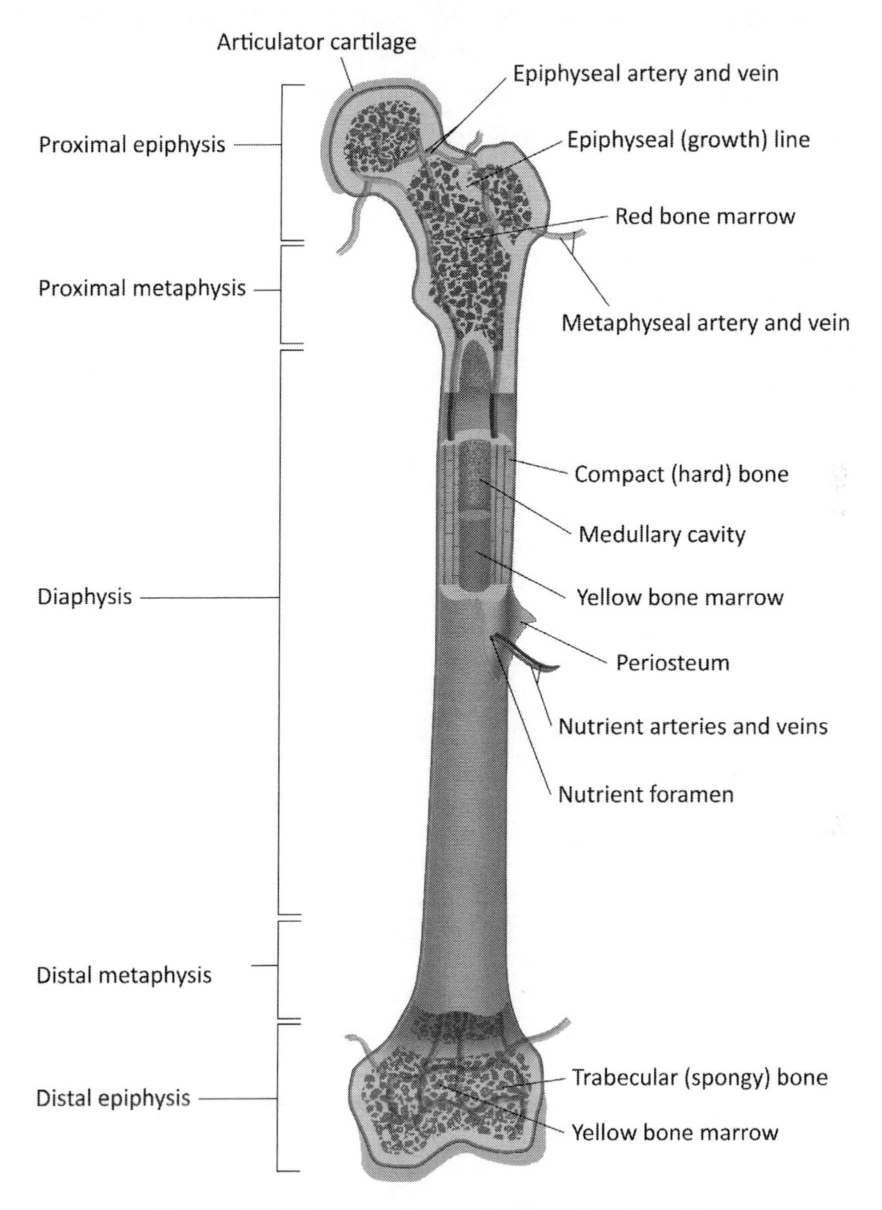

Figure 34: Femur schematic showing location of yellow and red bone marrow.

Lymph Nodes: These nodes serve to filter lymph fluid as T-cells and macrophages remove pathogens, dead cells and waste that have collected upstream. Hundreds of lymph nodes are located throughout the body to maximize its protection. When the body is fighting an infection your lymph nodes swell up, which is why doctors examine them—primarily those located in your neck—when you are sick. Details of the lymph network, how it interacts with the circulatory system, as well as the location of lymph nodes and lymphoid organs, is shown in Figures 31 through 33.

Mucosa-Associated Lymphoid Tissue (MALT): This tissue is found in submucosal membranes such as the breasts, eyes, gastrointestinal tract, lungs, salivary glands, skin and thyroid. B-cells and T-cells are present to remove pathogens, as well as M-cells in MALT located in the gastrointestinal tract.

Chapter 6

Fertility

Infertility is a possibility for some types of cancers and more likely for various types of cancer treatments. This can be an emotional topic, but if you are planning to have children it is important to understand the potential impact of your cancer and treatment on fertility and the steps that can be taken to increase your chances of having a child post-cancer.

Chemotherapy

Chemotherapy is the most likely cause of infertility in both men and women. In men, the potent agents will kill sperm since chemotherapy agents attack rapidly dividing cells, which includes sperm cells. This can be temporary unless the spermatogonial stem cells in the testes are damaged during treatment and they permanently can no longer produce healthy sperm cells. Damage to these stem cells will prevent them from dividing as part of the spermatocytogenesis process, which forms spermatocytes, followed by spermatids and finally spermatozoons, or motile sperm cells. Being treated at a younger age increases a man's chances of recovering to a normal level of fertility post-treatment.

For women, chemotherapy can damage their eggs. Although the historical belief has been that women produce all their eggs at birth—in excess of a million, dwindling to a few hundred thousand by puberty—new stem cell research suggests that women may continue producing eggs throughout their life. While this research is still ongoing, types of chemotherapy have been shown to cause infertility, whether by damaging the eggs themselves or the stem cells that make the eggs. As with men, having chemotherapy at a younger age decreases the odds of becoming sterile. Chemotherapy can also lead to premature menopause in women.

Women should not get pregnant during chemotherapy as the drugs can harm the fetus. Furthermore, women should wait six months after chemotherapy before conceiving. Impacts of chemotherapy can linger and there may be an increased risk of a miscarriage or genetic defect in the child. If already pregnant, surgical options instead of chemotherapy should be considered first as they will have the least impact on the fetus.

Chemotherapy during the first three months of fetus growth inside the womb is particularly advised against, as the baby's organs are still growing; however chemotherapy during all phases of pregnancy should be avoided. Make sure you find an oncologist who specializes in treating cancers in pregnant women before deciding on the best treatment options.

Drugs that increase the risk of infertility in both sexes (with generic names followed by trade names in parentheses) include:[22]

- Busulfan (Busulfex)
- Carboplatin (Paraplatin)
- Carmustine (BiCNU)
- Chlorambucil (Leukeran)
- Cisplatin (Platinol)
- Cyclophosphamide (Cytoxan)
- Ifosfamide (Ifex)
- Lomustine (Gleostine)
- Mechlorethamine (Mustargen)
- Melphalan (Alkeran)
- Procarbazine (Matulane)

Additionally, there are increased risks for men with the following drugs:

- Cytarabine (Cytosar-U)
- Dactinomycin (Cosmegen)

There are also increased risks for women with the following drugs:

- Dacarbazine (DTIC-Dome)
- Doxorubicin (Adriamycin)
- Temozolomide (Temodar)

22 Jakoi, "Homeostasis and Fluid Compartments."

Some common chemotherapy drugs that have a low likelihood of causing infertility in men and women include:

- Bleomycin (Blenoxane)
- Cytarabine (Cytosar-U)
- Daunorubicin (Cerubidine)
- Fludarabine (Fludara)
- Fluorouracil (Aducil)
- Methotrexate (Otrexup)
- Vinblastine (Alkaban-AQ)
- Vincristine (Oncovin)

Additional common drugs that have a low risk of infertility for men only include:

- 6-Mercaptopurine (Purinethol)
- Dacarbazine (DTIC-Dome)
- Doxorubicin (Adriamycin)
- Epirubicin (Ellence)
- Etoposide (Toposar)
- Mitoxantrone (Novantrone)
- Thioguanine (Tabloid)
- Thiotepa (Thioplex)

Common drugs with a low risk of infertility for women only include:

- Dactinomycin (Cosmegen)
- Gemcitabine (Gemzar)
- Idarubicin (Idarubicin)

These lists include some of the more common chemotherapy agents used. If your oncologist has prescribed a drug not on this list, consult your oncologist and additional literature about the likelihood of infertility.

Aside from chemotherapy, other cancer treatments can also cause infertility. These include radiation treatment, bone marrow or stem cell transplants, targeted immune therapies, surgery and hormone therapies.

Radiation

High-energy radiation purposely damages cancer cells and can unfortunately harm healthy cells, including cells involved in reproduction. In men, radiation on or near the testes can damage stem cells involved in sperm production and lead to permanent infertility. Testicular cancer and childhood leukemia are two cancers where high radiation to the testes may be involved. Nearby radiation to the abdomen is also a concern due to internal radiation scatter impacting the testes.

Radiation on or near the brain can impact the pituitary gland, which signals the testes to produce hormones for sperm production. Therefore radiation to this region of the body may impact fertility. For men with prostate cancer, brachytherapy is the preferred method for protecting fertility, as the implant reduces the amount of radiation exposure to the testes compared with external radiation.

For women, radiation near the uterus, ovaries and vagina pose the highest risks of impacting fertility. In addition to damaging the eggs themselves, the uterus can be constrained—not allowed to expand fully—during pregnancy, which can lead to a miscarriage, premature birth or low birth weight. Rays in nearby parts of the body such as the abdomen can be deflected and harm eggs as well, so be sure to discuss with your radiation oncologist how to minimize radiation to your female reproductive organs.

Brain radiation, specifically to the pituitary gland that controls ovary hormone release, can impact a woman's egg release from her ovaries. If pregnant, radiation near the fetus is a significant risk and should be avoided. Surgical options are the safest cancer treatment when a fetus is involved.

Surgery

Although not always an option for all cancer diagnoses (e.g. if the cancer has metastasized), surgery is generally the best option in terms of fertility. However, that should be a secondary consideration; the first consideration must be your health and choosing the best treatment to defeat your cancer. The exception is surgery near the reproductive organs. In men, this includes removing a testicle, or orchiectomy. Note that a bilateral orchiectomy, which is the removal of both testicles, is required in less than 5% of cases.

Prostate cancers that have spread to the testes may also require removal of one or both testes. If the prostate itself is cancerous, the prostate gland and seminal vesicles may require removal through a radical prostatectomy. This will leave the patient sterile, as he will no longer be able to produce semen. An additional potential side effect is the inability to have an erection due to nerve damage from surgery.

A cystectomy removes the bladder in addition to the prostate gland and seminal vesicles. Sperm can still be produced and men can have the feeling of an orgasm, yet since there is no longer a pathway for the sperm to reach the urinary tube, semen will not be produced. However if the sperm is still being produced but you lack semen, there are methods of harvesting the sperm to fertilize an egg in either a radical prostatectomy or cystectomy.

An additional surgery that can impact male fertility is removing lymph nodes in the pelvis, which may be required to treat testicular or colon cancers. This can also cause nerve damage that can impact erections and ejaculation.

In women, a hysterectomy—the removal of the uterus—is required for certain cancer treatments, leaving her unable to carry a child. An oophorectomy removes a woman's ovaries and accompanying eggs. This may be required for ovarian or cervical cancer. If one of the ovaries can be saved, this will allow the woman to become pregnant while also delaying the onset of menopause.

Another possible surgery for women with cervical cancer is a trachelectomy, where the cervix is removed but not the uterus, allowing the woman to bear children. Depending on the specific surgery, scarring may occur on the fallopian tubes, which can prevent eggs from properly being deployed and interacting with sperm.

ADDITIONAL TREATMENTS

Fertility can be impacted by additional treatments including bone marrow and stem cell transplants, hormone therapy and targeted immune therapies. Bone marrow and stem cell transplants require high doses of chemotherapy and sometimes radiation to the entire body before the transplant, which often leads to infertility in men and women. Hormone therapy can harm a woman's ability to release eggs as well as a man's ability to release sperm, both potentially leading to infertility.

Immune targeted therapies utilize the body's natural immune defense system to attack cancer cells. These are relatively new in usage; therefore, their impact on fertility is not well known. Discuss with your oncologist the potential impacts on infertility if he or she recommends these types of drugs as part of your treatment plan.

Cryogenic Bank

Going to a cryogenic bank is recommended prior to starting cancer treatment if you plan on having children. I must admit, showing up at a fertility clinic to provide a sample can be somewhat depressing and was not what I envisioned myself doing in my twenties. But you need to tell yourself that it is a smart decision. So I asked the appropriate questions about how much it costs, what the process was and other concerns.

I paid around $400 to store a sample and they wanted an additional $400 per year to maintain the specimen. These prices are like what I found searching the website as of 2020, i.e. $460 for a specimen and $395 for annual storage from one website.

For women, the cost of freezing their eggs is significantly higher. Although prices will vary, as of 2020 one clinic charged $10K to harvest and store eggs for a year, followed by a $500 annual storage fee and another $5K to fertilize and transfer to the uterus when the mother is ready. Insurance does not cover the costs so unfortunately this option will not be available for everyone due to its high cost.

Spending time discussing fertility with your oncologist is recommended to see if there are treatment options that will reduce the chances of infertility. For example, certain chemotherapy drugs are in a similar family as others in terms of their treatment effect but are less likely to damage reproductive cells.

My experience of storing a sample was not very pleasant because of the emotional and personal thoughts associated with the possibility of not being able to have kids. After getting diagnosed with cancer at such a young age, infertility was one more thing that I did not want to worry about.

After my consultation visit, I showed up to provide my sample; they give you a cup and send you into a room. They advise no ejaculations for a few days prior to your appointment in order to increase the sample size. Once the sample is received, they test the motility (how mobile your sperm is) of your sperm as well as count. Thankfully all my readings were healthy.

A few months after my treatment ended, I decided to have my sperm retested to see if I could avoid having to pay the yearly storage fees. Since I did have chemotherapy agents that put me at risk for infertility (specifically methotrexate), finding out sooner rather than later was preferred. Although my sperm motility was not quite as high as the pre-cancer sample, the overall motility and quantity were considered healthy.

What a big relief! Based on this result I decided not to store my sample. There were no near-term plans to have children and since my sperm appeared healthy, paying annual storage fees and having the constant reminder was not something I wanted to do.

How you deal with whether you want to go through this process and how long to keep your sample is a personal decision. However ensure that you understand all the pros and cons associated with your treatment before deciding. I encourage you to err on the side of caution, especially men, with the price being more affordable and the process less complicated.

Chapter 7

Caretakers

Having a strong support network is instrumental in your battle against cancer. Each person will have their own comfort zone in terms of whom to lean on and how many people they want around them to help.

For me there was a balance between wanting support and needing space. Throughout the initial treatment months, I shared my diagnosis with my family and a few close friends. I intentionally asked my work manager to keep the diagnosis discreet until I was willing to share, which I did later in my treatment.

Once I extended my experience to friends and co-workers, their support lifted my spirits and helped with my fight. My brother-in-law, Nick, ran Relay for Life and lit a candle on my behalf as he raised thousands of dollars to help fight cancer. Co-workers pitched in to buy me some Wii games since I had a lot of time at home. My friend Jack organized a lunch with some of my close friends to show support. All of this meant a lot and was important for my emotional state.

There are different types of caretakers and you need to build a team around you with various strengths. For instance, I primarily leaned on my brother to discuss the more medical aspects of the treatments when I needed to make decisions as he was a good sounding board and read through some of the material with me. He had a similar mindset as me, feeling learning about the cancer was empowering.

My dad was great to lighten the mood, so to speak, trying to take my mind off the cancer, which was difficult. His sense of humor definitely helped. Mom and my sister also flew out and provided loving support throughout the process that helped when I was down in the dumps.

My girlfriend Moy was my main caretaker and I would joke to her about going to a caretaker burnout support group. While I made light of this, there are support groups for caretakers since burnout is a real issue that can occur. There is a multitude of emotional stress and energy shared by the patient's support group that they should be prepared for. Moy went to every chemotherapy session I had and would bring me food when I was able to eat, spending many hours with me at the doctor's office. She also went to all my doctor's appointments to discuss treatment and was there for my many tests.

At home she prepared food and took pictures of my body, as I wanted to track the treatment impact. The pills alone turned out to require a lot of work that I did primarily myself, yet this is also an area where caretakers can help. Not all patients are willing—or able, depending on their condition—to spend the time and rigor to organize and stay on top of the gaggle of pills required for treatment, which is discussed in detail in the *Chemotherapy and Pills* chapter. Moy would pick up specific food and drinks meant to help combat the cancer and was a constant throughout the process, for which I am very grateful.

Make sure you put together a diverse team of doctors and family members to help you through this trying stage in your life. Consider who you want close to you and what strengths they bring. I recommend one person to rely on as your main caretaker, someone who will be with you throughout the process, going with you to appointments and treatments, providing transportation—you will not be able to drive after certain procedures—shopping for healthy food and helping with pill organization and reminders.

My family lives across the country and they were able to fly out multiple times to help me; however, Moy lived with me and was able to be that one consistent person throughout the process. In addition to that main person, find an individual who will research cancer information with you and help make treatment decisions. For me this was my brother and it was a tremendous help. Aside from these two individuals, surround yourself with family and friends to provide emotional support. Sharing your experience with them and receiving their love and encouragement will do wonders in your battle to beat cancer!

Chapter 8

Be Your Own Doctor

Cancer treatment is a process that you will have to treat like any other important challenge in your life, whether it be getting a degree, landing a job, finding a life partner, having kids, meeting fitness goals, achieving successes in sports, etc. Only in this case the stakes are the highest they have ever been as there is nothing more important than your health.

With any goal in life, you need confidence, calmness, concentration and perseverance, four attributes I learned from my high school calculus teacher Mr. Lawrence and keep with me to this day. Additionally, you need hope and visualization. Treat your cancer like you would any other important goal and work hard at it and plan accordingly. For me, this meant picking up every book or pamphlet I could get my hands on to educate myself about my illness. It meant researching what treatments were available and which were likely to be most effective for my condition. It meant making smart decisions about what I would eat, how I would exercise, which facilities to go to, who I wanted on my support team and how I would overcome my disease.

You are your biggest advocate. Now is the time to be selfish and think of yourself and your health, putting all your energy into getting healthy. Doctors have many patients and as much as they want to help you, they just do not have the time or resources to provide as much attention to you as you and your support network do.

Doctors are also human and make mistakes. It took me growing up and becoming educated to realize this. When you are young you look at doctors as these omnipotent humans who can do no wrong when in fact, yes, they can! They make mistakes. Hopefully not severe or they will lose their medical license, but just like in any profession there are more and less capable doctors and even the best make mistakes. With me it started from the beginning when I was not properly diagnosed after experiencing my first lymphoma symptoms. Then it continued with a blood clot in my arm and not being on blood thinners. I do not blame doctors, as they are intelligent people trying to help you, so that is not the premise. Rather you must not sit back and rely solely on doctors to make you better. You need to take ownership of your health and treatment more than anyone.

Arrange your pills and take them as directed. Ask lots of questions of your oncology team and ensure you—or someone from your support team—understand treatment options and side effects. Eat healthfully and know how your diet impacts your recovery. Go for second opinions if you are not sure about a treatment option. Report side effects to your oncology team. In short, be your own doctor; you are the most important person in this process!

Chapter 9

Chemotherapy and Pills

Although there are many promising studies and billions of dollars spent annually searching for a cure, unfortunately there is no cure for cancer as of this writing. Cancer treatment involves various forms including surgery, chemotherapy, radiation or a combination of the three. Other less common treatments are also available, including ablation and embolization. The former involves heating or freezing tumors to try and kill them while the latter attempts to block blood flow to the cancer.

Which type of treatment you receive depends on what part of the body your cancer originated in and its stage. Even within a certain cancer type there are a host of options that can be very specific to your diagnosis, which is why it is so important to choose a strong oncology team. For example, breast cancer can involve all three types of treatment depending on the size of the tumor and how much the cancer has spread. Since my Hodgkin's lymphoma was stage IV and had spread to my groin and neck, chemotherapy was used followed by radiation on the large tumor. Unless the tumor is localized, chemotherapy is the first form of treatment for all cancers.

Understanding the drugs going into your body and their side effects is important. While many patients and their caretakers may not realize it, there are usually options when it comes to what type of chemotherapy program you undertake. Oncologists may or may not explain or volunteer that information—mine did not. For example, Hodgkin's lymphoma has 33 different chemotherapy drugs approved for treatment and 14 individual plans for how to use them!

I highly recommend patients and their caretakers study the various types of chemotherapy available and read the statistics and side effects for each one. While one would hope that there was a singular approved approach for a given cancer, unfortunately any type of drug used for treatment always has pros and cons. Just as one might compare over-the-counter pain or allergy medications, one should treat their chemotherapy selection the same way, albeit on a much more important scale. I was unable to find a singular resource that had all the data available with success rates for various chemotherapy treatment plans and therefore had to do a significant amount of research in a short amount of time.

Providing that research for all types of cancers is outside the scope of this book, but I have listed all of the approved chemotherapy drugs for each cancer type along with drug combinations in Appendix B and have pointed readers to sources on success and side effect statistics for these treatments. Of course your oncology team will recommend either one or a few different options, but once again you need to be your own doctor and make sure that the final chosen treatment is the best for you.

I can definitively say that the extra research I took on saved three months of treatment time and statistically improved my chances of beating cancer. My oncologist recommended ABVD, the standard treatment for Hodgkin's lymphoma. It consists of concurrent treatment with the chemotherapy drugs doxorubicin (Adriamycin), bleomycin (Blenoxane), vinblastine (Alkaban-AQ) and dacarbazine (DTIC-Dome)[23]. However, upon reading *Living with Lymphoma* and doing independent research, I realized there were numerous chemotherapy treatments available and wanted to make sure I chose the best one. I was pleasantly surprised that when I brought this up with my oncologist, he agreed and left the decision up to me, printing out numerous studies at my request that I could take home and read to compare their efficacy.

My research concluded that one form of treatment, Stanford V, was relatively new and more aggressive than ABVD. Rather than the same

23 Note: ABVD uses the trade name of doxorubicin (i.e. Adriamycin) along with the generic names of the other three drugs to form the ABVD acronym.

four drugs every other week, Stanford V would rotate various drugs and provide treatment every week, effectively cutting my treatment time in half. Furthermore, Stanford V had a slightly better success rate than ABVD. The only drawback to Stanford V was that it was a more aggressive treatment program, meaning less time to recover between treatments and more stress on you both physically and mentally. Stanford V includes six chemotherapy agents and one corticosteroid: doxorubicin (Adriamycin), bleomycin (Blenoxane), vinblastine (Alkaban-AQ), mechlormethine (Mustargen), vincristine (Oncovin), etoposide (Toposar) and prednisone (Deltasone).

Being 28 and in great shape, along with possessing a fighter's spirit, I knew Stanford V was right for me. Shaving three months off my therapy was important to me both professionally (less time away from my career) and physically (as an athlete who enjoys competition). Now this decision will be different for every patient depending on their diagnosis, age, personality and other factors. For instance, an elderly patient who is retired has different considerations. Perhaps they would prefer a less taxing but prolonged treatment that would be easier on their body. In addition, if they are on a retirement pension or living off a 401K, there may be less urgency for them as opposed to a working professional. All these considerations and more should be analyzed when deciding your best treatment options while discussing them with your oncologist.

Another eye-opening part of the chemotherapy process was realizing that it effectively poisons your body and kills healthy cells along with cancerous cells. Research will eventually provide breakthrough treatments that target cancer cells only, but the present course of treatment carries numerous unwanted side effects. I was surprised and somewhat morbidly amused to read that one of the drugs I was taking was a derivative of the mustard gas used in World War I! That really hit home with just how nasty these drugs are; I was knowingly taking a drug used in chemical warfare to try and kill my cancer cells.

Chemotherapy Categories

Chemotherapy drugs can be divided into various categories and subcategories based on what or how they are derived.[24] For instance, many come from plants while a large portion are made synthetically in a laboratory.

Alkylating Agents

These drugs were among the first developed to treat cancers in the 1940s and work by inhibiting the DNA inside cancer cells from reproducing. This is done by adding an alkyl group to the DNA strand to confuse the DNA and prevent it from linking up properly during reproduction. Cancer cell DNA is delicate, which is why this method of chemical reaction is effective.

Alkylating agents are particularly successful against tumors and slow-growing cancers including bladder cancer, brain cancer, breast cancer, leukemia, lung cancer, lymphoma, melanoma, myelomas, neuroblastoma, ovarian cancer, retinoblastoma, sarcomas and testicular cancer. They are also attractive since they attack cancer cells in all phases of cell reproduction. Due to the adverse effect on bone marrow, prolonged dosages can have a cumulative impact and make patients more susceptible to leukemia, generally within five to ten years of treatment.

Alkylating agents are still the most common chemotherapy drug used today and can be broken up into six subsets.

Alkyl Sulfonates

Busulfan (Busulfex)

Taken by pill or intravenously (into the vein), busalfan (trade name Busulfex) is used as a conditioning agent to prepare the body for bone

24 "Types of Chemotherapy," *Chemocare.com*, https://chemocare.com/ chemotherapy /what-is-chemotherapy/types-of-chemotherapy.aspx;
"Types of Chemo Drugs," *American Cancer Society*, https://www.cancer.org/treatment /treatments-and-side-effects/treatment-types/chemotherapy/how-chemotherapy -drugs-work.html;
"Types of Chemotherapy Drugs," *Chemoth.com*, https://chemoth.com/types;
"Types of Chemotherapy," *University of Iowa Hospitals and Clinics*, https://uihc .org/health-topics/types-chemotherapy;
"Categories of Chemotherapy Agents," *Mesothelioma Web*, https://www .mesotheliomaweb.org/categories.htm;
"Types of Chemotherapy," *MedlinePlus*, https://medlineplus.gov/ency /patientinstructions/000910.htm;
"Types of Chemotherapy Drugs," *National Institute of Health*, https://training .seer.cancer.gov/treatment/chemotherapy/types.html.

marrow transplants. This drug is also used to treat CML and blood disorders such as myeloid metaplasia and polycythemia. Common side effects (>30% occurrence) include low RBC, WBC and platelet counts, nausea, vomiting, appetite loss, mouth sores, diarrhea and infertility. Less common side effects include skin discoloration, itching and rashes.

ETHYLENIMINES

HEXAMETHYLMELAMINE OR ALTRETAMINE (HEXALEN)

Taken via pills with food, hexamethylmelamine, also known as altretamine (trade name Hexalen), is used to treat ovarian cancer. Common side effects (>30% occurrence) include low RBC, WBC and platelet counts, nausea, vomiting and peripheral neuropathy (numbness and tingling in the extremities, i.e. the hands and feet). Less common side effects (<30% occurrence) include abdominal cramps, dizziness, insomnia, appetite loss, depression and irritability.

THIOTEPA (THIOPLEX)

Given intravenously, thiotepa (trade name Thioplex) is used to treat breast cancer, Hodgkin's lymphoma, non-Hodgkin's lymphoma and ovarian cancer. Common side effects (>30% occurrence) include hair loss and low RBC, WBC and platelet counts. Less common side effects (>30% occurrence) include nausea, vomiting, mouth sores, bladder irritation, skin irritation (darkening, rashes, redness, flaking and peeling) and allergic reactions (tightness of the throat, shortness of breath and skin hives).

HYDRAZINES AND TRIAZINES

DACARBAZINE (DTIC-DOME)

Given intravenously, dacarbazine (trade name DTIC-Dome) is used to treat fibrosarcomas, Hodgkin's lymphoma, islet cell carcinoma, medullary thyroid carcinoma, metastatic malignant melanoma, neuroblastoma, rhabdomyosarcoma and soft tissue sarcomas. Common side effects (>30% occurrence) include skin and vein irritation around the IV insertion site, low red/, WBC and platelet counts, appetite loss, nausea, vomiting and an increase in blood liver enzymes. Less common side effects (<30% occurrence) include flu-like symptoms, sun sensitivity, as well as numbness and tingling in the extremities.

PROCARBAZINE (MATULANE)

Taken via capsule, procarbazine (trade name Matulane) is used primarily to treat Hodgkin's lymphoma. It can also be used for brain tumors, lung cancer, melanoma, multiple myeloma and non-Hodgkin's lymphoma. Common side effects (>30% occurrence) include low WBC and platelet counts, nausea, vomiting and appetite loss. Less common side effects (<30% occurrence) include hair loss, diarrhea, constipation, mouth sores, flu-like symptoms, infertility, central neurotoxicity (dizziness, drowsiness and headaches) and allergic reactions (rashes, hives and flushing).

TEMOZOLOMIDE (TEMODAR)

This drug is taken via pill form with water on an empty stomach. Temozolomide (trade name Temodar) is used to treat specific brain cancers, including anaplastic astrocytoma and glioblastoma. Common side effects (>30% occurrence) include nausea, vomiting, headaches, fatigue and constipation. Less common side effects (<30% occurrence) include low RBC, WBC and platelet counts.

METAL SALTS

CARBOPLATIN (PARAPLATIN)

Carboplatin (trade name Paraplatin) is given either intravenously or via an intraperitoneal injection directly into the lining of the abdomen. This drug is used primarily to combat ovarian cancer, but can also treat bladder cancer, breast cancer, central nervous system or germ cell tumors, cervical cancer, endometrial cancer, esophageal cancer, head and neck cancer, lung cancer and osteogenic sarcoma. Carboplatin is also used as preparation for stem cell or bone marrow transplants.

Common side effects (>30% occurrence) include low RBC, WBC and platelet counts, nausea, vomiting, hair loss, fatigue, taste bud changes and abnormal magnesium levels. Less common side effects (<30% occurrence) include abdominal pains, diarrhea, constipation, a burning sensation or infection (or both) around the injection site, mouth sores, peripheral neuropathy, ototoxicity (hearing loss), abnormal blood electrolyte levels (specifically sodium, potassium and calcium), abnormal blood liver enzymes such as alkaline phosphate and aspartate aminotransferase, allergic reactions (itching, rashes, shortness of breath, dizziness) and cardiovascular events (heart attacks, strokes, and blood clots).

Cisplatin (Platinol, Platinol-AQ)

Cisplatin (also referred to as CDDP, with trade names of Platinol and Platinol-AQ) is administered intravenously and treats bladder cancer, breast cancer, cervical cancer, esophageal cancer, head and neck cancer, Hodgkin's lymphoma, lung cancer, melanoma, mesothelioma, multiple myeloma, neuroblastoma, non-Hodgkin's lymphoma, ovarian cancer, prostate cancer, sarcomas, stomach cancer and testicular cancer.

Common side effects (>30% occurrence) include low RBC, WBC and platelet counts, nausea, vomiting, reduced kidney function, ototoxicity (hearing loss) and low magnesium, calcium and potassium levels. Less common side effects (<30% occurrence) include peripheral neuropathy, appetite loss, hair loss, changes in liver function and taste bud changes.

Oxaliplatin (Eloxatin)

Taken intravenously, oxaliplatin (trade name Eloxatin) is used for metastasized colon or rectal cancer. Common side effects (>30% occurrence) include peripheral neuropathy, appetite loss, nausea, vomiting, diarrhea, fatigue and low RBC, WBC and platelet counts. Less common side effects (<30% occurrence) include pain, fevers, constipation, headaches, coughing, changes in liver function and allergic reactions (rashes, hives and difficulty swallowing or breathing).

Nitrogen Mustards

Chlorambucil (Leukeran)

Taken orally in tablet form, chlorambucil (trade name Leukeran) is used to fight breast cancer, choriocarcinoma, giant follicular lymphoma, Hodgkin's lymphoma, lymphocytic leukemia, lymphosarcoma, non-Hodgkin's lymphoma, ovarian cancer, polycythemia vera, testicular cancer, thrombocythemia and Waldenström's macroglobulinemia.

Common side effects (>30% occurrence) include bone marrow degeneration that increases susceptibility to future cancers and decreased RBC, WBC and platelet counts. Less common side effects (<30% occurrence) include nausea, vomiting, skin rashes and decreased liver function. Infertility and pulmonary toxicity occur in rare instances.

CYCLOPHOSPHAMIDE[25] (CYTOXAN)

Cyclophosphamide (trade name Cytoxan) can be taken in many forms, including orally as tablets with food, intravenously or by direct injection into the muscle, abdominal or lung linings. This chemotherapy agent treats ALL, AML, brain cancer, breast cancer, Burkitt's lymphoma, CLL, chronic CML, endometrial cancer, Ewing's sarcoma, Hodgkin's lymphoma, lung cancer, multiple myeloma, neuroblastoma, non-Hodgkin's lymphoma, ovarian cancer, retinoblastoma, rhabdomyosarcoma, T-cell lymphoma and testicular cancer.

Common side effects (>30% occurrence) include bone marrow suppression resulting in low RBC, WBC and platelet counts, nausea, vomiting, hair loss, hair color or texture change, poor appetite, darkening of the skin and nails, stomachaches and lethargy. Less common side effects (<30% occurrence) include diarrhea, hemorrhagic cystitis, mouth sores and bladder irritation.

IFOSFAMIDE (IFEX)

Administered intravenously, ifosfamide (trade name Ifex) is used to treat bladder cancer, cervical cancer, germ cell tumors, head and neck cancer, Hodgkin's lymphoma, lung cancer, non-Hodgkin's lymphoma, sarcomas and testicular cancer. A common side effect (~50% of patients) includes encephalopathy (brain dysfunction) symptoms spanning from mild (fatigue and concentration loss) to moderate (delirium and psychosis) and severe (nonconvulsive status epilepticus seizure and coma). These symptoms are particularly damaging to children, as their neurological development can be impacted.

Other common side effects (>30% occurrence) include low WBC and platelet counts, hair loss, nausea, vomiting, loss of appetite and blood in the urine. A less common side effect (<30% occurrence) is impairment of the peripheral nervous system, i.e. the nervous system outside of the brain and spinal cord.

MECHLORETHAMINE[26] (MUSTARGEN)

Given intravenously or topically for skin cancers, mechlorethamine

25 Used in my Stanford V treatment.

26 Used in my Stanford V treatment.

(trade name Mustargen) is a nitrogen mustard used to treat breast cancer, bronchogenic carcinoma, CML, Hodgkin's lymphoma, lung cancer, lymphosarcoma, mycosis fungoides (a common form of cutaneous T-cell lymphoma, which is a form of non-Hodgkin's lymphoma), other forms of non-Hodgkin's lymphoma, polycythemia vera and prostate cancer. Incidentally, this drug is also used in chemical warfare under the code name HN2. IV dosages are mixed with a saline solution to decrease vein irritation.

Common side effects (>30% occurrence) include decreased blood cell counts (RBCs, WBCs and platelets), hair loss, nausea, vomiting, mouth sores, vein darkening, skin irritation (dryness and redness) and loss of fertility. Less common side effects (<30% occurrence) include diarrhea, fevers, appetite loss, taste bud change, ear ringing and an increase in blood uric acid levels.

MELPHALAN (ALKERAN)

Generally taken in pill form on an empty stomach, melphalan (trade name Alkeran) can also be administered intravenously. Treatments span AL amyloidosis, breast cancer, malignant melanoma, multiple myeloma, neuroblastoma, ovarian cancer and rhabdomyosarcoma.

Common side effects (>30% occurrence) include nausea, vomiting and decreased RBC, WBC and platelet counts. Less common side effects (<30% occurrence) include allergic reactions, pulmonary fibrosis (lung tissue scarring), hair loss, interstitial pneumonitis, skin irritations (rashes, itching), irreversible bone marrow failure and cardiac arrest.

NITROSOUREAS

As opposed to most chemotherapy agents, nitrosoureas can cross the blood-brain barrier, which makes them useful for treating brain cancers.

CARMUSTINE (BICNU)

Carmustine (trade name BiCNU; also known as BCNU) is an orange-yellow liquid taken intravenously or applied topically, although a solid form of the drug can also be placed at the site of a removed tumor for a slow and controlled release. This drug is used for the treatment of brain cancers (astrocytoma, brainstem glioma, ependymoma, glioblastoma, medulloblastoma and metastatic brain tumors), colon cancer, cutaneous T-cell lymphoma, Hodgkin's lymphoma, lung cancer, melanoma, multiple myeloma and non-Hodgkin's lymphoma.

Common side effects (>30% occurrence) include nausea, vomiting, facial flushing, low WBC and platelet counts and pain around the injection site. Less common side effects (<30% occurrence) include dizziness, eye redness or blurring, low blood pressure, low RBC counts and decreased liver function.

Lomustine (CeeNU)

Taken in capsule form with plenty of liquid and on an empty stomach, lomustine (trade name CeeNU) is used to treat colon cancer, Hodgkin's lymphoma, lung cancer, melanoma and non-Hodgkin's lymphoma, as well as primary and metastatic brain tumors. Common side effects (>30% occurrence) include nausea, vomiting and low WBC, RBC and platelet counts. Less common side effects (<30% occurrence) include appetite loss, mouth sores and infertility.

Streptozocin (Zanosar)

Given intravenously, streptozocin (trade name Zanosar) treats carcinoid tumors and islet cell pancreatic cancer. This drug is a vesicant, i.e. a chemical that can cause severe reactions to tissues such as redness, swelling and blistering if secreted from the vein. Common side effects (>30% occurrence) include nausea, vomiting and decreased kidney function. Less common side effects (<30% occurrence) include low WBC, hemoglobin and platelet counts, low blood sugar and diarrhea.

Antimetabolites

Like alkylating agents, antimetabolites prevent cell division by disrupting DNA synthesis, although using a different technique. Antimetabolites mimic metabolites that are used for cell metabolism—antifolates mimic folate, for example; however, since the antimetabolite is not the actual metabolite, the cell metabolism is disrupted and the cell fails to divide. Specifically, antimetabolites prevent DNA replication in the s-phase of the cell-division cycle. This differs from alkylating agents in that the cancer drug attacks cells during a specific point within the cell cycle. While unfortunately impacting healthy cells as well, fast-dividing cancer cell replication will be reduced or halted.

ADENOSINE DEAMINASE INHIBITORS

CLADRIBINE (LEUSTATIN)

Cladribine (Leustatin) is given intravenously and used to treat AML, CLL, Langerhans cell histiocytosis (LCH), non-Hodgkin's lymphoma and Waldenström's macroglobulinemia. Common side effects (>30% occurrence) include low RBC and WBC counts, fevers and fatigue. Less common side effects (<30% occurrence) include nausea, infection, skin rash, appetite loss, headaches, low platelet counts, diarrhea, coughing, skin irritation around the infusion site and abnormal breathing.

FLUDARABINE (FLUDARA)

Taken intravenously, fludarabine (trade name Fludara) treats ALL, AML, CLL and non-Hodgkin's lymphoma. Common side effects (>30% occurrence) include low RBC, WBC and platelet counts, weakness, coughing, fevers, infection, nausea, vomiting and appetite loss. Less common side effects include peripheral neuropathy, chills, fatigue, joint and muscle pain, sweating, diarrhea, rashes, swelling, taste bud changes and shortness of breath.

NELARABINE (ARRANON)

Given intravenously, nelarabine (trade name Arranon) is used for fighting forms of ALL. Common side effects (>30% occurrence) include low RBC, WBC and platelet counts, nausea and fatigue. Less common side effects (<30% occurrence) include coughing, fevers, lethargy, diarrhea, constipation, dizziness, peripheral neuropathy, shortness of breath, headaches, hypoesthesia, swelling in the extremities, muscle aches, petechiae (small, red purple spots on the skin), joint pain and pleural effusion, the buildup of fluid in the pleural cavity due to metastasized cancer cells such as those from lymphoma, breast cancer, lung cancer or ovarian cancer.

PENTOSTATIN (NIPENT)

Pentostatin (trade name Nipent) is given intravenously and treats acute graft-versus-host disease, chronic graft-versus-host disease, CLL, cutaneous T-cell lymphoma and T-cell prolymphocytic leukemia. Common side effects (>30% occurrence) include low RBC, WBC and platelet counts, nausea, fevers, skin rash and fatigue. Less common side effects (<30% occurrence) include itching, coughing, muscle aches, chills, headaches, diarrhea, abdominal pain, anorexia, weakness, upper respiratory infection, elevated liver enzymes and shortness of breath.

FOLATE ANTAGONISTS

METHOTREXATE (OTREXUP, RASUVO, RHEUMATREX OR TREXALL)

Methotrexate (trade name Otrexup) is given either via intravenous infusion, intramuscular (into the muscle) injection, intraventricular (into the brain or heart) infusion or intrathecal (into the spinal canal) infusion. This drug is used to combat ALL, breast cancer, chorioadenoma, choriocarcinoma, gestational trophoblastic disease, mycosis fungoides, non-Hodgkin's lymphoma and osteosarcoma.

Common side effects (>30% occurrence) include low RBC, WBC and platelet counts, appetite loss and mouth sores. Less common side effects (<30% occurrence) include nausea, vomiting, kidney toxicity, rashes, diarrhea, skin darkening, infertility, hair loss and photosensitivity.

PEMETREXED (ALIMTA)

Pemetrexed (trade name Alimta) is given intravenously, treating malignant mesothelioma, non-small cell lung cancer and non-squamous. Common side effects (>30% occurrence) include low RBC and WBC counts, nausea, vomiting, fatigue, appetite loss, constipation, chest pains and shortness of breath. Less common side effects (<30% occurrence) are low platelet counts, flu-like symptoms, fevers, creatinine (breakdown of creatine phosphate found in muscle) increase, depression, rashes, peripheral neuropathy and mouth sores.

PYRIMIDINE ANTAGONISTS

CAPECITABINE (XELODA)

Taken as a pill with food, capecitabine (trade name Xeloda) treats breast cancer, colon cancer, esophageal cancer, fallopian tube cancer, gastric cancer, hepatobiliary cancer, neuroendocrine tumors, ovarian cancer, pancreatic cancer, peritoneal cancer and rectal cancer.

Common side effects (>30% occurrence) include low RBC and WBC counts, palmar-plantar erythrodysesthesia (PPE—also known as hand-foot syndrome—which causes skin irritation, rashes, pain and swelling), nausea, vomiting, diarrhea, increased liver enzymes, fatigue, abdominal pain and skin irritation (rashes and itching). Less common side effects (<30% occurrence) include low platelet counts, appetite loss, fevers, mouth sores, feet or ankle swelling, eye irritation, constipation, headaches, peripheral neuropathy, joint pain, muscle pain, bone pain and gastrointestinal motility disorders.

CYTARABINE (CYTOSAR-U)

Cytarabine (trade name Cytosar-U) is given either intravenously or via intrathecal infusion. This drug treats ALL, AML, APL, CML, Hodgkin's lymphoma and non-Hodgkin's lymphoma. Common side effects (>30% occurrence) include low RBC, WBC and platelet counts, nausea, vomiting, mouth sores, headaches and increased blood in liver function blood tests.

Less common side effects (<30% occurrence) include skin rash, diarrhea, appetite loss, flu-like symptoms, hair loss, eye pain, dizziness, fatigue, increased uric acid levels and PPE.

FLUOROURACIL (ADRUCIL)

Fluorouracil (trade name Adrucil) is given intravenously as an IV push injection, infusion or topically. Usage includes anal cancer, bladder cancer, breast cancer, cervical cancer, esophageal cancer, head and neck cancer, hepatobiliary cancers (originating in the liver, bile ducts and gallbladder), neuroendocrine tumors, pancreatic cancer, basal cell and squamous cell skin cancers, stomach cancer and thymic cancer.

Common side effects (>30% occurrence) include low RBC, WBC and platelet counts, nausea, vomiting, diarrhea, appetite loss, mouth sores, watery eyes, light sensitivity, change in taste buds and vein discoloration. Less common side effects (<30% occurrence) include skin irritation (cracking, dryness, peeling, darkening), hair thinning, nail discoloration and PPE.

FLOXURIDINE (FUDR)

Floxuridine (trade name FUDR) is given either intravenously or using intra-arterial methods (infusion into an artery, commonly into the hepatic artery going to the liver). This drug is used to treat colon, kidney and stomach cancers. Common side effects (>30% occurrence) include low RBC, WBC and platelet counts, mouth sores and diarrhea. Less common side effects (<30% occurrence) include poor appetite, nausea, vomiting, hair loss, elevated liver enzymes, PPE and stomach ulcers.

DECITABINE (DACOGEN)

Decitabine (trade name Dacogen) is given intravenously and is used to treat myelodysplastic syndrome (MDS), a type of bone cancer. Common side effects (>30% occurrence) include low RBC, WBC and platelet counts, fevers, nausea, vomiting, fatigue, coughing, diarrhea, constipation, high blood pressure (hyperglycemia) and petechiae, which are red dots on the skin due to bleeding. Less common side effects (<30% occurrence) include headaches, insomnia, chills, low magnesium, potassium, sodium and albumin, swelling, bruising, rashes, dizziness, appetite loss, sore throat, abdominal pain, heart murmurs, joint pain and muscle pain.

GEMCITABINE (GEMZAR)

Given intravenously, gemcitabine (trade name Gemzar) treats bladder cancer, metastatic breast cancer, non-small cell lung cancer, ovarian cancer, pancreatic cancer and soft-tissue sarcoma. Common side effects (>30% occurrence) include low RBC, WBC and platelet counts, flu-like symptoms, fevers, nausea, vomiting, fatigue, appetite loss, rashes, increased liver enzymes and bloody urine. Less common side effects (<30% occurrence) include hair loss, diarrhea, weakness, mouth sores, insomnia and shortness of breath.

PURINE ANTAGONISTS

MERCAPTOPURINE (PURINETHOL)

Given orally, mercaptopurine (trade name Purinethol) is used to fight ALL, APL, Crohn's disease, lymphoblastic lymphoma and ulcerative colitis. Common side effects (>30% occurrence) include low RBC, WBC and platelet counts and liver toxicity. Less common side effects (<30% occurrence) include nausea, vomiting, appetite loss, fevers, diarrhea, mouth sores, lethargy, loss of fertility, skin rash, darkening of the skin and difficulty breathing.

Thioguanine (Tabloid)

Thioguanine, trade name Tabloid, is taken orally and used to treat ALL and AML. Common side effects (>30% occurrence) include low RBC, WBC and platelet counts. Less common side effects (<30% occurrence) include nausea, appetite loss, mouth sores, swelling of the legs, hands or feet, pain swallowing and coughing up mucus.

Anti-Tumor Antibiotics

These drugs are made from *Streptomyces*, which are bacteria found in soil and fresh water. Natural products of these species are used in medicine to inhibit or prevent cell replication. Like the alkylating agents, these drugs attack cell replication in all phases of the cell cycle.

Anthracyclines

Daunorubicin (Cerubidine)

Given via intravenous injection, daunorubicin (trade name Cerubidine) treats ALL, AML and APL. Common side effects (>30% occurrence) include low RBC, WBC and platelet counts, nausea, vomiting, pain around the injection site, bloody urine, mouth sores and hair loss. Less common side effects (<30% occurrence) include skin or nail darkening, diarrhea and infertility.

Doxorubicin[27] (Adriamycin or Rubex)

Administered as an intravenous injection, doxorubicin (trade name Adriamycin or Rubex) treats ALL, AML, bone sarcoma, breast cancer, endometrial cancer, gastric cancer, head and neck cancer, Hodgkin's lymphoma, kidney cancer, liver cancer, multiple myeloma, neuroblastoma, non-Hodgkin's lymphoma, ovarian cancer, small-cell lung cancer, soft tissue sarcoma, thymoma, thyroid cancer, transitional cell bladder cancer, uterine sarcoma, Waldenström's macroglobulinemia and Wilms tumor. This drug is a vesicant if secreted from the vein.

Common side effects (>30% occurrence) include low RBC, WBC and platelet counts, nausea, vomiting, pain around the injection site and hair loss. Less common side effects (<30% occurrence) include skin or nail darkening, watery eyes, mouth sores, infertility and urine discoloration (red, orange, brown or pink from the medication).

27 Used in my Stanford V treatment.

Epirubicin (Ellence)

Taken via intravenous injection, epirubicin (trade name Ellence) is used to treat breast cancer and can also be used as a substitute for doxorubicin for some cancer treatments. This drug is a vesicant if secreted from the vein. Common side effects (>30% occurrence) include nausea, vomiting, discolored (reddish) urine, pain around the injection site, hair loss, mouth sores, fatigue and amenorrhea (loss of menstrual cycle). Less common side effects (<30%) include skin darkening, diarrhea, infection, infertility and conjunctivitis (pink eye).

Idarubicin (Idamycin)

Given as an intravenous injection, idarubicin (trade name Idamycin) fights ALL, AML, CML and MDS. This drug is a vesicant if secreted from the vein. Common side effects (>30% occurrence) are low RBC, WBC and platelet counts, pain around the injection site, nausea, vomiting, discolored urine from medication (red, orange, brown or pink), hair loss, mouth sores and diarrhea.

Mitoxantrone (Novantrone)

Mitoxantrone (trade name Novantrone) is given via intravenous injection and is used to treat AML, breast cancer, non-Hodgkin's lymphoma and advanced prostate cancer. This drug is an irritant that can cause vein inflammation as well as tissue damage if the chemical escapes the vein.

Common side effects (>30% occurrence) include low RBC, WBC and platelet counts, nausea, vomiting, liver functionality and fevers. Less common side effects (<30% occurrence) are hair loss, diarrhea, weakness, mouth sores, low blood pressure, discoloration of the eyes or urine (blue or green), heart arrhythmia and abnormal heart electrocardiogram.

Chromomycins

Dactinomycin (Cosmegen)

Dactinomycin (trade name Cosmegen) is administered intravenously and is classified as a vesicant. It is used to treat childhood rhabdomyosarcoma, Ewing's sarcoma, gestational trophoblastic neoplasia, locally recurrent solid tumors, metastatic testicular tumors (nonseminomatous), osteosarcoma, ovarian cancer, soft tissue sarcoma and Wilms tumor.

Common side effects (>30% occurrence) include low RBC, WBC and platelet counts, hair loss, nausea, vomiting, fatigue, mouth sores, liver problems, diarrhea, loss of fertility and an increased risk of infection. Less common side effects (<30% occurrence) include loss of appetite, skin peeling, acne, skin reactions to sunlight and darkening of the skin.

PLICAMYCIN (MITHRACIN)[28]

Plicamycin (trade name Mithracin) is administered intravenously and used to treat testicular cancer. Side effects include nausea, vomiting, diarrhea, appetite loss, mouth sores, fevers, fatigue, headaches, depression, skin rash, facial flushing and discomfort.

OTHER ANTI-TUMOR ANTIBIOTICS

BLEOMYCIN[29] (BLENOXANE)

Bleomycin (trade name Blenoxane) is given either by intravenous infusion or intrapleural injection (into the pleural cavity of the lungs, i.e. the fluid-filled region between a lung's visceral and parietal pleural membranes). Treatments are intended for Hodgkin's lymphoma, melanoma, non-Hodgkin's lymphoma, ovarian cancer, sarcoma, squamous cell skin cancer and testicular cancer. Bleomycin is also useful in treating malignant pleural effusion.

Common side effects (>30% occurrence) include hair loss, fevers and chills, skin irritation (peeling, darkening, redness, thickening, stretch marks) and nail thickening or bending. Less common side effects (<30% occurrence) include nausea, vomiting, appetite loss, pneumonitis (lung inflammation), phlebitis (vein inflammation), mouth sores and radiation recall (skin reaction when chemotherapy is administered after radiation treatment).

MITOMYCIN (MUTAMYCIN)

Given either by injection or intravenously, mitomycin (trade name Mutamycin) is a vesicant used to treat anal, bladder, breast, cervical, colorectal, head and neck, non-small cell lung, pancreatic and stomach cancer.

28 "Mithracin," *RxList*, https://www.rxlist.com/mithracin-side-effects-drug-center. htm.

29 Used in my Stanford V treatment.

Common side effects (>30% occurrence) include low RBC, WBC and platelet counts, fatigue, mouth sores and appetite loss. Less common side effects (<30% occurrence) include nausea, vomiting, diarrhea, hair loss and bladder inflammation.

CORTICOSTEROIDS

Although having some direct cancer-treating effects such as treating the cancer itself, reducing inflammation and the immune response,[30] and reducing sickness during chemotherapy, corticosteroids are primarily used to treat other chemotherapy and radiation drug side effects. Common corticosteroids naturally produced in humans include cortisol, cortisone, corticosterone and aldosterone. Corticosteroids are produced in the adrenal glands, located on top of the kidney as shown in Figure 35. While the medulla regions of the adrenal glands produce adrenaline, noradrenaline and small amounts of dopamine, the cortex produces androgens (testosterone, dehydroepiandrosterone, androstenedione, androsterone, androstenediol and dihydrotestosterone) as well as corticosteroids (glucocorticoids and mineralocorticoids).

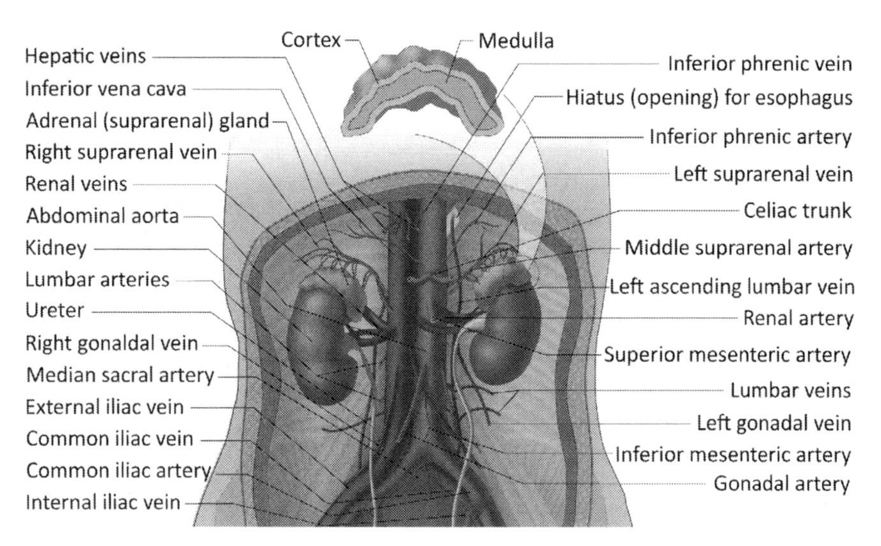

Figure 35: Adrenal gland location.

30 W.Y. Almawi and O.K. Melemedjian, "Molecular Mechanisms of Glucocorticoid Antiproliferative Effects: Antagonism of Transcription Factor Activity by glucocorticoid Receptor," *Journal of Leukocyte Biology* 71, no. 1 (Jan. 2002): 9–15, https://jlb.onlinelibrary.wiley.com/doi/pdf/10.1189/jlb.71.1.9.

Of these two subsets of corticosteroids, glucocorticoids include cortisol and cortisone, which have anti-inflammatory, immunosuppression (reduced immune system), vasoconstrictive (blood vessel restriction) and anti-proliferative (cell-multiplying) effects. Additionally, glucocorticoids impact carbohydrate, protein and fat metabolism, which is why they cause an increase in appetite and are used to combat weight loss side effects of some chemotherapy drugs. The main mineralocorticoid is aldosterone, which regulates the concentration of electrolytes and water balance within the body via the kidneys.

Glucocorticoids are synthetically reproduced in laboratories to enhance the body's naturally occurring cortisol and cortisone (corticosterone has both glucocorticoid and mineralocorticoid traits but is minor in humans and much more prominent in various animals). The four main drugs produced and used in cancer treatment are dexamethasone, hydrocortisone, methylprednisolone and prednisone.[31]

Dexamethasone (Decadron)

Dexamethasone (trade name Decadron) is primarily given as a pill taken with food after meals, although it can also be administered from intravenous infusion, topically, or with eye drops. Cancer treatments include leukemia, lymphoma, myeloma and to cope with cancer side effects such as nausea, vomiting and appetite loss. Dexamethasone is also used to treat adrenal insufficiency, allergic reactions, asthma, autoimmune diseases, skin and lung conditions, and as an anti-inflammatory. In addition to specifically treating eye inflammation, dexamethasone is unique from the other cancer treatment corticosteroids in that it is used specifically to reduce swelling from brain and spinal tumors.

Common side effects (>30% occurrence) include irritability, increased appetite (when not being used for this purpose), insomnia, heartburn, ankle and feet swelling, muscle weakness, increased blood sugar, increased blood pressure and inhibited wound healing. Less common side effects (<30% occurrence) include headaches, dizziness and mood swings. Long-term usage can lead to cataract conditions and bone thinning.

---

31 Laurence Fardet, "What is Cortisone," *Henri Mondor Hospital,* https://cortisone-info.com/General-Information/What-is-Cortisone;
Mallatt, *Human Anatomy*, 7th Edition, 535–536;
"Dexamethasone, Hydrocortisone, Methylprednisolone and Prednisone," *Drugsite Trust,* https://www.drugs.com.

HYDROCORTISONE (ALA-CORT, HYDROCORTONE PHOSPHATE, SOLU-CORTEF, HYDROCORT ACETATE, LANACORT, CORTEF OR WESTCORT)

Taken in pill form with food, injected intramuscularly or intravenously, applied topically or with eye drops, hydrocortisone (trade name Ala-Cort, Hydrocortone Phosphate, Solu-Cortef, Hydrocort Acetate, Lanacort, Cortef or Westcort) is used to treat leukemia, lymphoma and multiple myeloma. Cancer side effects also treated include nausea, vomiting and appetite loss. Additionally, hydrocortisone is used to treat asthma, autoimmune diseases, skin and lung conditions, allergic reactions, adrenal insufficiency, eye conditions (as an anti-inflammatory) and thyroid complications.

Common side effects (>30% occurrence) include irritability, mood swings, increased appetite (if not being used for this purpose), insomnia, upset stomach, increased blood sugar, increased blood pressure and inhibited wound healing. Less common side effects (<30% occurrence) include headaches, dizziness, muscle weakness, aches, heartburn, ankle and feet swelling, bloating in the face and trunk, and thinning in the arms and legs. Long-term usage can lead to cataract conditions, bone thinning and stunted growth in children.

METHYLPREDNISOLONE (DURALONE, MEDROL, MEDRALONE, M-PREDNISOL OR SOLU-MEDROL)

Methylprednisolone is taken either in pill form with food, injected intramuscularly or injected intravenously. Methylprednisolone (trade name Duralone, Medrol, Medralone, M-Prednisol or Solu-Medrol) treats leukemia, lymphoma and myeloma. Cancer side effects also treated include nausea, vomiting and appetite loss. Additionally, methylprednisolone is used to treat asthma, autoimmune diseases, skin and lung conditions, allergic reactions, adrenal insufficiency, inflammation and graft-versus-host disease after a bone marrow transplant.

Common side effects (>30% occurrence) include irritability, increased appetite (if not being used for this purpose), insomnia, heartburn, increased blood sugar, increased blood pressure, nausea, vomiting, ankle and feet swelling, muscle weakness and inhibited wound healing. Less common side effects (<30% occurrence) include headaches, dizziness, mood swings and cataract conditions (with long-term usage).

PREDNISONE[32] (DELTASONE)

Given orally via pills with food, prednisone (trade name Deltasone) treats leukemia, lymphoma and myeloma. Prednisone is also used to cope with cancer side effects including nausea, vomiting and appetite loss. Additionally, this drug is an anti-inflammatory made to treat asthma, autoimmune diseases, skin and lung conditions, allergic reactions and adrenal insufficiency.

Common side effects (>30% occurrence) include irritability, increased appetite (if not being used for this purpose), insomnia, heartburn, increased blood sugar, increased blood pressure, nausea, vomiting, ankle and feet swelling, muscle weakness and inhibited wound healing. Less common side effects (<30% occurrence) include headaches, dizziness, mood swings and cataract conditions (with long-term usage).

MISCELLANEOUS ANTINEOPLASTICS

There are many useful types of chemotherapy drugs that are unique and do not fall into a broader category; these miscellaneous drugs are combined into an antineoplastics[33] category and include the following:

- ***Ribonucleotide Reductase Inhibitor:*** Hydroxyurea (Hydrea).
- ***Adrenocortical Steroid Inhibitor:*** Mitotane (Lysodren).
- ***Enzymes:*** Asparaginase (Elspar) and pegaspargase (Oncaspar).
- ***Antimicrotubule Agent:*** Estramustine (Emcyt).
- ***Retinoids:*** Bexarotene (Targretin), isotretinoin (Accutane) and tretinoin (Vesanoid).
- ***Monoclonal Antibody:*** Alemtuzumab (Campath), bevacizumab (Avastin), cetuximab (Erbitux), gemtuzumab (Mylotarg), ozogamicin (Besponsa), ipilimumab (Yervoy), ofatumumab (Arzerra), panitumumab (Vectibix), pembrolizumab (Keytruda), ramucirumab (CYRAMZA), rituximab (Rituxan) and trastuzumab (Herceptin).
- ***Cytokines:*** Aldesleukin (Proleukin).

32 Prednisone was used to supplement the chemicals in my Stanford V treatment.

33 An antineoplastic is an agent that prevents, halts or inhibits the formation of a neoplasm, i.e. a tumor.

- ***MTOR Inhibitors:*** Everolimus (Afinitor), ridaforolimus (still in testing, formerly known as deforolimus), sirolimus (Rapamune) and temsirolimus (Torisel).

Additional information on the uses of these drugs can be found in Appendix B. This is a supplemental chemotherapy resource that organizes chemical agents not only by type, but also by form of cancer. Patients can identify which chemotherapy agents are typically used to treat their cancer and conduct additional research in conjunction with their oncologist on what the best treatment plan is. For side effects and administration, readers are referred to Chemocare.com.

PLANT ALKALOIDS

Alkaloids are naturally occurring compounds that come as byproducts of animals, plants, bacteria and fungi. Primarily containing carbon, hydrogen and nitrogen, these compounds also can contain oxygen, sulfur and less frequently bromine, chlorine and phosphorus.

There are four main categories of plant alkaloids[34] used as chemotherapy drugs: camptothecin analogs, podophyllotoxins, taxanes and vinca alkaloids. The latter two are considered antimicrotubule agents, which block cell division by interfering with cell microtubules that transport chromosomes during the mitosis (type of cell division) stage of the cell cycle. Similarly, the former two can be grouped together based on their function and are classified as topoisomerase inhibitors. This means the drugs interfere with topoisomerases, enzymes that help ravel and unravel DNA strands by breaking them, thereby preventing entanglement.

Camptothecin analogs come from the bark of the Asian happy tree (*Camptotheca acuminata*), which is native to China and Tibet. Podophyllotoxins are produced from the mayapple plant (*Podophyllum*

34 Remarkably, the National Cancer Institute has screened some 35,000 plant species for potential anticancer properties; just as remarkably, about 3,000 of them have demonstrated reproducible anticancer activity. The use of plant alkaloids in chemotherapy drugs has been the focus in recent years of numerous scholarly articles;
Avni G. Desai, Ghulam N. Qazi, Ramesh K. Ganju et al., "Medicinal Plants and Cancer Chemoprevention," *Current Drug Metabolism* 9, no. 7 (Sept. 2008): 581–591;
Elham Safarzadeh, Siamak Sandoghchian Shotorbani and Behzad Baradaran, "Herbal Medicine as Inducers of Apoptosis in Cancer Treatment," *Advanced Pharmaceutical Bulletin* 4, suppl. 1 (October, 2014): 421–427.

peltatum) and can be found in the eastern United States and southeastern Canada. Taxanes come from the bark of the Pacific yew tree (*Taxus brevifolia*), primarily found in the United States Pacific Northwest. Vinca alkaloids are produced from the Madagascar rosy periwinkle plant (*Catharanthus roseus*), shown in Figure 36, which is found mainly in the southern region of Mexico, specifically in Champoton, Campeche and Merida.

Figure 36: Madagascar rosy periwinkle plant. [35]

Camptothecin Analogs (Topoisomerase I Inhibitors)

Irinotecan (Camptosar)

Given intravenously, irinotecan (trade name Camptosar) treats metastatic colon and rectal cancer. This drug is an irritant that can cause vein inflammation as well as tissue damage if the chemical escapes the vein. Common side effects (>30% occurrence) include low RBC and WBC counts, diarrhea, nausea, vomiting, hair loss, appetite degradation, fevers, weight loss and weakness. Less common side effects (<30% occurrence) include coughing, headaches, insomnia, shortness of breath, constipation,

35 Joydeep, "Catharanthus roseus flower," *Wikimedia Commons*, https://commons.wikimedia.org/wiki/File:Catharanthus_roseus24_08_2012_(1).JPG.

dehydration, chills, rashes, face flushing, heartburn, mouth sores and swelling of the feet and ankles.

TOPOTECAN (HYCAMTIN)

Topotecan (trade name Hycamtin) can be taken either intravenously or as a capsule with water and with or without food. Primarily used to treat ovarian cancer, topotecan is also used for small-cell lung cancer. Common side effects (<30% occurrence) include nausea, vomiting, hair loss and diarrhea. Less common side effects (<30% occurrence) include constipation, fatigue, fevers, bone pain, abdominal pain, appetite loss, weakness, mouth sores, rashes, coughing, headaches and shortness of breath.

PODOPHYLLOTOXINS (TOPOISOMERASE II INHIBITORS)

ETOPOSIDE[36] (TOPOSAR, VEPESID OR ETOPOPHOS)

Etoposide (trade name Toposar, VePesid or Etopophos) can be taken either intravenously or by tablet. This drug is used to treat bladder cancer, brain tumors, Ewing's sarcoma, Hodgkin's lymphoma, Kaposi's sarcoma, lung cancer, mycosis fungoides, non-Hodgkin's lymphoma, prostate cancer, rhabdomyosarcoma, stomach cancer, uterine cancer and Wilms tumor. Etoposide can also be used in high doses as part of bone marrow transplant procedures. This drug is an irritant that can cause vein inflammation as well as tissue damage if the chemical escapes the vein.

Common side effects (>30% occurrence) include low RBC, WBC and platelet counts, hair loss, infertility, nausea, vomiting, chemotherapy-induced menopause and low blood pressure. Less common side effects (<30% occurrence) include appetite reduction, diarrhea, mouth sores and radiation recall (a severe skin reaction that occurs when radiation is given too soon after chemotherapy).

TENIPOSIDE (VUMON)

Given intravenously, teniposide (trade name Vumon) is used for ALL, which occurs primarily in children. This drug is an irritant that can cause vein inflammation as well as tissue damage if the chemical escapes the vein. Common side effects (>30% occurrence) include low RBC, WBC and platelet counts, diarrhea and mouth sores. Less common side effects (<30% occurrence) include nausea, vomiting and infection.

36 Used in my Stanford V treatment.

TAXANES (ANTIMICROTUBULE AGENTS)

DOCETAXEL (TAXOTERE)

Given intravenously, docetaxel (trade name Taxotere) treats breast cancer, head and neck cancer, metastatic prostate cancer and non-small cell lung cancer. This drug is also under investigation for bladder cancer, melanoma, ovarian cancer, pancreatic cancer, small-cell lung cancer and soft tissue sarcoma. Common side effects (>30% occurrence) include low RBC, WBC counts, peripheral neuropathy, fluid retention, weight gain, nausea, vomiting, hair loss, diarrhea, mouth sores, infection, fatigue, weakness and nail discoloration.

PACLITAXEL (TAXOL OR ONXAL)

Paclitaxel (trade name Taxol or Onxal) is given via IV push injection or IV infusion. Cancers treated include bladder cancer, breast cancer, esophageal cancer, Kaposi's sarcoma, lung cancer, melanoma, ovarian cancer and prostate cancer. This drug is an irritant that can cause vein inflammation as well as tissue damage if the chemical escapes the vein.

Common side effects (>30% occurrence) include low RBC, WBC and platelet counts, peripheral neuropathy, hair loss, joint pain, muscle pain, mouth sores, diarrhea and fevers. Less common side effects (<30% occurrence) include changes in liver function, feet or ankle swelling, low blood pressure, nail discoloration and skin darkening.

VINCA ALKALOIDS (ANTIMICROTUBULE AGENTS)

VINBLASTINE[37] (ALKABAN-AQ OR VELBAN)

Vinblastine (trade name Alkaban-AQ or Velban) is given intravenously either via injection push or infusion and is used to fight bladder cancer, breast cancer, choriocarcinoma, germ cell tumors, fibromatosis, head and neck cancers, Hodgkin's lymphoma, Kaposi's sarcoma, non-small cell lung cancer, melanoma, mycosis fungoides (T-cell lymphoma), non-Hodgkin's lymphoma, soft tissue sarcoma and testicular cancer. As a vesicant, vincristine irritates the skin if it escapes the vein near the IV injection site, causing tissue damage such as redness, swelling and blistering.

---

37 Used in my Stanford V treatment.

Common side effects (>30% occurrence) are low RBC, WBC and platelet counts, fatigue, weakness and injection site reactions. Less common side effects (<30% occurrence) include nausea, vomiting, peripheral neuropathy, fevers, diarrhea, constipation, appetite loss, hair loss, hearing loss, headaches, mouth sores, taste bud change, depression, jaw pain, bone pain, tumor pain, muscle pain, joint pain, shortness of breath and high blood pressure.

VINCRISTINE[38] (ONCOVIN OR VINCASAR PTS)

Given intravenously, vincristine (trade name Oncovin or Vincasar Pts) is used to treat ALL, AML, brain tumors, CLL, CML, Hodgkin's lymphoma, multiple myeloma, neuroblastoma, non-Hodgkin's lymphoma, rhabdomyosarcoma, thyroid cancer and Wilms tumor. As a vesicant, vincristine irritates the skin if it escapes the vein near the IV injection site, causing tissue damage such as redness, swelling and blistering. Hair loss is a common side effect (>30% occurrence) for this drug. Less common side effects (<30% occurrence) include constipation and low RBC, WBC and platelet counts.

VINORELBINE (NAVELBINE)

Given intravenously either with a push injection or infusion, vinorelbine (trade name Navelbine) primarily treats non-small cell lung cancer. This drug may also be used for breast cancer, Hodgkin's lymphoma and ovarian cancer. Vinorelbine is a vesicant that can cause severe tissue reactions if secreted from the vein.

Common side effects (>30% occurrence) include low RBC, WBC counts, nausea, vomiting, muscle weakness and constipation. Less common side effects (<30% occurrence) include peripheral neuropathy, hair loss, low platelet counts, diarrhea and injection site pain.

TOPOISOMERASE INHIBITORS

These are drugs that block the action of topoisomerase, an enzyme involved in the overwinding or underwinding of DNA.

They fall under two categories:

Topoisomerase I Inhibitors: Irinotecan (Camptosar) and topotecan (Hycamtin).

38 Used in my Stanford V treatment.

Topoisomerase II Inhibitors: Daunorubicin (Cerubidine), doxorubicin (Adriamycin), etoposide (Toposar), mitoxantrone (Novantrone) and teniposide (Vumon).

Note that etoposide and teniposide were also listed under plant alkaloids while daunorubicin, doxorubicin and mitoxantrone were covered under anti-tumor antibodies; thus these chemicals all fall under more than one category. Readers are once again encouraged to utilize Appendix B/ Chemotherapy Agents and Chemocare.com to find information on chemicals above not already covered in terms of side effects and administration.

PILLS

In addition to chemotherapy and radiation, many pills will be prescribed to offset the various side effects. While every chemotherapy agent has unique side effects, many share similarities with other agents as well as radiation side effects. These include nausea, kidney or liver issues, constipation, blood clots and low blood cell counts. Some pills will be prescribed before your treatment starts to combat known side effects, while others will be added as needed to mitigate additional side effects during treatment.

The specific pills I took for the Stanford V Hodgkin's lymphoma chemotherapy treatment, as well as for my follow-on radiation treatment, are shown in Table 2. Six classes of pills were prescribed as part of the standard treatment plan, including prednisone, Septra DS, acyclovir, ketoconazole, Zantac and Colace. Prednisone is an integral part of the Stanford V treatment because it reduces inflammation as well as the risk of allergic reactions and nausea. The remaining drugs used are listed in Table 2 and prevent infections, an upset stomach and constipation. Additional pills are taken as needed to fight nausea, lessen kidney side effects and for blood-thinning purposes to combat blood clotting.

Understanding the frequency and dosage of your pills is a rigorous process that requires your or your caretaker's full attention. While chemotherapy and radiation are administered at your medical facility, picking up, understanding and taking your medication is your team's responsibility. My oncologist gave me a printout of the Stanford V initial treatment plan with frequencies and dosages. When additional drugs were added during my treatment, I placed them in a numbered list along with instructions to stay on track. I strongly advise getting a pill organizer for dividing up the medication you will be taking multiple times a day over your treatment. Having to buy one of those at 28 years of age was a combination of comical and depressing! However, it proved invaluable for organizing my pill ingestion. Table 2 shows that in addition to the seven chemicals administered to me during treatment, I took 14 different types of pills. This amounted to dozens of pills taken throughout the day, many with special instructions such as with or without food, restrictions on taking with another pill, ability to cause drowsiness, etc. Organizing these notes is imperative and should be done by either you or your caretaker. I chose to have everything on one sheet of paper with shorthand for a drug's use, how often to take it and how much to take along with any special instructions.

Table 2: Summary of pills taken for my Stanford V Hodgkin's lymphoma chemotherapy treatment and radiation regimen.

Drug	Frequency	Dosage	Description
Stanford V Supplemental Pills			
Prednisone	qod	40 mg	Nine tablets at once (dosage based on height and weight). Do not lie down for 30 minutes after. Best taken with food in the morning.
TMP-SMX (Septra DS)	bid	160 - 800 mg	Antibiotic to prevent infection while on prednisone.
Acyclovir	tid	200 mg	Anti-viral medication to supplement low WBC counts.
Ketoconazole	qd	200 mg	Anti-fungal medication.
Ranitidine (Zantac)	bid	150 mg	Antihistamine that prevents acid buildup in the stomach (causing upset stomach and heartburn) associated with prednisone. Take separately from ketoconazole.
Docusate Sodium (Colace)	qd	250 mg	Prevents constipation during Stanford V treatment.

Drug	Frequency	Dosage	Description
Take as Prescribed by or After Consulting with Your Oncologist			
Ondansetron (Zofran)	bid	8 mg	Prevents nausea.
Allopurinol	qd	300 mg	Reduces uric acid to help the kidneys.
Lorazepam	qd to q4h	1 mg	Prevents nausea.
Prochlorperazine (Compazine)*	q6h	10 mg	Prevents nausea.
Multivitamin	qd	1 tablet	Supplements diet.
Calcium with Vitamin D	qd	1 tablet	Supplements diet.
Neupogen	qd	5mcg/kg	Man-made protein used to increase WBC.
Coumadin	As directed	As pre-scribed	Blood thinner.
Emend	qd	125 mg	Prevents nausea.
Milk of Magnesia*	As needed	400 mg	Stool softener.

Key			
qod	Every other day	qd	Once/day
bid	Twice/day	q4h	Every 4 hours
tid	Three times/day	*	I did not take

Chapter 10

Blood Work

Throughout your cancer experience, blood tests will be taken. This is important early on when diagnosing your cancer, during the treatment itself and after the treatment has ended. Your oncologists will be tracking important measurables to determine the efficacy of the treatment and ensure you are improving. Being the engineer that I am, I rigorously plotted and tracked my blood test results in Excel. Not only was this interesting, but more importantly I wanted to understand what was happening to my body and be certain that my oncology team did not miss any red flags. While each patient's blood testing may differ depending on the cancer type, there are common categories to all patients that your oncologists will track.

Complete Blood Count (CBC)

Table 3 shows my CBC samples taken before, during and after chemotherapy and radiation treatments. This is the most comprehensive blood test your oncologist will give you to monitor your blood cell levels. Each column (as well as associated data in Figures 37 through 43) has a minimum and maximum value, providing a range of normal, healthy values. WBC counts are in thousands per microliter while RBC counts are in millions per microliter. (Your body has approximately one thousand times more RBCs than WBCs.) Having low WBCs can lead to infection while low RBCs can cause anemia—low oxygen in the bloodstream that results in fatigue, lethargy and pallor. HGB is hemoglobin, introduced earlier as a protein inside RBCs that carries oxygen (measured in grams per deciliter, where 1 dL = 100 mL). Hematocrit (HCT) is the volume percentage of RBCs in blood. This can be estimated by tripling the HGB g/dL value and dropping the units.

MCV stands for mean corpuscular volume—or mean cell volume since a corpuscle is a cell—which is the average volume of an RBC inside your body measured in femtoliters, or 10^{-15} liters. RDW is RBC distribution width, which is a measure of the distribution of red cell volume. For example, 10% means there is a 10% difference in volume between the smallest and largest RBCs. In conjunction with MCV, RDW helps diagnose anemia. When both MCV and RDW are high, this is a sign of folate and vitamin B12 deficiency. When MCV is normal but RDW is high, this is a sign of low iron. Finally, platelet count, measured in thousands per microliter, is a measure of your body's ability to clot wounds. Low counts can lead to excessive bleeding while high counts can cause dangerous blood clots.

Interpreting the data, several conclusions can be made. Blood diagnostics are relatively normal before cancer treatment apart from HGB, which is slightly below normal. Some cancers can suppress RBC production, particularly blood cancers including leukemia, lymphoma and myeloma. These cancers attack the bone marrow and reduce blood cell production. While having low blood cell counts may be a sign of a blood cancer, most cancers will have less of an impact on blood counts and all cancers will require a biopsy for a confirmed diagnosis.

Chemotherapy itself causes an even more drastic decline in RBC production,[39] as evidenced by the continued decrease in HGB, HCT and RBC values once my chemotherapy commenced on December 3, 2009. Note that in Tables 3 and 4, dates in green are weeks that included chemotherapy treatment while weeks in red included radiation treatment (exact treatment dates are provided in Tables 5 through 8 in the *Mental Health* chapter). Also, values highlighted in red are above a healthy range while those in blue are below. My decrease in RBC count continued through radiation treatments, which also reduced cell production in the bone marrow. Reductions are particularly dramatic when radiation impacts the bones in your leg, chest, abdomen or pelvis.[40] My radiation was centered on my chest near my heart and left lung, where the tumor had expanded and displaced my organs.

39 "Your Blood, Bone Marrow and Cancer Drugs," *Cancer Research UK*, https://www.cancerresearchuk.org/about-cancer/cancer-in-general/treatment/cancer-drugs/side-effects/your-blood-and-bone-marrow.

40 "Radiotherapy Effects on Your Blood," *Cancer Research UK*, https://www.cancerresearchuk.org/about-cancer/cancer-in-general/treatment/radiotherapy/side-effects/general-radiotherapy/blood.

Table 3: Complete blood count results for my Hodgkin's lymphoma treatment.

Date	WBCs	RBCs	HGB	Hematocrit	MCV	RDW	Platelets
[-]	[K/μL]	[M/μL]	[g/dL]	[%]	[fL]	[%]	[K/μL]
Min.	3.5	4.1	13	39	80	11.9	140
Max.	12.2	5.7	17	51	100	14.3	400
11/23/09	6.8	4.32	12.7	40.1	93	14.2	324
11/30/09	9.7	4.1	**12.3**	**37.2**	91	**14.4**	366
12/8/09	8	4.13	**12.1**	**37**	90	13.6	222
12/15/09	1.7	4.1	**11.9**	**38**	93	13.9	318
12/17/09	3.5	**3.84**	**11.6**	**35.1**	92	**14.6**	303
12/23/09	3.6	**4.08**	**12.2**	**36.7**	90	**14.8**	200
12/30/09	**50.6**	**3.9**	**11.6**	**36.2**	93	**16.1**	292
1/6/10	5.8	**3.88**	**11.5**	**33.7**	87	**16.4**	130
1/13/10	**14.8**	**3.83**	**11.4**	**35.3**	92	**18.2**	**429**
1/20/10	10.7	**3.53**	**10.7**	**33.3**	94	**20**	186
1/27/10	**26.6**	**3.63**	**11.1**	**34.8**	96	**21.4**	379
2/3/10	4.7	**3.6**	**10.9**	**33.9**	94	**21.9**	182
2/10/10	11.1	**3.73**	**11.7**	**33.1**	97	**23.8**	358
2/17/10	8.8	**3.22**	**10.4**	**30.6**	99	**23.2**	183
2/24/10	5.4	**3.73**	**11.6**	**35.7**	96	**23.6**	386
3/3/10	4.6	**3.78**	**12.1**	**37**	98	**21.6**	277
3/11/10	6	**3.87**	**12.5**	39.4	**102**	**19.1**	267
3/17/10	**3.1**	**3.78**	**12.2**	**37.7**	100	**16.2**	210
3/24/10	3.6	3.95	13.1	39.3	100	14.2	160
3/31/10	**3**	4.3	14.1	42.1	98	13.3	141
4/7/10	**3**	4.1	13.7	39.3	96	12.8	191
4/14/10	**3.1**	4.38	14.3	42.3	97	12.7	182
4/21/10	**3.2**	4.38	14.2	41.7	95	12.5	162
6/1/10	5.1	4.66	15.1	43.6	94	12.6	162

Key: The first grouping of dates in bold are weeks with chemotherapy while the second bold grouping are weeks with radiation. Bold data falls outside of the recommended range listed at the top of the table.

MCV and RDW both increased during my treatment, indicating I was suffering from anemia. This is caused by the chemotherapy agents killing off healthy RBCs along with the unwanted cancerous cells. The effect contributed to the overall lethargy I experienced from the reduced oxygen circulation in my bloodstream. Although radiation can contribute to anemia, this happens when the radiation attacks your bone and reduces the ability of your marrow to produce RBCs. My treatment generally avoided my bone marrow, which is why the RBCs began to recover after chemotherapy, as seen by my blood counts returning to normal.

Platelet counts (PLT) were volatile throughout my treatment, making conclusions more difficult. These oscillations went away after chemotherapy ended, indicating that the oscillations were caused by the chemical agents. Cyclophosphamide in particular has been noted to cause oscillations in platelet counts.[41] Finally, WBC counts exhibited some interesting behavior as well, which is discussed in the *WBC Differential* section.

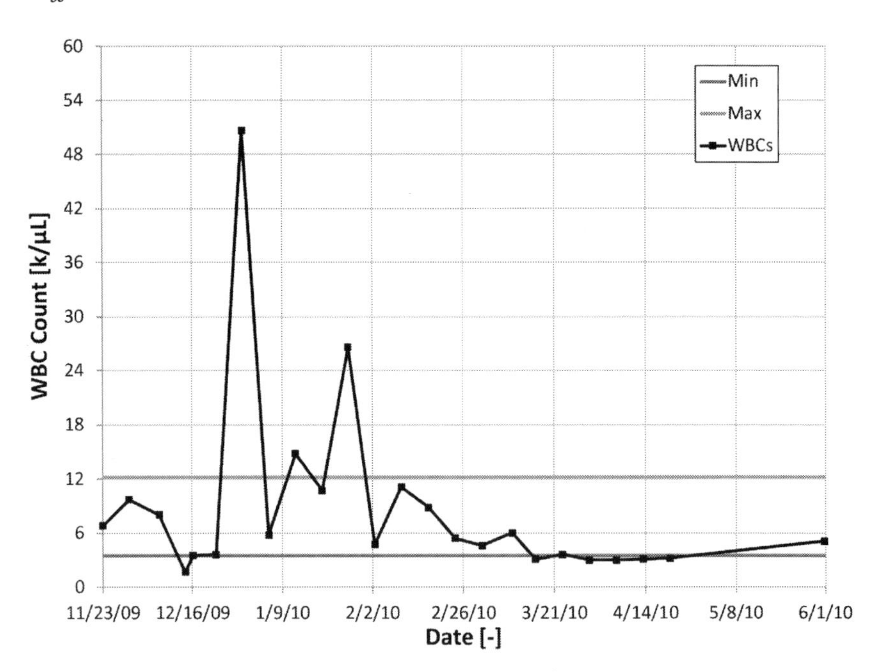

Figure 37: WBC counts during and after my treatment.

41 Anne Beuter, Roderick Edwards and Michele S. Titcombe. *Nonlinear Dynamics in Physiology and Medicine,* (Berlin/Heidelberg: Springer-Verlag, 2003), 242–245.

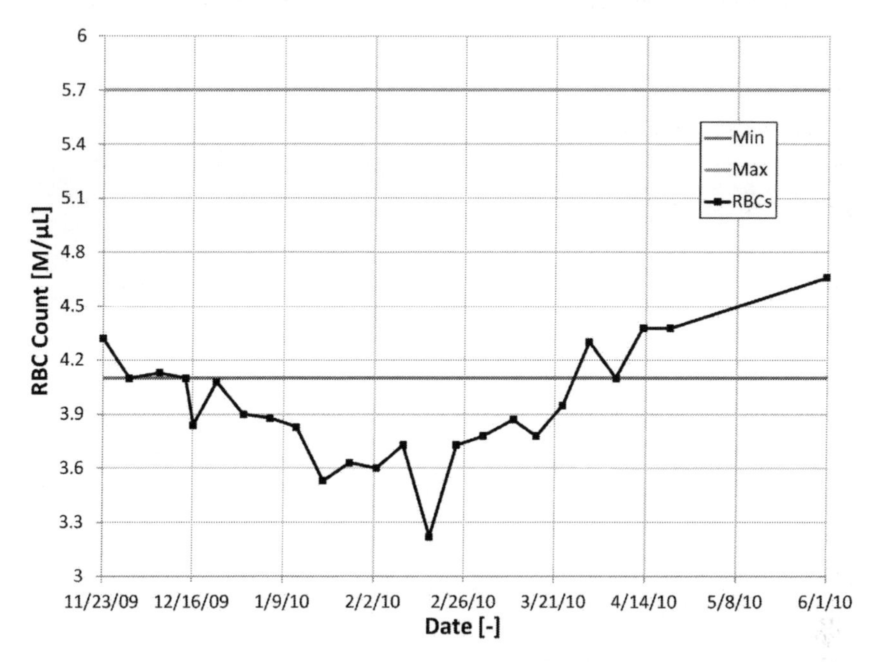

Figure 38: RBC counts during and after my treatment.

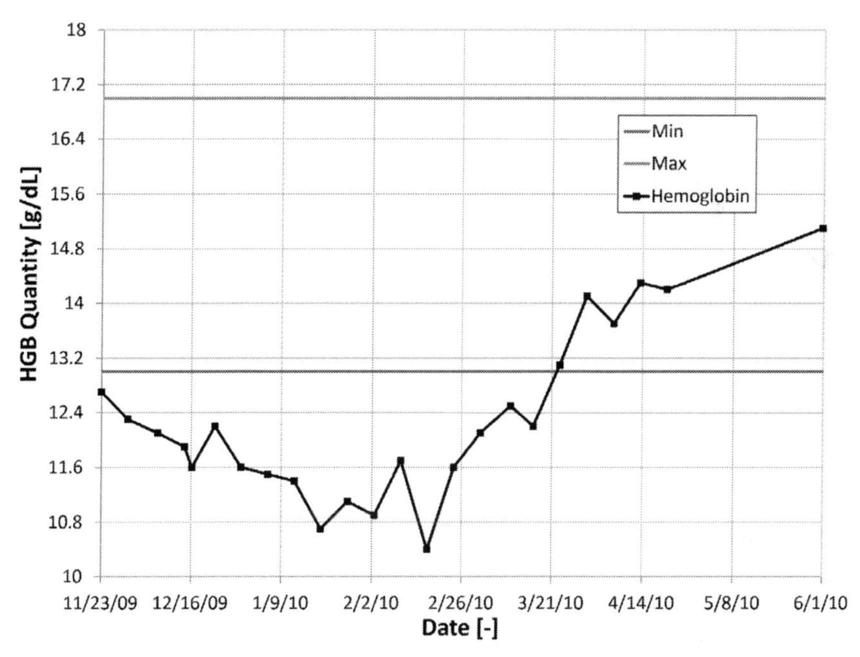

Figure 39: Hemoglobin quantity during and after my treatment.

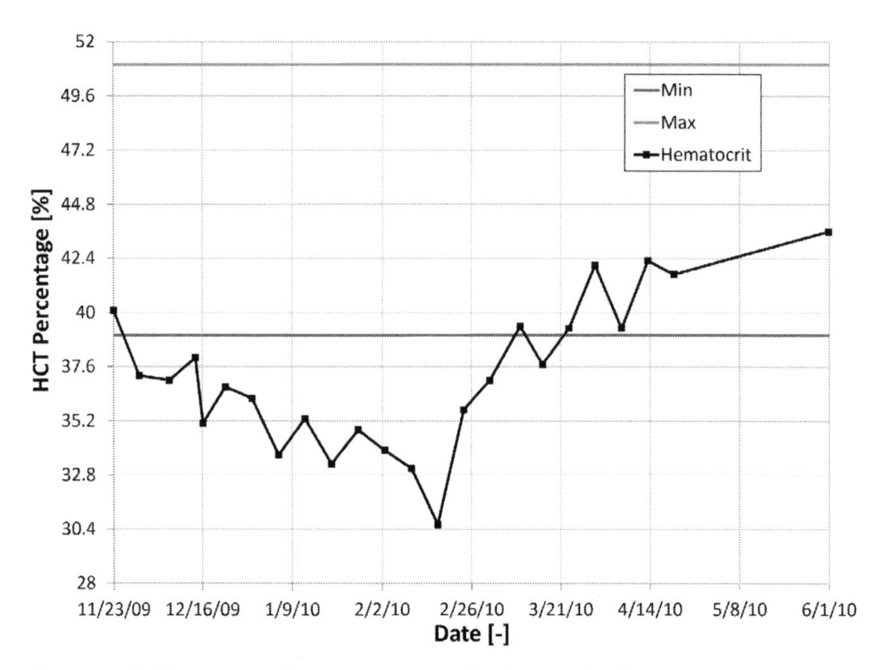

Figure 40: Hematocrit percentage during and after my treatment.

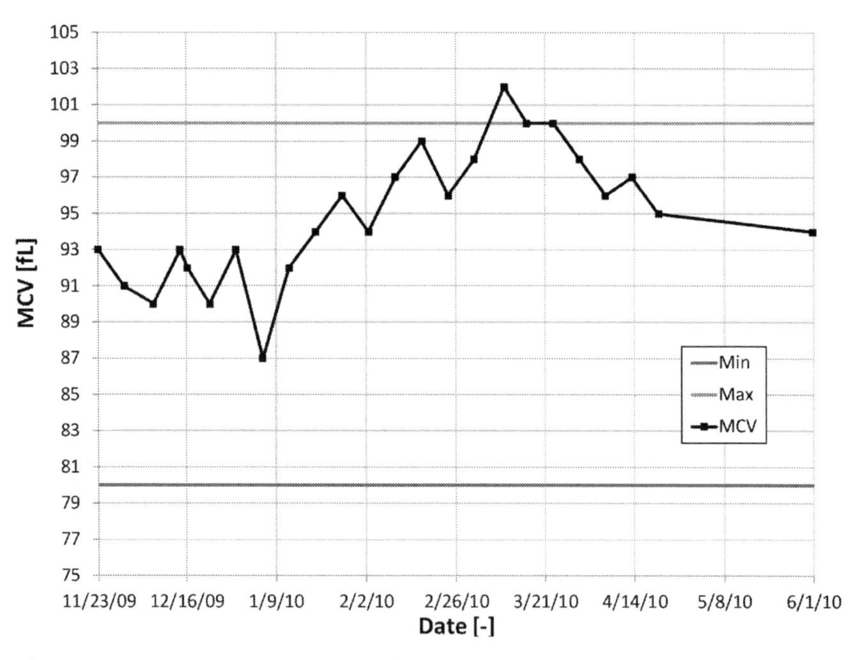

Figure 41: Mean corpuscular volume during and after my treatment.

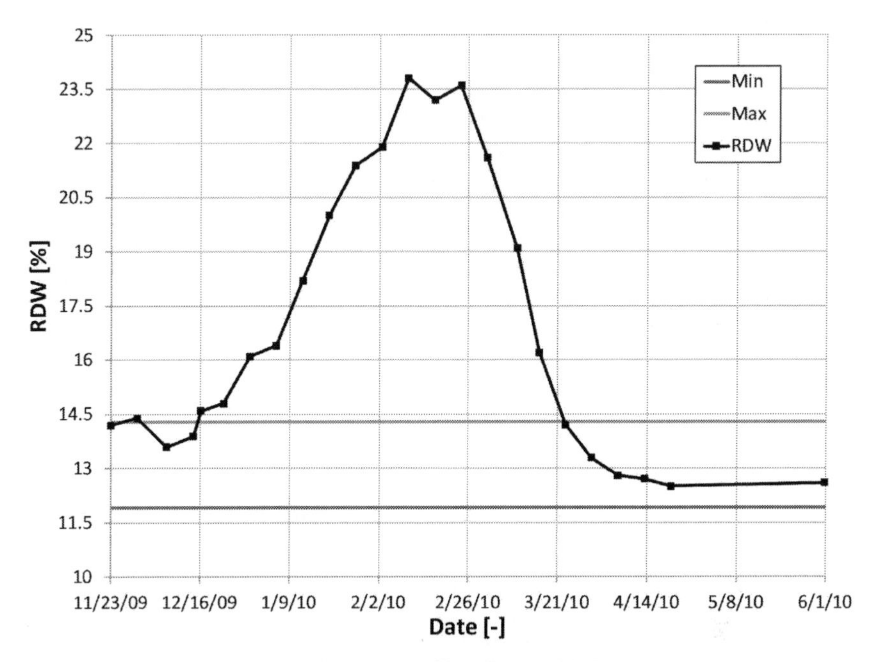

Figure 42: RBC distribution width (RDW) during and after my treatment.

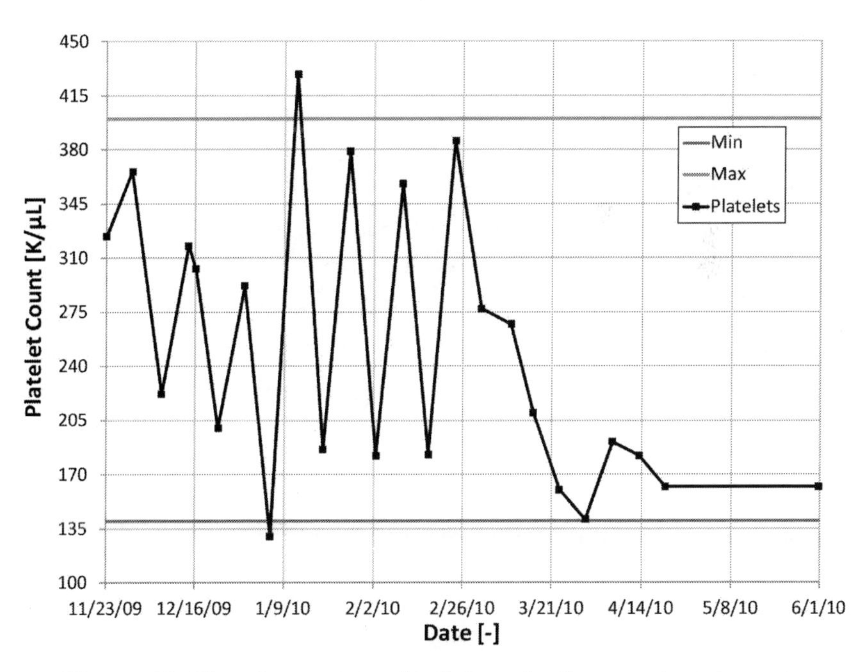

Figure 43: Platelet counts during and after my treatment.

WBC Differential

WBC differential looks at the distribution of WBCs throughout your treatment. Counts were more volatile throughout my treatment, as can be seen by the data in Table 4 and Figure 44. Before treatment, WBC counts were relatively normal with all WBC types within their standard distribution. While WBC counts decreased after chemotherapy began, the impact was more delayed compared to RBC counts, which immediately began to fall off. Since the first treatment consisted of three drugs that cause low blood cell counts (doxorubicin, mechlorethamine and vinblastine), the larger drop-off after the second treatment was most likely a cumulative effect of these drugs plus the vincristine taken during the second treatment on December 10.

There were three noticeable WBC spikes during my treatments occurring on December 30, January 13 and January 27, with the December 30 treatment particularly causing a drastic increase. Blood samples were generally taken the day before treatment, which allowed about six days of recovery between the chemotherapy sessions and the blood tests. This recovery period, coupled with the body's fight against the tumor, can account for the increases in WBC counts. However, the data is not consistent; on some weeks the WBC counts remained in the low or normal range even with the recovery time. With two WBC effects countering one another—an increase in WBCs to fight the cancer and a decrease in WBCs due to the chemotherapy drugs—predicting the data becomes more difficult compared with the RBC data. The fact that I had a blood cancer further complicates the predictions because lymphoma produces additional lymphocytes.

Finally, variables such as the time of day, time since last treatment, water and food intake, rest, type of chemotherapy drugs taken—Stanford V varies the type every week—and laboratory error also impact the data. The most reasonable conclusion to be drawn is that these competing variables caused the oscillatory nature of the WBC data, with periods of increased WBC counts from the lymphoma itself and a decrease in WBC counts due to the drugs.

Looking a bit deeper into the distribution of WBCs, Table 4 and Figure 44 show the breakdown between the five main types of WBCs: neutrophils, lymphocytes, monocytes, eosinophils and basophils. This distribution may vary depending on your cancer type. Once again, normal ranges are displayed in the table, with values below normal in blue and above normal in red. Basophils and eosinophils were within their normal range before, during and after treatment. These two types of WBCs are associated with inflammation and the eosinophils in particular were higher before and near the beginning of my chemotherapy treatment, likely from fighting the lymphoma.

Monocytes and lymphocytes followed a similar pattern, with elevated levels near the beginning of treatment and reduced values during the middle and end of treatment. Neutrophils were generally elevated throughout the cancer treatment with the exception of a large dip in December that coincided with large increases in lymphocyte and monocyte counts. This variability makes it difficult to draw conclusions.

Although a different combination of drugs was taken in the middle of December versus the end, the same two drugs, bleomycin and vincristine, were taken again on January 7 and follow-up blood tests did not show the same low neutrophil and high lymphocyte/monocyte concentrations. Also, the distribution was fairly normal before chemotherapy began, suggesting that the cancer itself was not having a large impact on these concentrations.

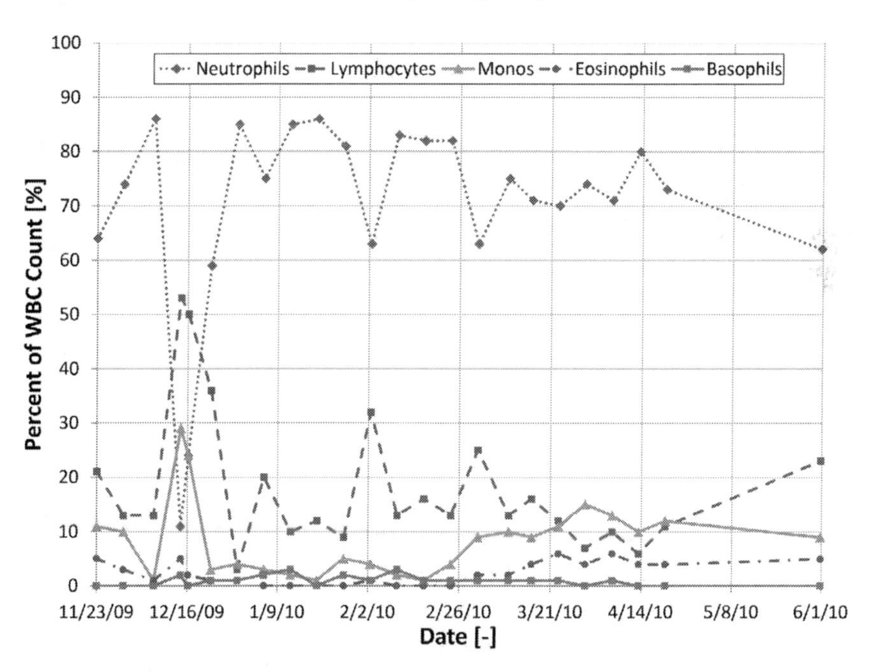

Figure 44: White blood cell count distribution during and after my treatment.

Table 4: White blood cell results for my Hodgkin's lymphoma treatment

Date	Neu-tro-phils	Lym-pho-cytes	Monos	Eosin-ophils	Baso-phils	Meta_ myelo-cytes	Myelo-cytes	Total	Neu-tro-phils
[-]	[%]	[%]	[%]	[%]	[%]	[%]	[%]	[%]	K/uL
Min.	41	13	5	0	0	N/A	N/A	59	2.1
Max.	79	44	14	6	2	N/A	N/A	145	7.4
11/23/09	64	21	11	5	0	N/A	N/A	101	4.3
11/30/09	74	13	10	3	0	N/A	N/A	100	7.1
12/8/09	**86**	13	1	1	0	N/A	N/A	101	6.9
12/15/09	**11**	**53**	**29**	5	2	N/A	N/A	100	N/A
12/17/09	**24**	**50**	**24**	2	0	N/A	N/A	100	N/A
12/23/09	59	36	3	1	1	N/A	N/A	100	2.1
12/30/09	85	**3**	**4**	N/A	1	3	4	100	N/A
1/6/10	75	20	3	0	2	0	0	100	4.3
1/13/10	**85**	**10**	**2**	0	3	0	0	100	**12.6**
1/20/10	**86**	**12**	**1**	0	0	0	0	99	**9.2**
1/27/10	**81**	**9**	5	0	2	2	1	100	N/A
2/3/10	63	32	**4**	1	1	0	0	101	3
2/10/10	**83**	13	**2**	0	3	0	0	101	**9.1**
2/17/10	**82**	16	**1**	0	1	0	0	100	7.2
2/24/10	**82**	13	**4**	0	1	0	0	100	4.4
3/3/10	63	25	9	2	1	0	0	100	N/A
3/11/10	75	13	10	2	1	0	0	101	4.5
3/17/10	71	16	9	4	1	0	0	101	2.2
3/24/10	70	**12**	11	6	1	0	0	100	2.5
3/31/10	74	**7**	**15**	4	0	0	0	100	N/A
4/7/10	71	**10**	13	6	1	0	0	101	2.1
4/14/10	**80**	**6**	10	4	0	0	0	100	N/A
4/21/10	73	**11**	12	4	0	0	0	100	N/A
6/1/10	62	23	9	5	0	0	0	99	3.2

Key: The first grouping of dates in bold are weeks with chemotherapy while the second bold grouping are weeks with radiation. Bold data falls outside of the recommended range listed at the top of the table.

This further validates the conclusion that cancer and the impact of chemotherapy and radiation have many variables that influence WBC distribution. As a patient, you should track your values and point out anomalies such as spikes in data to your oncologist, making sure there are no issues with your treatment.

Additional Blood Tests

On top of your CBC and WBC distribution, your oncologist may order additional tests. Mine are discussed in the following sections.

Total Bilirubin

When hemoglobin in old RBCs breaks down, it turns into the organic molecule bilirubin. Bilirubin circulates in the bloodstream before reaching the liver, where it is converted into one of the components making up bile. This biotransformation—a chemical modification of an organism's molecule—occurs when bilirubin conjugates (the process of making a molecule water-soluble) with glucuronic acid via the enzyme UDP-glucuronyltransferase.[42] Now termed direct bilirubin, as opposed to the indirect non-soluble version, the majority of this orange-yellow liquid gets mixed with similar bile pigments.[43] These include the brownish stercobilin, which gives feces its color, as well as the green biliverdin and the bilirubin and biliverdin derivatives urobilinogen, urobilin, bilicyanin and bilifuscin.[44]

The remaining bile constituents are mostly water, with smaller contributions from salts, cholesterol, fatty acids, lecithin, sodium, potassium, calcium, chlorine and bicarbonate ions. Bile formed in the liver gets stored in the gallbladder, where it is excreted via bile ducts into the small intestine when needed for digestion. Eventually bile waste gets passed out of the body through stool.

Excess bilirubin in the bloodstream is an additional sign of liver failure. As one can see by the trends of these blood tests, liver function

42 UDP stands for uridine diphosphate, a nucleotide diphosphate.

43 Christine Case-Lo, "Bilirubin Blood Test," *Healthline,* https://www.healthline.com/health/bilirubin-blood#risks.

44 "Bile Pigment," *The Free Dictionary by Farlex,* https://medical-dictionaryx.thefreedictionary.com/bile+pigment.

is monitored closely throughout treatment. This is particularly important during chemotherapy due to the toxic nature of the chemicals entering the bloodstream and the potential damage they can do as your liver tries to process and purge them. Excess levels of bilirubin can cause jaundice, which leads to a yellowing of the skin and sclera—the whites of the eyes.

Examining the total bilirubin levels (TBILI) from my blood work (Figure 45), they fluctuated back and forth during chemotherapy but remained within standard values. Values actually went up slightly toward the end of my treatment while I was under radiation.

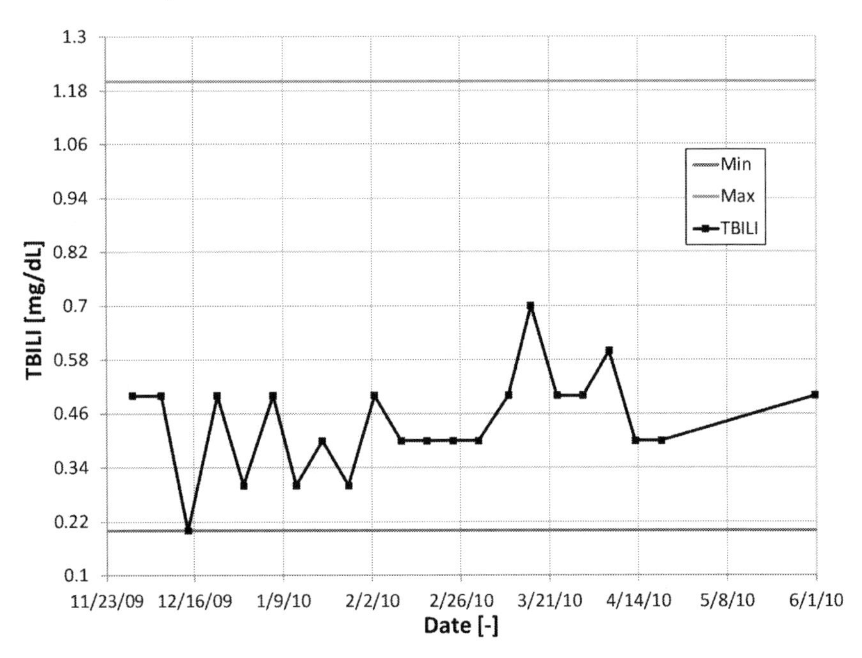

Figure 45: Total bilirubin levels during and after my treatment.

This suggests that TBILI is not as strong an indicator of liver function as other to be discussed tests (i.e. alanine aminotransferase (ALT), alkaline phosphatase (ALP) and lactate dehydrogenase (LDH)), although more patient samples would need to be analyzed to strengthen this conclusion. Additionally, the fact that the levels go up is interesting. Most of my tumor broke down during radiation treatment, which coincided with the rise in values for ALT, ALP and LDH as the liver had to eliminate the excess dead cells. Higher values during radiation may be a statistical anomaly or a real trend; however, additional data is required to draw solid conclusions.

Prothrombin Time (PT)

Prothrombin time (PT) shows how long it takes your blood to clot, measured in seconds. An additional way to represent this is using the PT international normalized ratio (INR), which is the same data but presented as a ratio of the patient's value to a normal value. The standard range of PT at my laboratory was 8.2–10.6 seconds, but ranges are dependent on how the test is done (e.g. a range of 10–12 seconds may be considered normal at a different laboratory). A PT INR reading of < 1 means your blood is thicker than average while a PT INR ratio > 1 means it is thinner than average (a value of 0.8–1.2 is considered normal and healthy). While on warfarin or a similar blood-thinning medication, doctors want your PT INR to be between 2 and 3. Figure 46 plots my values throughout treatment, although more data exists after my chemotherapy began since a blood clot in my right biceps caused me to have to go on blood thinners and add PT as well as PT INR to my blood work. This is discussed in more detail in the *Side Effects* chapter.

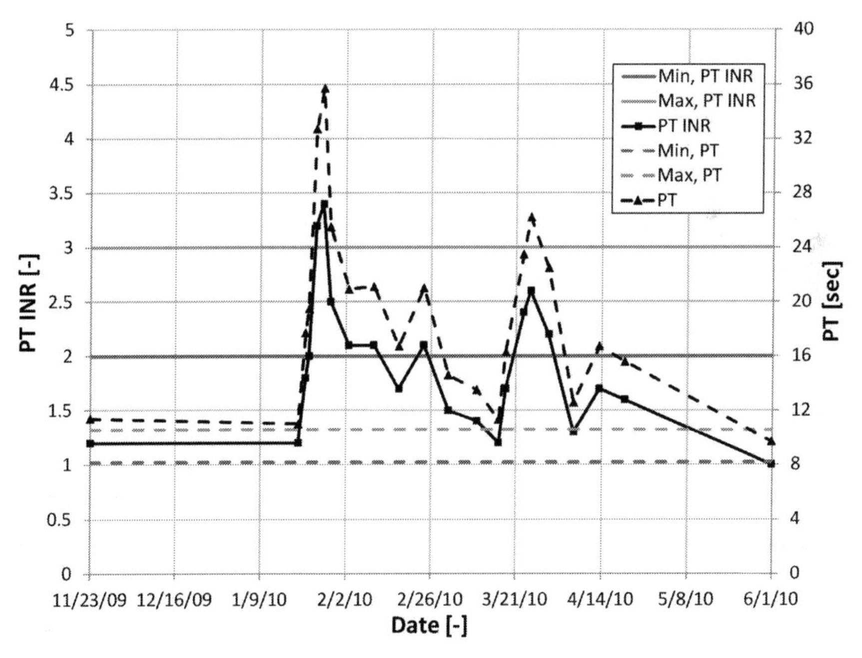

Figure 46: Blood thickness during and after my treatment (lower values = thicker blood).

Creatinine and Calculated Glomerular Filtration Rate

These tests are used to measure kidney function, which can become impaired from certain types of chemotherapy treatment (see the *Chemotherapy and Pills* chapter for drugs that have this side effect). Creatinine is a waste byproduct of creatine, which is a molecule either produced by the body or ingested via food or supplement that is important in the ATP muscle energy process described in the *Vitamins* section. Removed by the body through the kidneys, excess creatinine is a sign of kidney failure. Healthy values are <1.34 mg/dL.

While my levels remained healthy throughout my cancer treatment, as shown in Figure 47, I ended up getting a kidney stone in early 2014 when blood tests showed values of 1.42 mg/dL. This resulted in me having robotic surgery to cut a section of my ureter out that was too narrow, reattaching it to my kidney with a larger opening. My surgeon suggested the condition was congenital, although with my past chemotherapy treatment and not having any kidney issues until age 32, the toxic chemicals ingested into my body when I was 28 may have contributed to the condition.

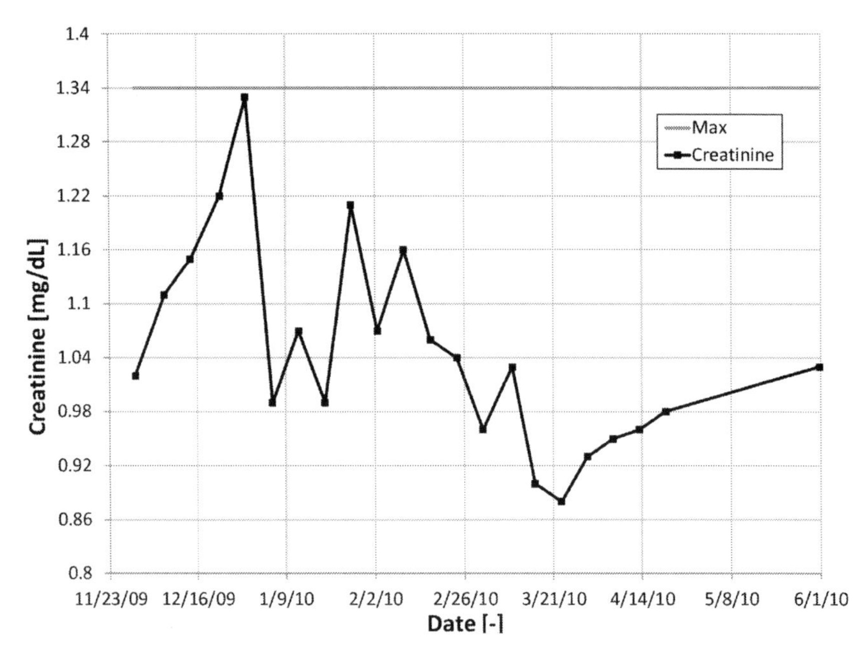

Figure 47: Creatinine values during and after my treatment.

Glomerular filtration rate is an additional method to measure kidney function, with values >60 mL/min considered healthy. Specifically, this test measures the amount of fluid filtered through the renal (kidney) glomerular capillaries into the Bowman's capsule per unit time. The equation is urine concentration multiplied by urine flow divided by plasma concentration.

LACTATE DEHYDROGENASE (LDH)

LDH is an enzyme in the body involved in energy conversion, transforming lactic acid into pyruvic acid and vice-versa as part of the Krebs cycle. LDH is produced in body tissues including the lungs, heart, brain, kidneys, liver and muscles, as well as in blood cells. When these tissues are damaged, excess LDH is produced. Therefore blood cancers—along with liver, lung, kidney, brain and heart cancers—are particularly coupled with elevated LDH levels.

During my treatment, LDH levels had a delayed spike a few weeks into chemotherapy (Figure 48). This was likely due to the time necessary for the chemotherapy drugs to have an impact on killing both the tumor and healthy cells. Values were elevated throughout the chemotherapy

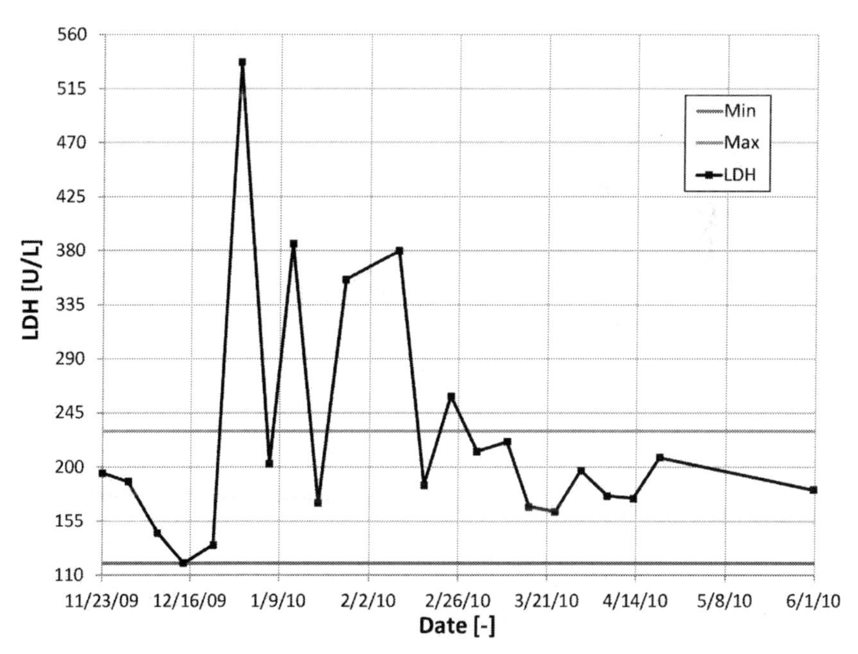

Figure 48: LDH levels during and after my treatment.

treatment, often exceeding the 230 U/L healthy limit, with 130 U/L being the healthy lower limit. Levels began dropping during radiation, indicating that although cells continued dying, most of the cell deaths occurred earlier in the treatment from chemotherapy. This is consistent with PET and CT scans, which showed the lion's share of my tumor shrinkage occurring before the onset of radiation treatment.

URIC ACID

Uric acid is produced by the body breaking down purines, which are organic nitrogen-containing molecules found in almost all foods you eat. Purines are important in DNA/RNA synthesis and play a role in other body biomolecules including ATP, guanosine-5'-triphosphate[45] (GTP), cyclic adenosine monophosphate[46] (AMP), coenzyme A and nicotinamide adenine dinucleotide (NAD). Uric acid gets filtered through the kidneys and is released via urine. Excess uric acid can lead to gout or kidney stones.

Cell death also produces purines, which is why uric acid is monitored closely during certain cancer treatments, most notably lymphoma and leukemia. The large size of my tumor, roughly the volume of a football, had my oncologist paying attention to my uric acid levels. As seen by the plot in Figure 49, values rose slightly during treatment but stayed within the healthy range of 2 and 8.5 mg/dL.

Diet also plays a role in these levels, where foods high in purine content should be avoided. These foods include meats (especially organ meats such as liver, brain and kidneys), certain seafood (anchovies, herring, mackerel, sardines and scallops), gravy, beer and wine.[47] To avoid high uric acid levels during treatments, large quantities of meat are not advised, although healthy portions of meat[48] or a substitute protein source are important to maintain your protein intake.

45 Other names include guanylyl imidodiphosphate and guanosine triphosphate.

46 Other names include cAMP, cyclic AMP and 3',5'-cyclic adenosine monophosphate.

47 Amber Gabbey and Rachel Nall, "Uric Acid and the Uric Acid Blood Test," *Healthline*, https://www.healthline.com/health/uric-acid-blood.

48 For example, 6 oz of meat daily is considered a healthy amount of protein for most adults, although portions should be adjusted based on size, fitness goals and recommended percentage of body fat (PBF) for men and women. See *Fatty Acids* section for recommended PBF ranges.

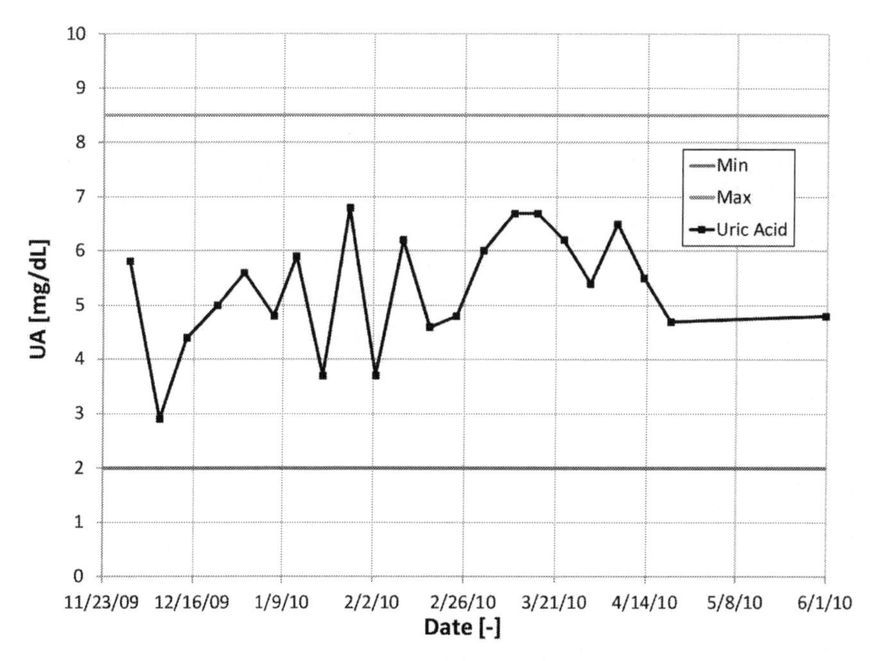

Figure 49: Uric acid levels during and after my treatment.

Alanine Aminotransferase (ALT) and Aspartate Aminotransferase (AST)

Liver health is measured by ALT and AST levels, which are both enzymes found in RBCs. These enzymes are also found in the liver, heart, kidneys, brain and skeletal muscles. ALT and AST are transaminase enzymes, molecules that catalyze reactions between amino acids and α-keto acids as part of the Krebs cycle[49] and glycolysis, both of which convert stored energy into usable free energy.

High levels of these enzymes are a sign of liver damage, which unless the cancer is on the liver itself, can be caused by chemotherapy agents.[50] If the patient has preexisting liver issues caused by conditions such as alcoholism, their levels will start higher but still increase. As seen by the data from my blood tests in Figures 50 and 51, both ALT and AST had an average rise with my first treatment on December 3 and final treatment on February 18.

49 Note the Krebs cycle is also referred to as the citric acid cycle or the tricarboxylic acid cycle.

50 Eric Roth, "Chemo Effects on the Liver," *Livestrong Foundation,* https://www.livestrong.com/article/22807-chemo-effects-liver.

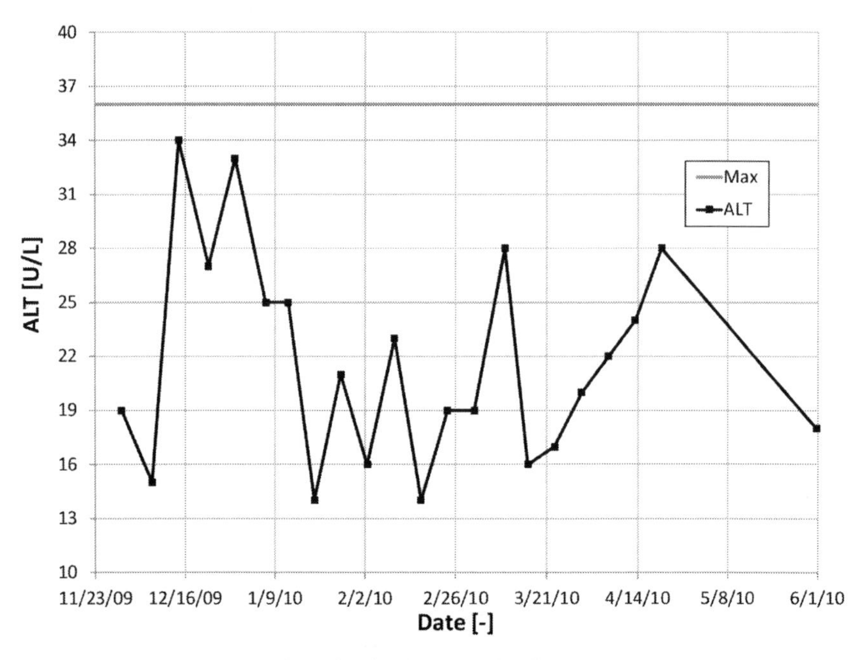

Figure 50: ALT levels during and after my treatment.

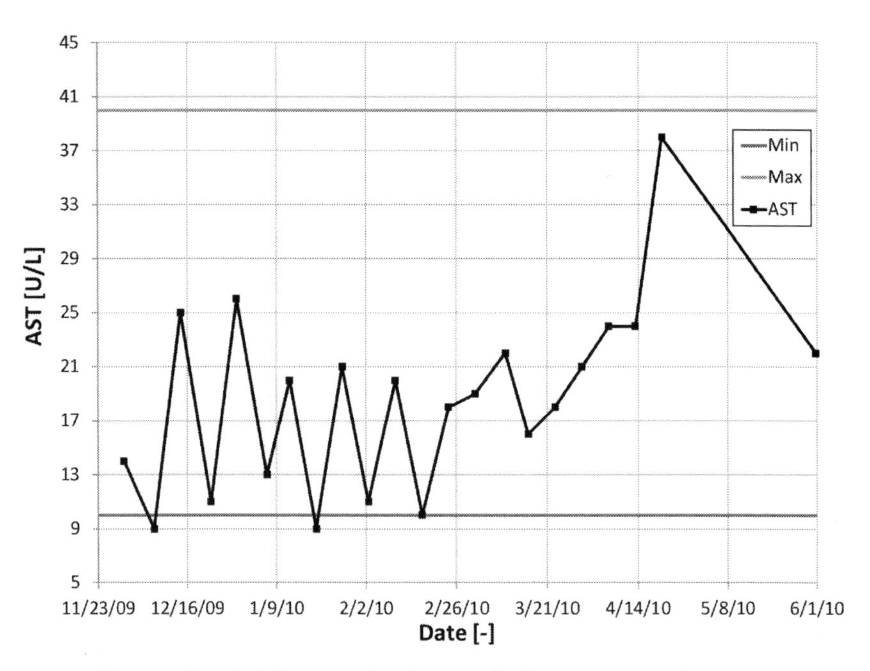

Figure 51: AST levels during and after my treatment.

While ALT values had a pronounced spike, my AST values were more cyclical and harder to draw conclusions from. Additionally, none of the chemotherapy agents I took had liver enzyme increase as a side effect; however, the data clearly shows impacts to AST due to chemotherapy as my liver had to work harder to process the poisonous chemotherapy agents. Both ALT and AST levels continued to increase after chemotherapy and throughout radiation. It is not clear whether this was due to prolonged effects of the chemotherapy or if the radiation itself had an impact.

With my last blood work on June 1—my final radiation treatment was on April 9—ALT levels were back to pre-chemotherapy values while AST values were reduced from the previous readings but not back to pre-chemotherapy values. In general, ALT levels showed more correlation with cancer treatment than AST levels. Note that while levels increased and got close to maximum safe values, they never exceeded their thresholds of 36 units per liter[51] (U/L) for ALT and 40 U/L for AST (with 10 U/L as the lower healthy limit for AST).

Alkaline Phosphatase (ALP)

Alkaline phosphatase (ALP) is a hydrolase enzyme—an enzyme that catalyzes (accelerates) the hydrolysis, or breakdown, of a chemical bond—that removes phosphates from various molecules within the body, including proteins, nucleotides and alkaloids. ALP is present throughout the body, most prominently in the liver, kidneys, intestines, bones, WBCs and, specifically for pregnant women, the placenta. Aside from elevated levels due to cancers impacting these regions of the body, ALP is another barometer to track liver function during cancer treatment.

My blood levels (Figure 52) had a large ALP spike between mid-December and mid-January, beyond the healthy range of 37–117 U/L, with elevated levels up until the end of chemotherapy treatment in February. These increased values correlate well with the other liver-functionality blood tests (i.e. ALT and LDH), demonstrating that my liver was overstressed throughout my chemotherapy treatment. While the chemotherapy drugs and cell death continued throughout the treatment, this is the portion where cell death was at its peak.

51 U is an abbreviation for IU, or international unit. This is a reference value that varies for each substance. For instance, 1 IU of insulin is equal to 0.0347 mg, or a 28.8 IU/mg conversion factor.

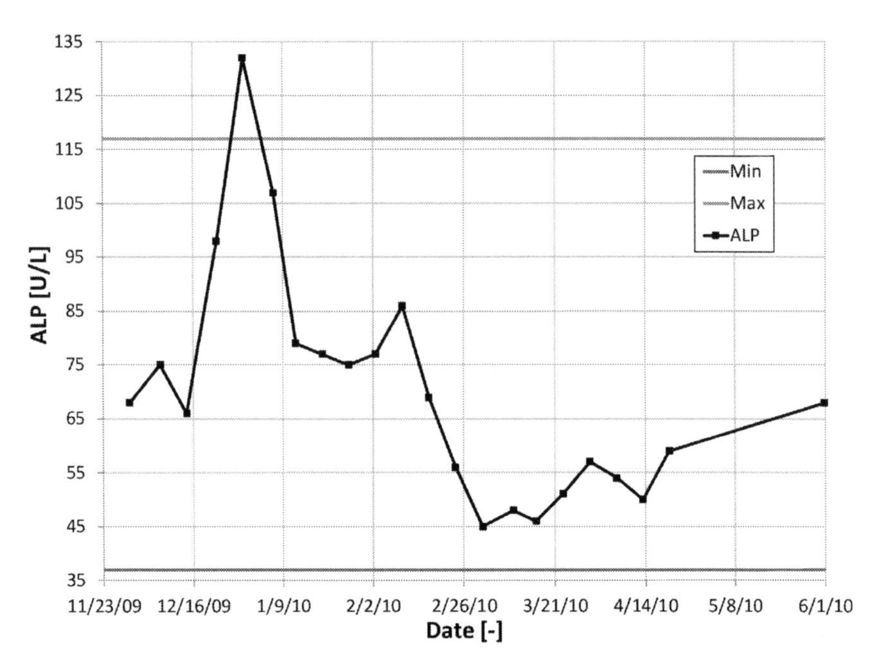

Figure 52: ALP (also referred to as ALKP) levels during and after my treatment.

Chapter 11

Radiation

Radiation therapy, along with chemotherapy and surgery, is one of the three main techniques used to fight cancer. Either one or a combination of the three is used depending on factors such as the patient's cancer type, stage, tumor size, tumor location, health and age. What type of radiation, how much dosage is given, the dosage frequency and the radiation footprint on the body are also tailored based on these factors and the radiation oncologist's expertise.

Your oncologist will recommend a treatment course after you complete your various imaging and diagnostic tests. For tumors that are shown to be local in nature, surgery is often the first choice, as it likely will be the least strenuous and detrimental to your body. An example is breast cancer, where localized surgery is often used to remove sections where cancer is present and has failed to metastasize. When metastasis is shown or feared, chemotherapy is commonly used since it works on cancer cells throughout the body that either would require too much surgery or are difficult to completely identify for surgical purposes.

Radiation is generally used to supplement chemotherapy by killing cancer cells and reducing the likelihood of a recurrence. Exactly which combination of therapies is very patient-specific. Your oncology team, including your primary oncologist and radiation oncologist, will consult and recommend a course of action. As stressed throughout this book, arming yourself with knowledge and asking questions will only improve your treatment and recovery.

There are three main types of radiation: external beam radiation therapy (EBRT), internal radiation therapy (also known as brachytherapy) and systemic radiation therapy.

External Beam Radiation Therapy (EBRT)

Intensity Modulated Radiation Therapy (IMRT)

This form of therapy generally uses a complex machine called a linear accelerator (LINAC), shown in Figure 53. Like a CT scan, electrons are produced by an electron gun where they are pulled off a hot tungsten filament cathode via a positively charged anode. Energy is required to accelerate these electrons, which is done via a magnetron (Figure 54). The magnetron creates a microwave, likewise using a heated cathode and anode combination; however, rather than the stripped electrons reaching the anode, they are reflected using magnets into a circular pattern. Machine-resonant cavities in the anode generate microwave radiation via transferred energy from the circulating electrons with a frequency dependent on the cavity size (Figure 54). This microwave radiation gets directed into the accelerator waveguide through a transfer waveguide, where it combines with electrons from the electron gun and accelerates them toward the other end of the waveguide.

The accelerator waveguide has both steering and focusing coils (Figure 55). Electrons from the electron gun first encounter the steering coils, which are magnets that deflect electrons toward the center axis of the waveguide; steering coils are also used to direct the beam once it exits the waveguide. Upon entering the main region of the accelerator waveguide, electrons encounter a series of copper electrode irises, which divide the waveguide into multiple cylindrical cavities. These cavity lengths are tailored to the particle of choice—in this case an electron—and designed such that their spacing and potential allow for the microwave's oscillating electric potential field to accelerate the electrons. Starting with a velocity on the order of 40% the speed of light, these electrons speed up to 99.9% the speed of light by the time they exit the waveguide. Focusing coils help maintain the electrons along the center of the waveguide axis during acceleration.

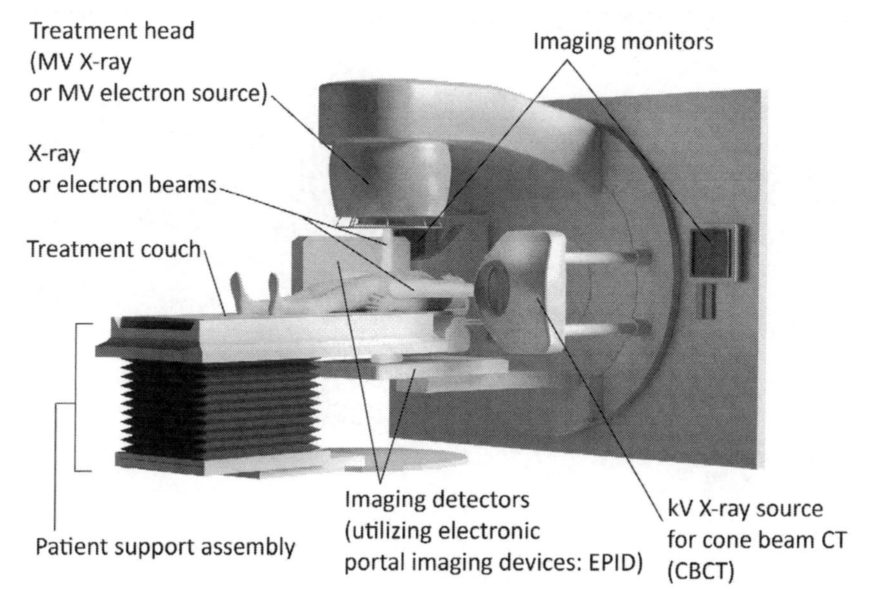

Figure 53: Front view of LINAC with patient being treated.

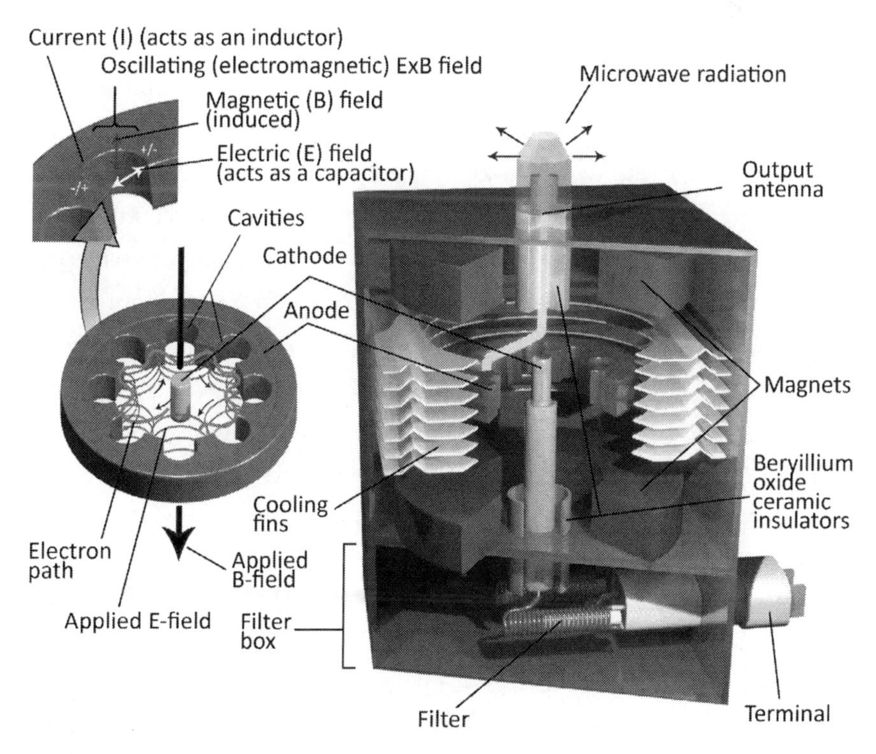

Figure 54: How a magnetron produces microwave radiation.

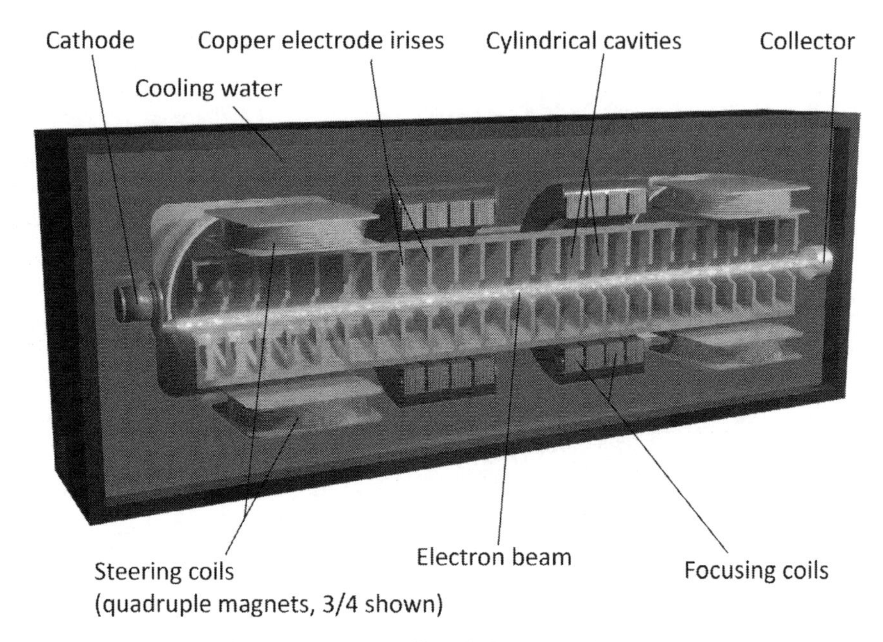

Figure 55: Details of the accelerator waveguide portion of a LINAC.

Electrons are then deflected 270° via quadrupole magnets, which are a form of achromatic bending. Achromatic bending is the opposite of chromatic bending, in which electron trajectories get segregated based on their speed. When electrons exit the waveguide they are not moving at the same speed, so the magnets are designed to get thicker with radial distance from the central beam axis. This causes faster particles to travel farther and by the time the electrons bend 270° they are going in the same direction once again with an exit-beam diameter of around 1 mm. Note that another variation of achromatic bending is achieved by three magnets in unison where the electrons are bent twice at 22.5° in opposite directions—think of a skier negotiating turns—followed by a final bend of greater than 90° to achieve the same final direction as the 270° turn magnet design.

Electrons then enter the treatment head, where the electron beam gets converted into a tailored X-ray beam specific to the location and shape of the patient's tumor. To convert the electrons into X-rays they collide with tungsten, which produces X-rays as described in the Computed Tomography (CT) Scan section via *characteristic X-ray* or *Bremsstrahlung X-ray production*. These X-rays enter a collimator, where only X-rays in the direction of the patient are allowed through. A flattening filter made of metals such as aluminum, tungsten or lead enhances the beam energy isotropy by absorbing more X-rays near the center than the perimeter, resulting in a near-uniform energy density. X-rays then enter ionization chambers for beam diagnostics and control. Ionization chambers are filled with a gas that is readily ionizable with an anode on one end and a cathode on the other. When incident radiation ionizes the gas, positive ions and electrons are produced and migrate toward their respectively charged electrodes. The current produced during this process can be measured by an electrometer, which can detect low currents. Specifically, three ionization chambers are utilized. The first chamber is for diagnostics and shuts off when the programmed dose radiation amount has been reached. While the second is merely a backup of the first, the third chamber is used for more precise monitoring as it compares the beam's profile with the programmed profile via a beam-quality function.

A multi-leaf collimator made of tungsten allows for final tuning of the beam shape by moving multiple metal leaves, also referred to as jaws, to get the desired shape for the patient's tumor. A lamp, lasers and mirrors are also located in the treatment head to help with aligning the patient's body. Due to the powers involved, a water pump cooling system is used to cool the magnetron, waveguide, beam transport system and collimator.

The patient sits on a couch—really a metal slab that can rotate around a pillar—that has four degrees of freedom, moving left-right and forward-back along the plane of the couch's surface, up-down (toward and away from the floor) and rotating around the beam's axis, i.e. yawing. Note that degrees of freedom depend on the specific LINAC design, with some companies including pitch and roll for all six degrees of freedom. Three degrees of freedom were used for my setup; there was no yawing, pitching or rolling.

A schematic of a LINAC with its full engineering detail is shown in Figure 56, along with three-dimensional representations in Figures 57 and 58.

One undesired effect of the LINAC is that neutrons and protons can be produced by X-rays interacting with material in the treatment head, where neutrons contribute to the patient's dosage while protons generally get absorbed within the treatment head. Additional neutrons are undesirable, as they increase the patient's prescribed radiation dosage and are more likely to hit healthy tissue, since unlike for X-rays, neutrons are not calibrated with penetration depth.[52] Electron beams—which can also be administered to the patient instead of X-rays—have the same issue. Neutrons can be produced from the treatment head's absorption of electrons; however, the reaction probability and resulting neutrons is much lower when in electron beam mode.

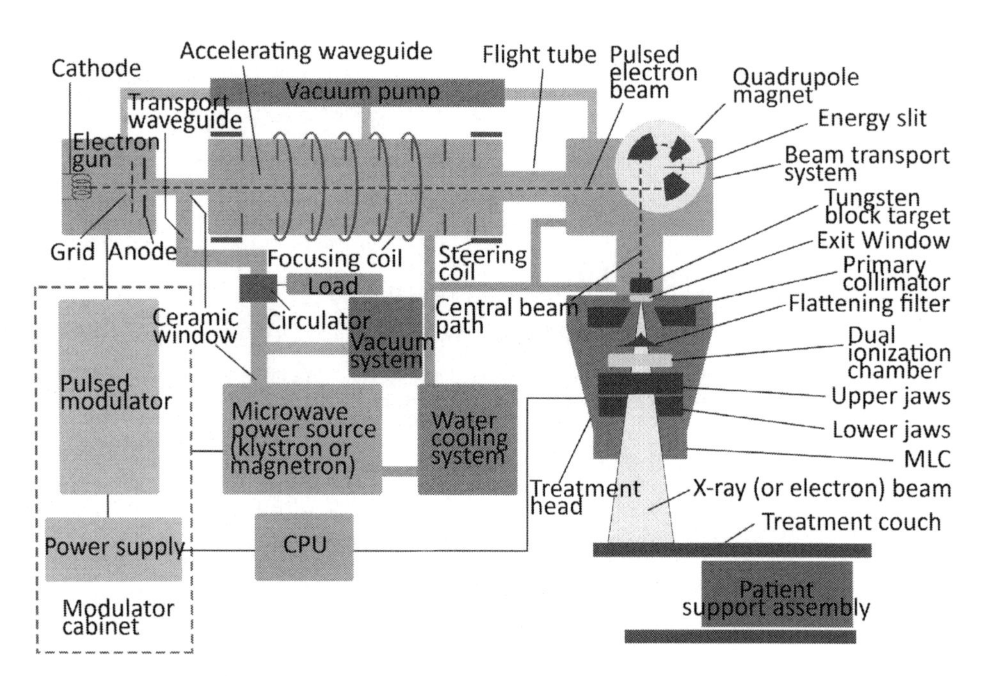

Figure 56: Detailed hardware overview for a LINAC.

Activated material in the treatment head can also expose technicians and medical personnel to radiation during maintenance, yet due to the short half-life, this can be avoided by waiting a few hours after the end of the last treatment for the radiation to dissipate. In order to protect doctors and staff from radiation, the LINAC is housed in a room with thick concrete walls.

52 Isra Israngkul-Na-Ayuthaya, Sivalee Suriyapee, and Phongpheath Pengvanich. "Evaluation of Equivalent Dose from Neutrons and Activation Products from a 15-MV X-ray LINAC," *Journal of Radiation Research* 56, no. 6 (Aug. 2015): 919–926, https://www.ncbi.nlm.nih.gov/pmc/articles/PMC4628218.

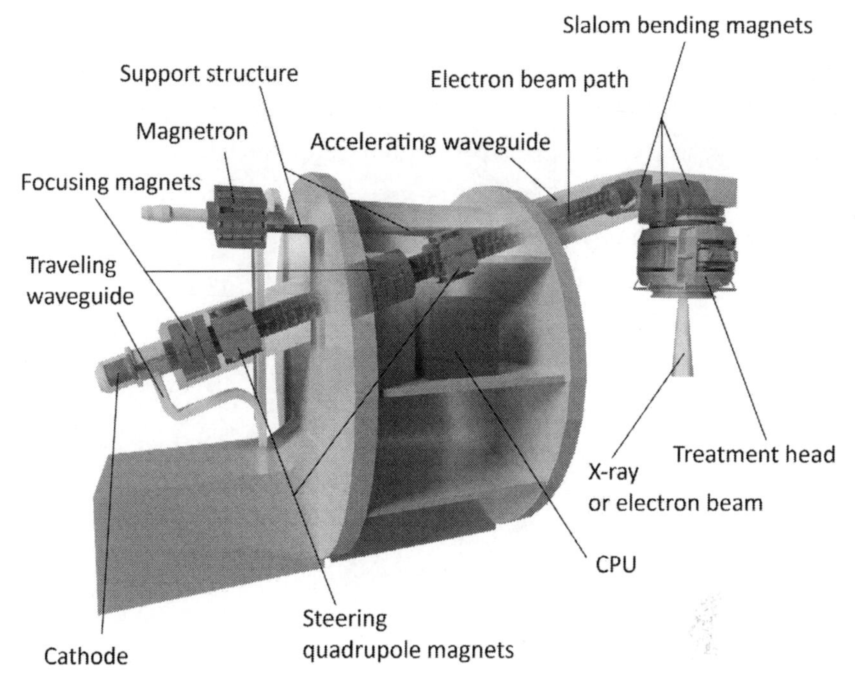

Figure 57: Side view 1 of LINAC during operation.

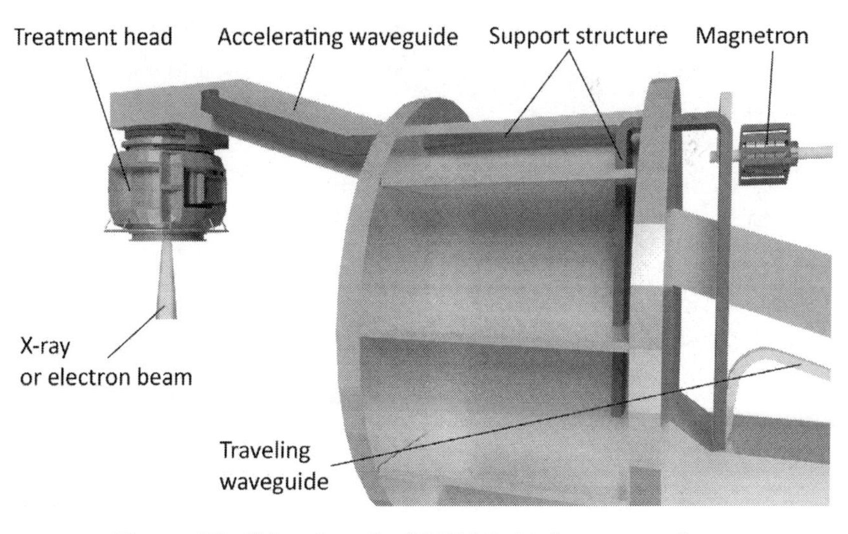

Figure 58: Side view 2 of LINAC during operation.

Setup

In order to properly align patients for treatment and to produce repeatable positions, which is very important for targeting the tumor and limiting the amount of radiation on healthy cells, a combination of laser alignment, markers, stickers, tattoos and a plastic conforming mask are used. During my initial setup, a flat piece of plastic was heated and stretched across my face as it cooled to retain shape; during treatment, this mask was secured over my face in order to properly position my head and prevent movement. Three tattoos—small green pin pricks, unfortunately not cool tattoos—were needled into my skin, one on each shoulder and the third on top of my chest to make a straight line. Although additional tattoos going down my chest were discussed, I successfully lobbied for stickers to avoid a tattooed cross of dots on my chest.

Radiation therapists administer these initial setups and will operate the LINAC throughout your treatment. Additionally, a medical radiation physicist and dosimetrist work with your radiation oncologist to plan your treatment. The physicist oversees the LINAC, making sure that it is operating properly. This person also oversees the dosimetrist, who is responsible for calculating your radiation dosage to ensure the tumor gets enough radiation without overexposing healthy tissues. While I did not meet my medical radiation therapist or dosimetrist, they were involved behind the scenes planning my therapy. Most importantly you must choose a radiation oncologist you are comfortable with and confident in, which fortunately I was able to.

Treatment

During treatment, the gantry is programmed to move in an arcing pattern that maximizes the cross-section of the tumor while minimizing the dosage to nearby organs. Unfortunately my tumor was wrapped around my heart, actually displacing it downward toward my stomach and collapsing my lungs, which made the use of the gantry and accuracy of the LINAC especially important. To this day, I wonder if the radiation received by my heart will have any long-term consequences. I received 3,600 cGy to my mediastinum using an IMRT photon plan at 180 cGy per fraction. My hope is that by exercising as much as I do and eating healthy, I will be able to offset any damage done by my treatment.

Treatment duration, dosage and length are dependent on your cancer, the size of tumor, etc. I had twenty-five sessions broken up into five days a week with the weekends off. Since my chemotherapy had reduced the tumor significantly, I asked whether the radiation therapy was necessary due to the risk of radiating my heart and lungs. My radiation oncologist cited studies showing that radiation treatment reduces the chance of a recurrence. This is why choosing a cancer facility is so important. Not only should you check the facilities and LINAC technology, but you must also have confidence in your oncology team to recommend the best treatment for your diagnosis.

3-D Conformal Radiation Therapy (CRT)

3-D CRT was developed before IMRT and is less accurate at targeting a tumor, as shown in Figure 59. One uniform dosage flux is used in 3-D CRT versus a varied flux for IMRT, allowing the latter to conform better to the tumor and reduce the amount of healthy tissue harmed by radiation. A planner for 3-D CRT will manually optimize beam parameters such as the number of beams, directionality, shape and intensity with a final uniform flux dosage computed from these inputs. IMRT does the opposite: a total dose is entered and the beams are optimized based on the dose and provided 3-D imaging (from a PET or CT scan) for a varying spatial flux dosage. While 3-D CRT takes less time due to the lower complexity, IMRT can provide higher dosage to the tumor and is more likely to be recommended by your radiation oncologist.[53]

Hadron Therapy

Hadron therapy refers to any type of EBRT that uses neutrons, protons or positive ions rather than X-rays to target a tumor. These are all hadron particles, i.e. particles made of quarks held together by the strong nuclear force as opposed to, say, an electron, which is a lepton that combines with hadrons via the weak nuclear force. Hadron particles are sped up to velocities >60% the speed of light to generate ionizing radiation that

53 "3D Radiation Versus IMRT Versus Proton Therapy," *OncoLink*. https://www.oncolink.org/frequently-asked-questions/cancer-treatments/radiation-therapy/3d-radiation-versus-imrt-versus-proton-therapy;
Erjona Bakiu, Ervis Telhaj, Elvisa Kozma et al., "Comparison of 3D CRT and IMRT Treatment Plans," *Acta Informatica Medica* 21, no. 3 (Sept. 2013): 211-212, https://www.ncbi.nlm.nih.gov/pmc/articles/PMC3804479.

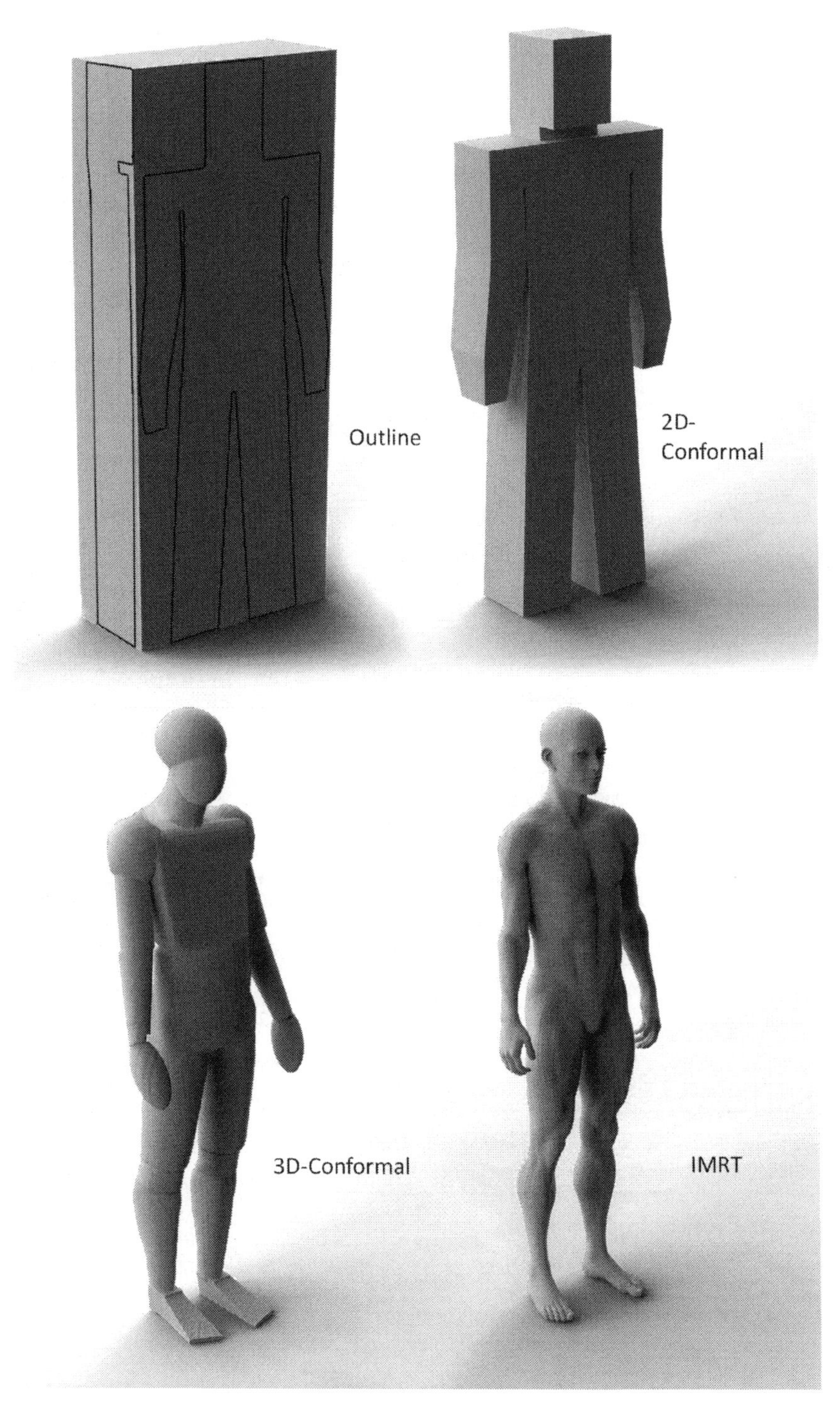

Figure 59: Figure showing the conceptual difference and evolution of 3D CRT versus IMRT.

has enough energy to elicit electron removal on their target molecules. This ionization damages the cell's DNA, which controls functions such as the ability to divide and replicate. Since cancer cells generally have more difficulty repairing damage than healthy cells, the net ionizing effect serves to destroy the tumor.

The most common type of hadron therapy used today is proton therapy. Proponents of proton therapy point to its capability of limiting unwanted radiation from interacting with healthy tissue after encountering the tumor, effectively restricting exit dosage. Protons can be programmed with specific velocities and energies that allow their primary energies to be imparted on the tumor itself rather than healthy tissue. Less-common forms of hadron therapy include fast-neutron therapy, heavy-ion therapy and carbon-ion therapy.

Proton therapy can be delivered using various devices. When using IMRT machines, proton therapy is referred to as intensity modulated proton therapy (IMPT). Proton therapy is also used in stereotactic radiosurgery (SRS), stereotactic radiotherapy (SRT), stereotactic body radiation therapy (SBRT), volumetric modulated arc therapy (VMAT) and brachytherapy.

Electron Beam Radiotherapy (EBR)

Like hadron therapy, EBR uses particles (electrons) rather than photons to treat cancer. Electrons also prevent exit dosage from damaging healthy skin tissue, as the electron stops at the cancer location. Due to their lower energy, electrons do not penetrate as far as hadron particles, which is why this therapy is used for cancer on or near the surface of the skin (generally above 6 cm in depth). Cancers treated include skin, breast, head and neck cancers. As mentioned in the *Intensity Modulated Radiation Therapy (IMRT)* section, LINACs can be configured to administer EBR.[54]

Total Body Irradiation (TBI)

TBI is used for patients preparing for a bone marrow transplant. In addition to killing cancer cells, this treatment suppresses the body's immune system to increase the chances of new bone marrow cells being accepted. Space is created for the new bone marrow cells by the death of cancer cells—and healthy bone marrow cells, unfortunately.

54 "Electron Beam Therapy," *SlideShare*, https://www.slideshare.net/ajjitchandran/electron-beam-therapy.

A rather rudimentary TBI stand (Figure 60) is used to support the patient—who is generally in an upright position while receiving treatment—although devices exist for lying down as well. This stand consists of a metal frame with a seat for support, handgrips and blocking metal to prevent X-ray radiation to the lungs.

Total lymphoid irradiation (TLI) and total marrow irradiation (TMI) are more specific forms of TBI in which radiation is focused on the lymph system and bone marrow, respectively.[55]

Image-Guided Radiation Therapy (IGRT)

IGRT is a radiation therapy technique that uses imaging to increase the accuracy of administered radiation. This technology is particularly useful for tumors located near organs that move considerably while breathing, such as the heart and lungs.[56]

IGRT is used in conjunction with other therapies and is therefore more of a technique than a therapy in and of itself. For instance, IGRT is used with IMRT, which in turn requires a LINAC for treatment.

Proton therapy, which uses a cyclotron or synchrotron to administer treatment, can also use IGRT. SBRT devices such as the CyberKnife use IGRT as well. All these devices have built-in equipment that utilize imaging technologies such as X-rays, CT scans, MRIs and ultrasounds.

Volumetric Modulated Arc Therapy (VMAT)

Introduced conceptually by Dr. Cedric Yu in 1995 and implemented after the turn of the century, VMAT is an advanced form of IMRT or SRS/SRT/SBRT that forgoes the need for CT scans prior to treatment, as it uses IGRT for more accurate radiation delivery. Radiation parameters can be adjusted in real-time and the total amount of time spent compared to traditional imaging is greatly reduced (1 to 2 minutes versus 10 to 20

---

55 Carson Wills, Sheen Cherian, Jacob Yousef et al., "Total Body Irradiation: A Practical Review," *Applied Radiation Oncology* 5, no. 2 (June 2016): 11–17, https://appliedradiationoncology.com/articles/total-body-irradiation-a-practical-review.

56 Kavitha Srinvasan, Mohammad Mohammadi and Justin Shepherd, "Applications of linac-mounted kilovoltage Cone-beam Computed Tomography in modern radiation therapy: A review," *Polish Journal of Radiology* 79, (July 2014): 181-193, https://www.ncbi.nlm.nih.gov/pmc/articles/PMC4085117;
Chris Shaw, *Cone Beam Computed Tomography*, (Boca Raton: Taylor & Francis Group, 2014), 224-226.

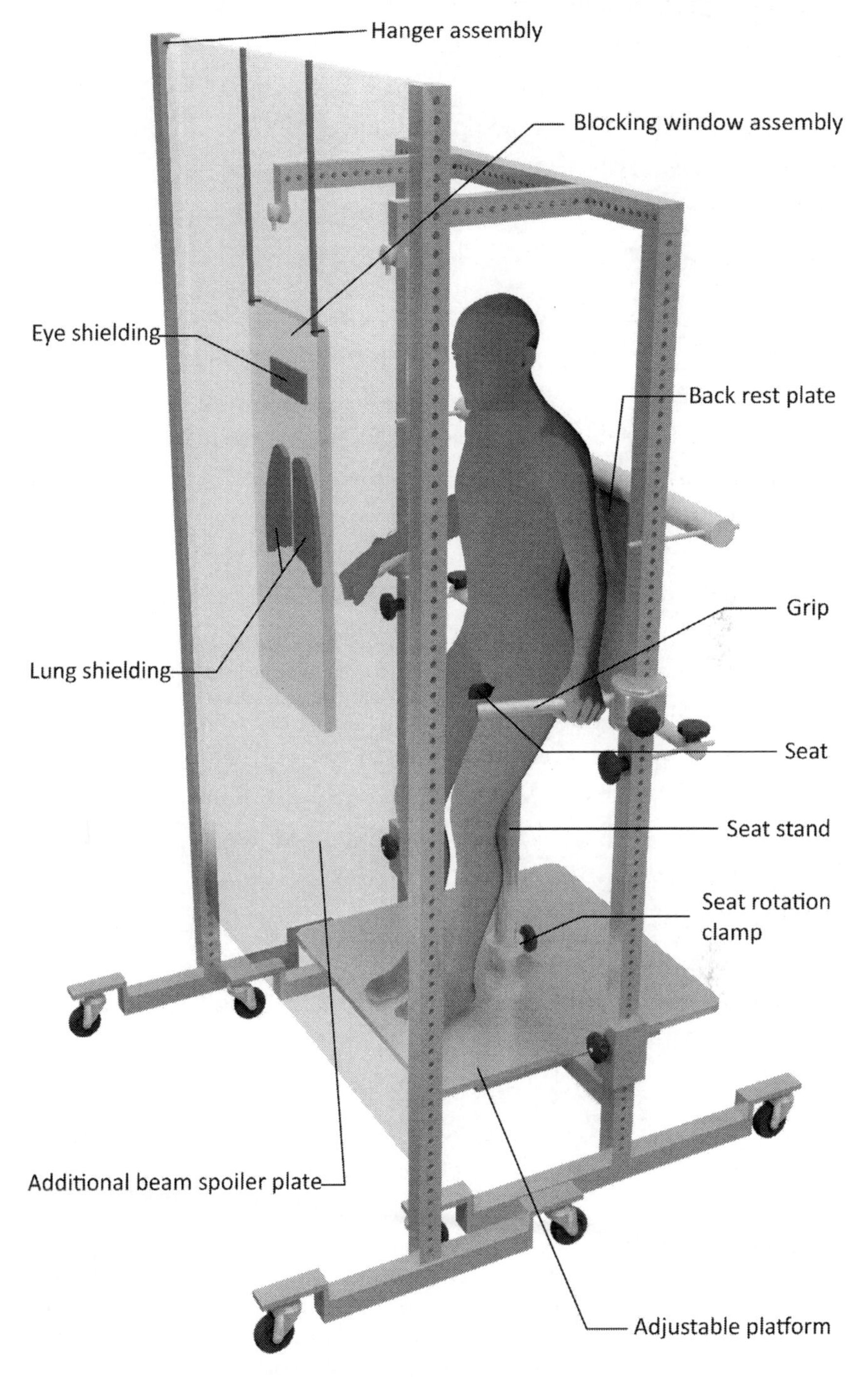

Figure 60: Apparatus for TBI treatment.

minutes). Rather than multiple repeated gantry rotations, VMAT delivers its entire dosage in one 360° rotation. By increasing the fidelity of radiation delivered, VMAT reduces the damage to healthy tissue compared to traditional IMRT. VMAT is referred to under different names depending on the company, such as RapidArc Radiotherapy by Varian, PreciseBeam Infinity by Elekta and SmartArc by Pinnacle.[57]

INTRAOPERATIVE RADIATION THERAPY (IORT)

IORT enables physicians to target tumors with radiation during surgery. After the tumor is removed, the leftover tumor bed is targeted to kill any leftover cancer cells. By performing the radiation treatment during surgery, the physician can better locate the region of interest and limit radiation exposure to healthy tissue. A protective metal disc is normally placed under the tumor bed before treatment to protect healthy tissue behind the incident site. Depending on the cancer type and specifics, IORT may be the only radiation treatment performed or it may precede more traditional ERT such as IMRT. Some of the more common cancers treated using this technique include breast and prostate cancer, in addition to cancers of the central nervous system, head, neck, lung, esophagus, rectum, stomach, pancreas, gynecological and soft tissue sarcoma.

Aside from being convenient—one session versus potentially weeks of IMRT—IORT can nip the cancer in the bud by not allowing regrowth time between surgery and IMRT sessions. Critical organs can also be better protected by IORT with the physician visually directing the treatment location. When combined with IMRT, IORT is referred to as a *focused boost of radiation* because it is more localized and a higher dosage. IORT is administered by a LINAC, which can include a traditional design used for IMRT or an SBS device such as a CyberKnife.

---

57 Enzhuo M. Quan , Xiaoqiang Li, Yupeng Li et al., "A Comprehensive Comparison of IMRT and VMAT Plan Quality for Prostate Cancer Treatment," *International Journal of Radiation Oncology Biology Physics* 83, no. 4 (July 2012): 1169–1178, https://www.ncbi.nlm.nih.gov/pmc/articles/PMC3805837;
Daliang Cao, "Volumetric Modulated Arc Therapy (VMAT): The Future of IMRT?," *American Association of Physicists in Medicine,* https://www.aapm.org/meetings/amos2/pdf/34-8076-8479-796.pdf;
"VMAT, IGRT and IMRT," *Terk Oncology Center for Prostate Care,* https://www.floridaprostate.com/vmatigrtimrt.html;
"VMAT: RapidArc," *Varian,* https://www.varian.com/oncology/treatment-techniques/external-beam-radiation/vmat.

STEREOTACTIC RADIATION

STEREOTACTIC RADIOSURGERY (SRS) AND STEREOTACTIC RADIOTHERAPY (SRT)

SRS is not actually surgery but rather a form of radiation therapy where a focused beam homes in on a tumor to destroy it during one session, replacing invasive therapy. SRS can be broken down into three categories based on the type of radiation administered: LINAC machine X-ray therapy, Gamma Knife gamma-ray therapy and proton beam therapy (often shortened to proton therapy). SRT refers to stereotactic radiotherapy, also known as fractional stereotactic radiotherapy, which uses SRS over multiple sessions (generally two to five) to destroy a tumor. Stereotactic refers to surgical techniques that rely on three-dimensional locating in space. SRS and SRT focus on cancers of the central nervous system, namely the brain and spine.

LINEAR ACCELERATOR X-RAY THERAPY

LINACs can be modified to perform SRS rather than IMRT. Various companies have produced LINACs for this purpose, with some of the more popular being Elekta's Axesse, Accuray's CyberKnife and Varian Medical Systems's and BrainLAB's Novalis Tx. Additional devices include Varian's Edge, Trilogy, Clinac iX and TrueBeam, ViewRay's MRIdian, Accuray's TomoTherapy and Elekta's Versa HD. Integra has also introduced X-Knife software to be used in SRS/SRT/SBRT. This is an ever-evolving field with new machines coming out regularly. A few of the more common devices are discussed below.[58]

Axesse: Built by Elekta, the Elekta Axesse, as shown in Figure 61, provides advanced VMAT and IGRT, producing more efficient and accurate therapy. In addition to the 360-degree rotating gantry, a six degrees of freedom bench allows for tumors to be treated throughout the body. The head and neck are immobilized throughout the process, using a device with a dental imprint similar to a mouthguard.

58 "Stereotactic Radiosurgery (SRS) and Stereotactic Body Radiotherapy (SBRT)," *RadiologyInfo.org,* https://www.radiologyinfo.org/en/info.cfm?pg=stereotactic; "Treatment Options, Pros and Cons," *Colorado CyberKnife,* https://www.radiosurgery.net/treatment-options; "Stereotactic Radiation Therapy," *RT Answers,* https://www.rtanswers.org/How-does-radiation-therapy-work/Stereotactic-Radiation-Therapy.

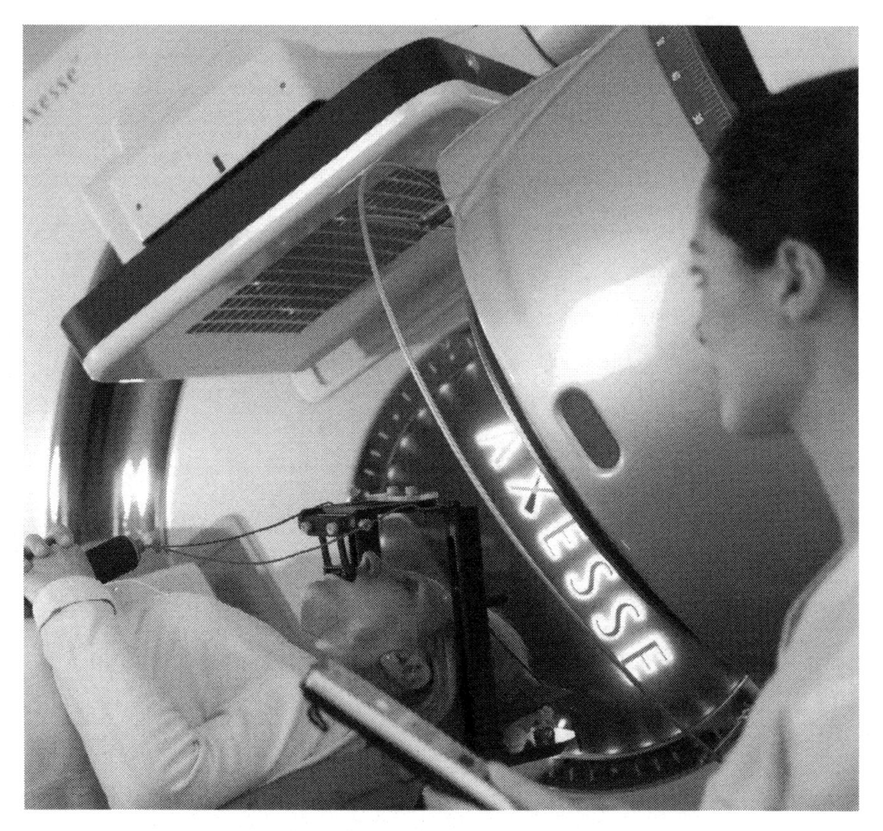

Figure 61: SRS treatment using Elekta's Axesse.

CyberKnife: This is one of the most advanced forms of radiation therapy, whereby a linear accelerator is placed at the end of a robotic arm. With this increased dexterity, the CyberKnife can attack the tumor from various angles and use IGRT to image it in real time. By moving with the tumor, such as when the patient breathes, the administered radiation will interact less with healthy tissues than traditional IMRT methods. The IGRT feature of CyberKnife also allows for more efficiency between treatments since CT or other desired scans can be taken during therapy rather than between treatments.

Invented by John Adler of Stanford University and Peter and Russell Schonberg of Schonberg Research Corporation in 1990, the CyberKnife (Figure 62) is made and sold by Accuray, with headquarters in Sunnyvale, California. Initially CyberKnife used a Japanese-developed robot (Fanuc), but newer models now use the German KUKA KR 240 robot.

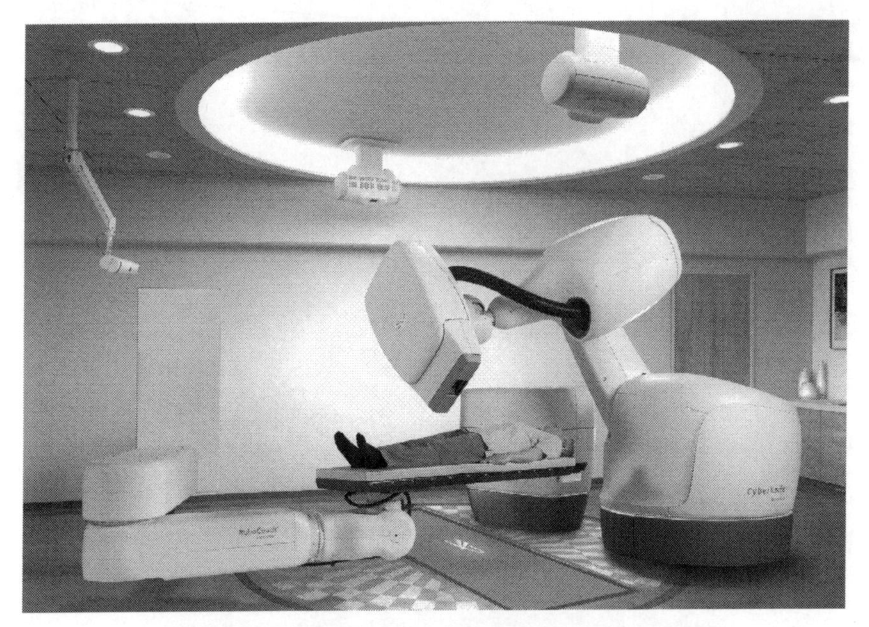

Figure 62: Schematic of Accuray's CyberKnife machine used for advanced SRS.[59]

Novalis Tx: The Novalis Tx was a collaboration between Varian Medical Systems and BrainLAB. Technology such as RapidArc was provided by the former while the latter provided treatment-planning software. Various monitors implement IGRT, which allows for real-time treatment monitoring and radiation positioning. While this system is commonly used for radiation treatment, it has evolved into newer models such as Varian's Edge Radiosurgery System (Figure 63) and TrueBeam Radiotherapy System.

GAMMA KNIFE GAMMA-RAY THERAPY

Gamma Knife therapy was developed in 1967 at the Swedish KarolinskaInstitute by Lars Leksell and Borje Larsson; Dr. Leksell would go on to found the Swedish company Elekta in order to sell his inventions. The Gamma Knife was brought over to the United States in 1979 via an agreement between Dr. Leksell and the University of California at Los Angeles's Dr. Robert Wheeler Rand.

Still used today, the Gamma Knife, shown in Figure 64, is primarily employed as a non-invasive alternative to the surgical removal of brain tumors. Gamma radiation is used on brain tumors up to 4 cm in size and is generally administered during one session, although multiple sessions may occur, such as for the treatment of non-cancerous trigeminal neuralgia.

59 Image used with permission from Accuray Incorporated.

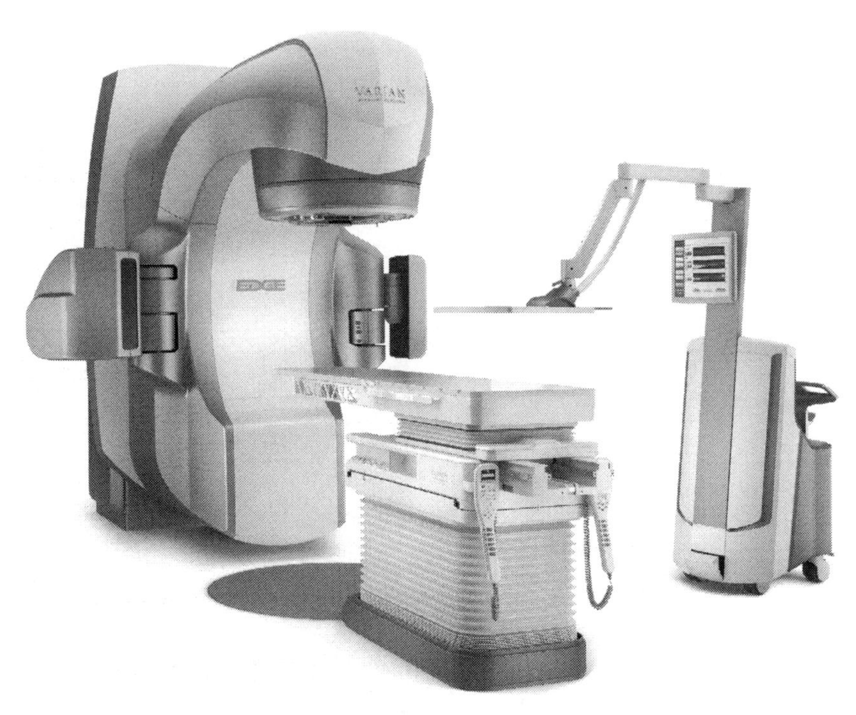

Figure 63: Schematic of Varian's Edge Radiosurgery System machine used for advanced SRS.

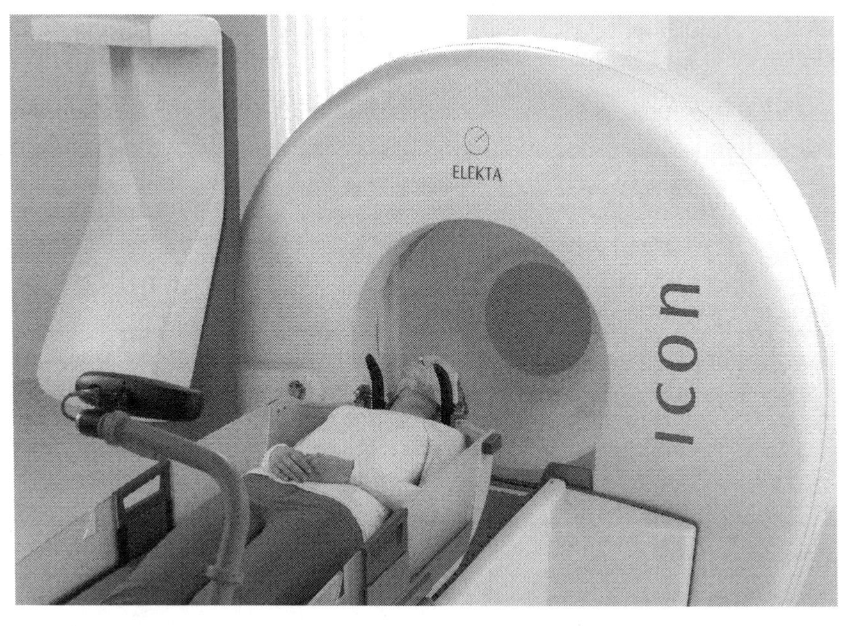

Figure 64: Schematic of Elekta's Leksell Gamma Knife Icon machine used for gamma-ray brain tumor therapy.

PROTON THERAPY

Discussed in other sections, LINACs and proton therapy can also be configured to deliver SRS/SRT/SBRT versus IMRT or IMPT.[60]

STEREOTACTIC BODY RADIATION THERAPY (SBRT)

SBRT is the same as SRT but is used for cancers outside of the central nervous system. SRT and SBRT are generally used for very small tumors.

ACCUBOOST

AccuBoost, made by AccuBoost, is a form of breast cancer radiation treatment, shown in Figure 65. For women diagnosed with early stage breast cancer, many will choose to preserve their breast in an approach known as breast conserving therapy (BCT) and will receive a lumpectomy, a procedure where a portion of breast cancer tissue is surgically removed. Following surgery, either radiation or radiation combined with chemotherapy is used to sterilize leftover cancer cells and reduce the likelihood of cancer cells returning.

There are two primary options for radiation treatments that patients will be offered. The first is referred to as whole breast irradiation (WBI) and consists of six to seven weeks of daily radiation to the entire breast using a LINAC. This is typically followed by supplemental radiation using AccuBoost applicators. These are paddles that deliver concentrated radiation doses—or boosts—in and around the lumpectomy cavity, where a recurrence is most likely. The second option is called accelerated partial breast irradiation (APBI) and involves the delivery of radiation via AccuBoost applicators only to the area around the original site of the disease. This treatment is completed in five to ten days.

Studies are ongoing comparing the efficacy of both treatment options. One study[61] determined that while WBI had fewer recurrences than APBI (71 versus 90 for 2,109 and 2,107 women, respectively), the recurrence in

60 Eric Shinohara, "Radiation Therapy: Which Type is Right for Me?," *OncoLink*, https://www.oncolink.org/cancer-treatment/radiation/overview/radiation-therapy-which-type-is-right-for-me.

61 FA Vicini, RS Cecchini and JR White et al., "Primary results of NSABP B-39/RTOG 0413 (NRG Oncology): A randomized phase III study of conventional whole breast irradiation (WBI) versus partial breast irradiation (PBI) for women with stage 0, I, or II breast cancer," *NSABP Oral Session* 4, https://www.breastcancer.org/research-news/apbr-vs-wbr-for-lower-recurrence-risk-after-lx.

the same breast was within 1%. However, since the recurrence rates were statistically significant, researchers were unable to declare the two treatment options equivalent in their success. Please discuss each treatment option with your oncology team and make the best decision for you.

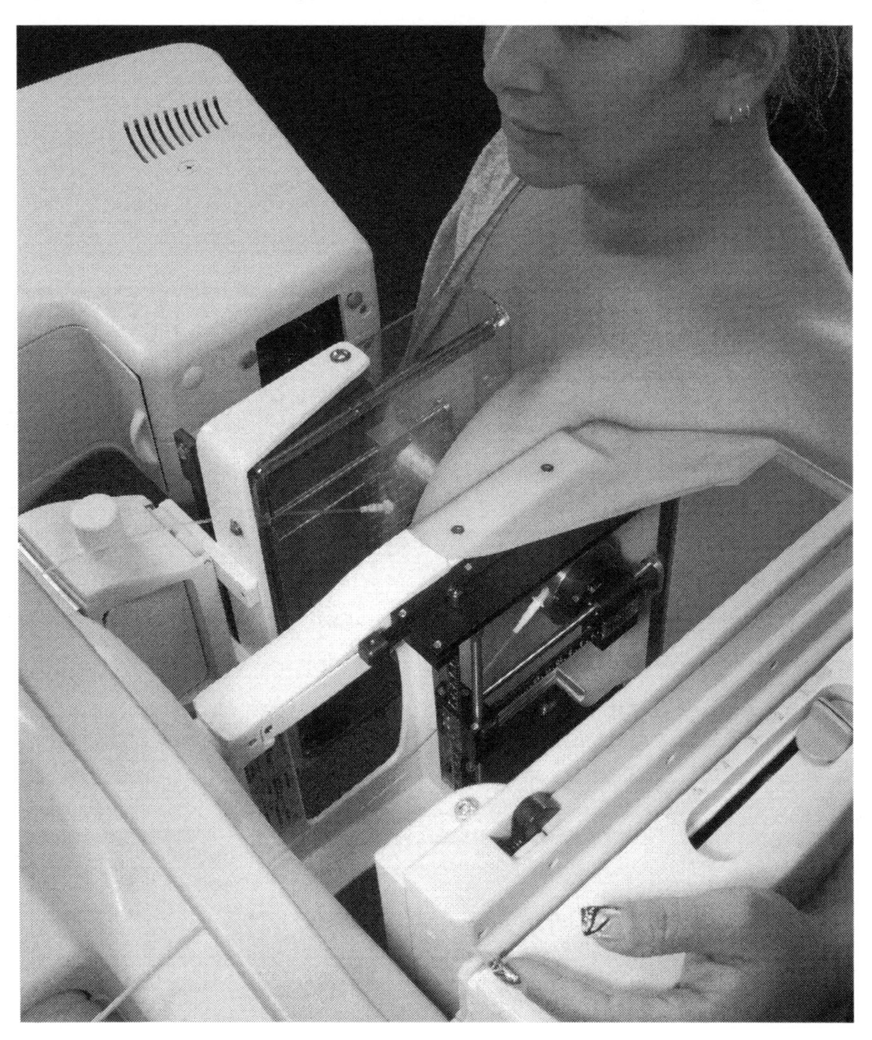

Figure 65: AccuBoost machine used for breast cancer radiation.

INTERNAL RADIATION THERAPY (BRACHYTHERAPY)

Internal radiation therapy, or brachytherapy, places a radioactive source inside the body to kill off cancer cells. There are various subcategories for this therapy. Radioactive sources are placed either in a body cavity such as the rectum or uterus, referred to as intracavitary, or as an implant near the tumor, referred to as interstitial. Additionally, internal radiation therapy can be grouped by radiation dose. This treatment is commonly used for prostate cancer.

HIGH-DOSE RATE BRACHYTHERAPY

As the name implies, high-dose rate (HDR) therapy uses a highly radioactive source for a short period of time, around a few minutes. Treatments may be repeated daily over a period of days to weeks. A metal or plastic applicator tube is used to insert the radioactive source and either left in between treatments or removed and reinserted before each treatment. Patients will stay at the hospital during treatment with restricted visitation since the patient's body will be giving off radiation. Children and pregnant women are particularly restricted.

LOW-DOSE RATE BRACHYTHERAPY

Low-dose rate (LDR) brachytherapy is similar to the HDR version but with a weaker radioactive source that stays in the body for longer periods of time, up to a few days. Generally, the applicator is removed and the patient allowed to return home. With this dosage the radiation is often restricted to the tumor and nearby tissue; however, as a precaution, patients are advised to restrict or limit visitors, especially children and pregnant women right after the radioactive substance is administered. Permanent implants are also used, which are on the order of the size of rice and do not typically cause discomfort.

SYSTEMIC RADIATION THERAPY

This therapy uses radioactive drugs called radiopharmaceuticals that are taken orally as pills, given intravenously or administered via instillation, which is when a substance is poured or injected drop by drop. One example is using the radioactive iodine isotope I-131 to treat thyroid cancers. Another example is using radioactive strontium (Sr-89) to treat cancers of the bone. Radioactive phosphorus is also a commonly used radiopharmaceutical.

RADIOIMMUNOTHERAPY (RIT)

RIT is the combination of two types of therapy: internal radiation therapy and immunotherapy. Immunotherapy is when the body's immune system is intentionally triggered for a therapy; this also applies to the process of getting vaccines or allergy shots. Specifically for cancer treatment, monoclonal antibodies are manufactured in a laboratory. These antibodies are made to mimic naturally occurring antibodies inside your body, but they are tailored to adhere to a specific cancer cell. Monoclonal antibodies are combined with a radioactive tracer that gives off radiation and when injected inside the body, destroys cancerous tumors.[62]

Hyperthermia

Hyperthermia is a nascent cancer treatment that heats cancer cells to high temperatures—up to 113°F depending on the location—to kill them. This therapy is generally used in conjunction with other treatments such as chemotherapy and radiation. Local hyperthermia focuses primarily on the tumor itself, while regional or whole-body hyperthermia treats a large area of tissue or the entire body, respectively. Whole-body treatment can reach temperatures of up to 108°F. Hyperthermia treatment is still under study and not yet widely available.

62 "Radiation Oncology," *Stanford Health Care,* https://stanfordhealthcare.org/medical-clinics/radiation-oncology.html;
"Radiation Therapy Basics," *American Cancer Society,* https://www.cancer.org/treatment/treatments-and-side-effects/treatment-types/radiation/radiation-therapy-guide/what-is-radiation-therapy.html;
"Radiation Therapy to Treat Cancer," *National Cancer Institute,* https://www.cancer.gov/about-cancer/treatment/types/radiation-therapy/radiation-fact-sheet;
"Targeted Therapy to Treat Cancer," *National Cancer Institute,* https://www.cancer.gov/about-cancer/treatment/types/targeted-therapies.

11/15/2008: With my dad roughly one year before being diagnosed with cancer.

9/8/09: At Capitola Beach a few months before my diagnosis, with no awareness of my tumor.

9/20/09: Here with Moy at my friend Yuta's wedding two months before diagnosis.

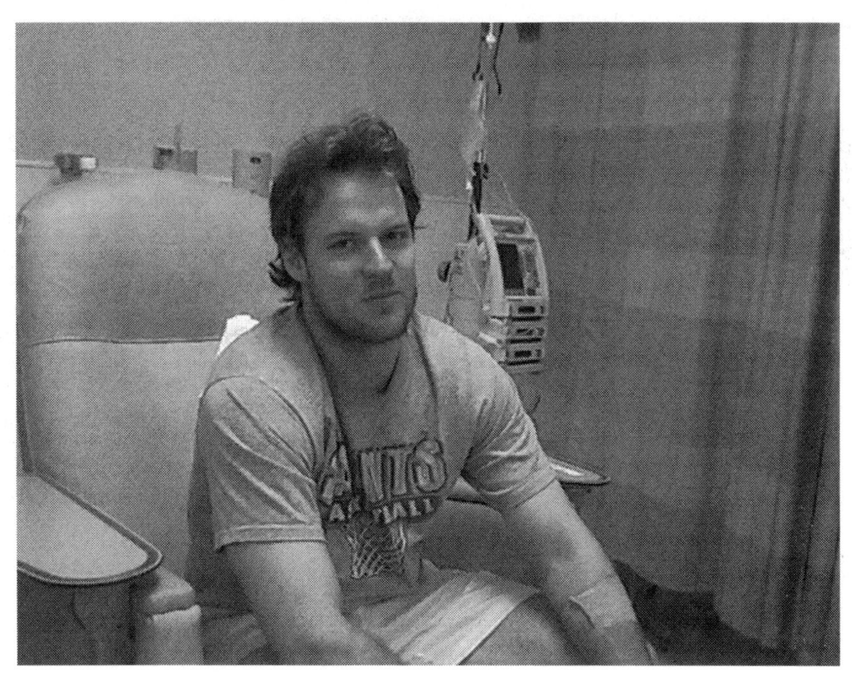

12/3/09: Having my first chemo treatment with my mom and Moy for support.

12/30/09: With my mom, Moy, aunt, uncle and cousins at Fisherman's Wharf in San Francisco.

1/7/10: My brother Marc shaved his head with me to show unity.

1/16/10: Having a visit from my sister Allison and cousin Ben lifted my spirits.

2/14/10: Getting out for social activities when up for it can help you mentally (pictured here watching a childhood idol, Pete Sampras).

3/7/10: Here I am pictured at Tiburon for a weekend break before my first radiation treatment.

3/7/10: Having my mom's support throughout my battle was a big help.

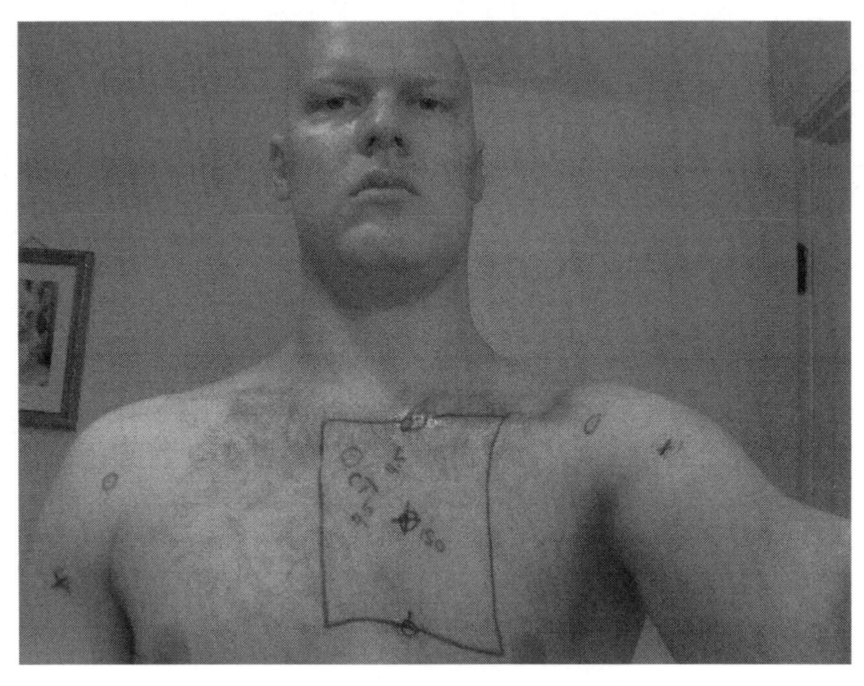

3/10/10: Radiation markings and three tattoo dots were used to align the radiation beam.

3/13/10: While tiring, a weekend at Indian Wells between chemo and radiation was helpful.

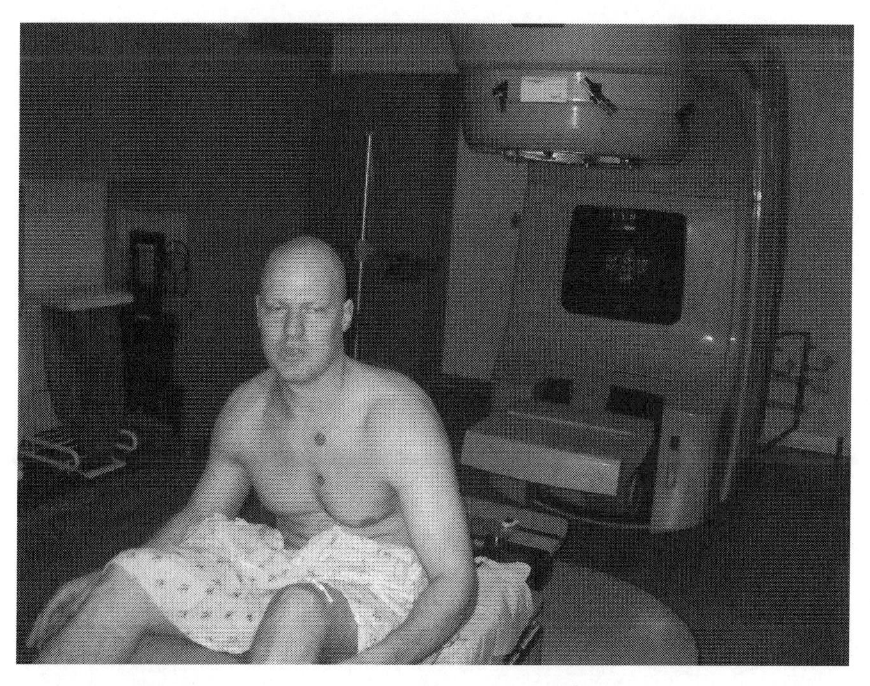

3/27/10: Here I am looking exhausted during a radiation session.

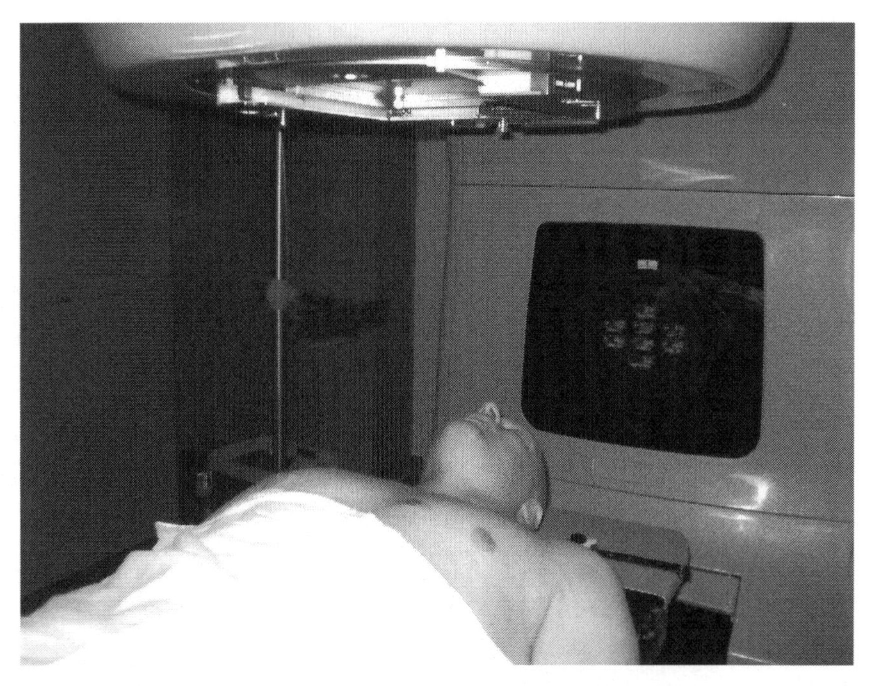

3/27/10: You can see the treatment head lined up over my tumor, near my heart and lungs.

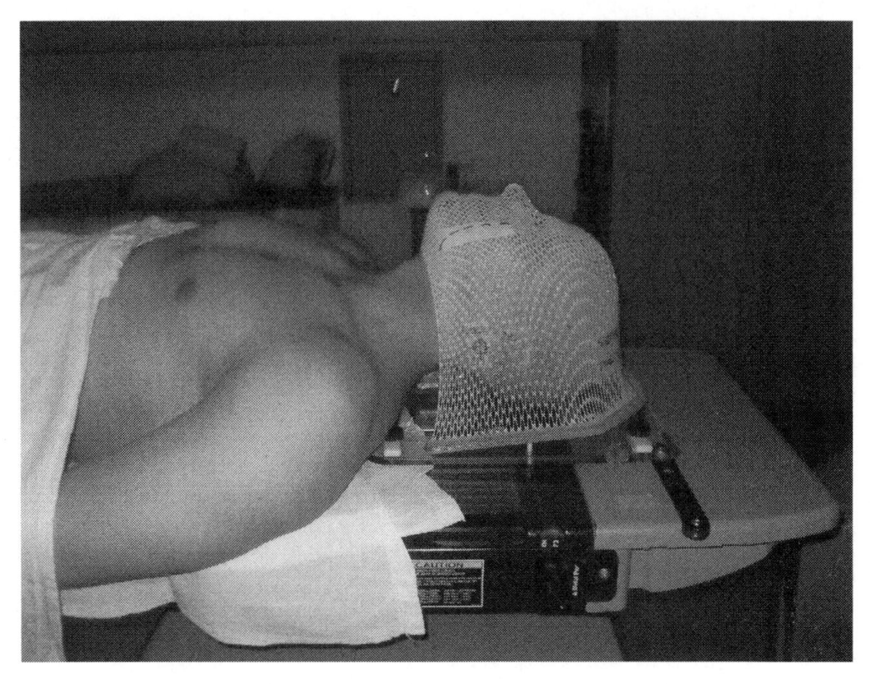

3/27/10: Radiation sessions required me to be very still to limit irradiation of my vital organs.

3/27/10: During radiation treatment, a custom mask secured my head to keep me still.

5/31/10: Enjoying time with Moy at our favorite beach (Capitola) six weeks post-treatment.

8/6/10: This picture with my dad is special to me, taken a few months after my treatment. Notice my hair is lighter and curlier due to the chemo, which eventually went back to normal.

10/11/10: I worked hard to get back into shape, shown here ready to climb a wall during our cruise.

10/11/10: Celebrating victory with Moy in the Bahamas.

11/16/10: Celebrating a team tennis title roughly half a year into remission.

7/29/10: Maverick and Callie provided comfort throughout my cancer journey.

12/28/10: Showing Moy my old stomping grounds at U of M, eight months after treatment.

7/2/17: Roughly seven years after my cancer treatment, enjoying family at our Shepard reunion.

7/3/17: Pictured here with my dad and Uncle Joe, two important figures in my life.

Chapter 12

Surgery

Surgery is often required either as a standalone treatment or in conjunction with other therapies such as chemotherapy and radiation. There are many types of surgeries and methodologies for grouping them. In order to discuss some of the most prominent surgeries relevant to cancer, this book groups them by methodology. The primary focus is on curative surgeries meant to eradicate the cancer. Other surgeries grouped by purpose include preventative surgery, reconstructive surgery, staging surgery such as a biopsy, diagnostic surgery that locates cancer, palliative surgery to reduce pain and supportive surgery, which helps other non-surgical cancer treatments work more effectively.[63]

Open Surgery

Open surgery involves making a large incision to gain access inside the body—generally for organs in the case of cancer treatment. For a discussion on the various types of open surgeries, it is convenient to divide the body into three regions: the head, chest and abdomen.

Abdomen

The primary surgeries of the abdomen address stomach cancer, although treatment of other organs such as the bladder, esophagus, small and large intestines, pancreas, spleen and uterus can also require open abdominal surgery. Stomach cancer is generally treated with some form of gastrectomy, i.e. removal of the stomach.

63 "Surgery," *MD Anderson Cancer Center,* https://www.mdanderson.org/treatment-options/surgery.html.

Subtotal (Partial) Gastrectomy

Partial removal of the stomach is referred to as a *subtotal* or *partial gastrectomy.* Most commonly this involves removing the lower stomach and reattaching the upper stomach to the small intestine. Less commonly the upper stomach is removed, where the lower stomach is reattached to the esophagus to complete your digestive tract.

Lymph nodes are also removed to help with staging, in addition to all or part of nearby organs wherein the cancer has spread. While a partial gastrectomy may be performed laparoscopically, your surgical oncologist will generally recommend open surgery in order to ensure that all the cancer is removed.

The patient will be under general anesthesia during surgery, which lasts a few hours. Recovery time takes two to three months with your diet and exercise routines restricted until you can function properly, albeit with some continuous dietary restrictions. For instance you will eat smaller, more frequent meals throughout the day, avoiding high-fiber foods while eating foods high in calcium, iron and vitamins C and D; you will also take vitamin supplements.

Total Gastrectomy

If the cancer has spread to both the upper and lower portions of the stomach, a total (also known as complete) gastrectomy will be performed where the entire stomach is removed. The esophagus is connected directly to your small intestine upon removal of your stomach.

Challenges with a total gastrectomy are similar to a partial, only amplified. Your diet and meal sizes are even more restricted after surgery while your weight decreases more rapidly. Although you will have to alter your diet for the rest of your life, you can live healthily without a stomach.

A particular form of stomach cancer that leads to a total gastrectomy is diffuse gastric cancer, which is hereditary and results from a mutation of the CDH1 gene. As a preventative measure the entire stomach is recommended for removal to prevent the cancer from spreading, since metastasis often leads to death.[64]

64 James Turner, "How to Live Without a Stomach," *No Stomach for Cancer,* https://www.nostomachforcancer.org/news/blog/167-how-to-live-without-a-stomach.

GASTRIC BYPASS

In the event that a tumor is unresectable (unremovable), a gastric bypass surgery may be performed. If the tumor gets too large and blocks the path from the lower stomach to the small intestine, you will be unable to excrete your waste. With a gastric bypass, the small intestine is attached to both the lower and upper portion of the stomach, allowing another entry into the small intestine—hence bypassing the obstructed pathway. Incidentally, this surgery is often done for extreme obesity, as it often leads to a severe reduction in body weight.[65]

BRAIN

There are various ways to classify types of brain surgery for cancer. The most straightforward is based on location in the brain. Within the brain there are three main regions: the brainstem, cerebellum and cerebrum as shown in Figure 66. The cerebrum can be further subdivided into the frontal, parietal, occipital and temporal lobes. Specific procedures for these locations are discussed in this section.

It is important to understand the skull, as sections will be drilled into it and pieces removed during brain surgery. There are twenty-two bones that make up the skull, with eight protecting the brain and fourteen forming your face. Bones in your skull are connected by sutures, consisting of dense, fibrous connective tissue that forms a solid joint between bones. The separation between bones is not smooth but rather jagged, like what is depicted in Figure 67.

CRANIOTOMY

In order to access parts of the brain, either a hole needs to be drilled for a minimally invasive craniotomy (i.e. keyhole surgery) or a piece of the skull must be removed (i.e. a traditional craniotomy). Which method is used depends on the location and size of the tumor. These procedures are named after either the location of the brain worked on or the type of procedure performed. For example, surgeries based on location include frontal, temporal, parietal and occipital craniotomies (Figure 68) or some combination thereof such as a pterional craniotomy, which is performed at the junction of the frontal, parietal, sphenoid and temporal bones.

65 "Surgery for Stomach Cancer," *American Cancer Society,* https://www.cancer.org/cancer/stomach-cancer/treating/types-of-surgery.html.

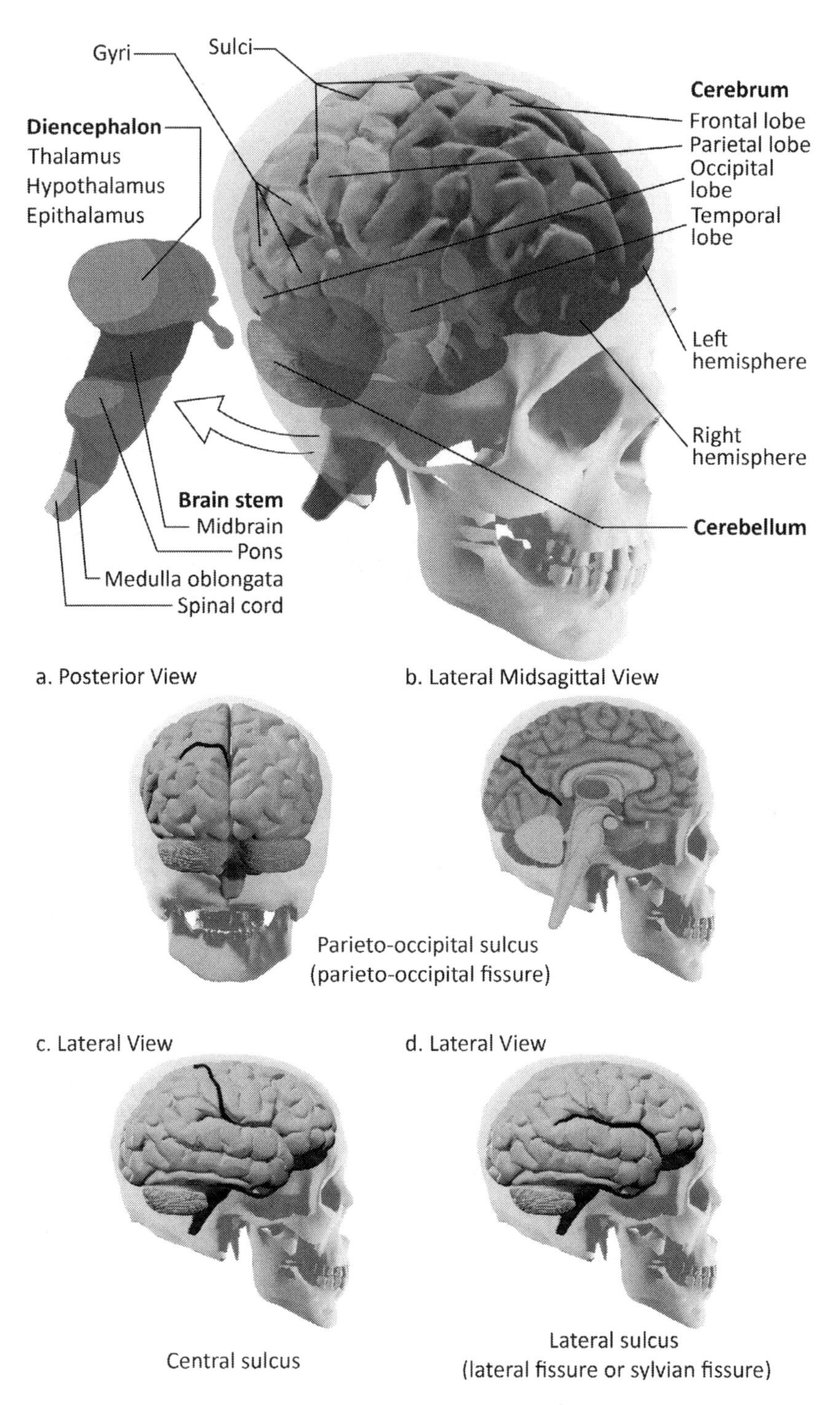

Figure 66: Major compartments of the brain.

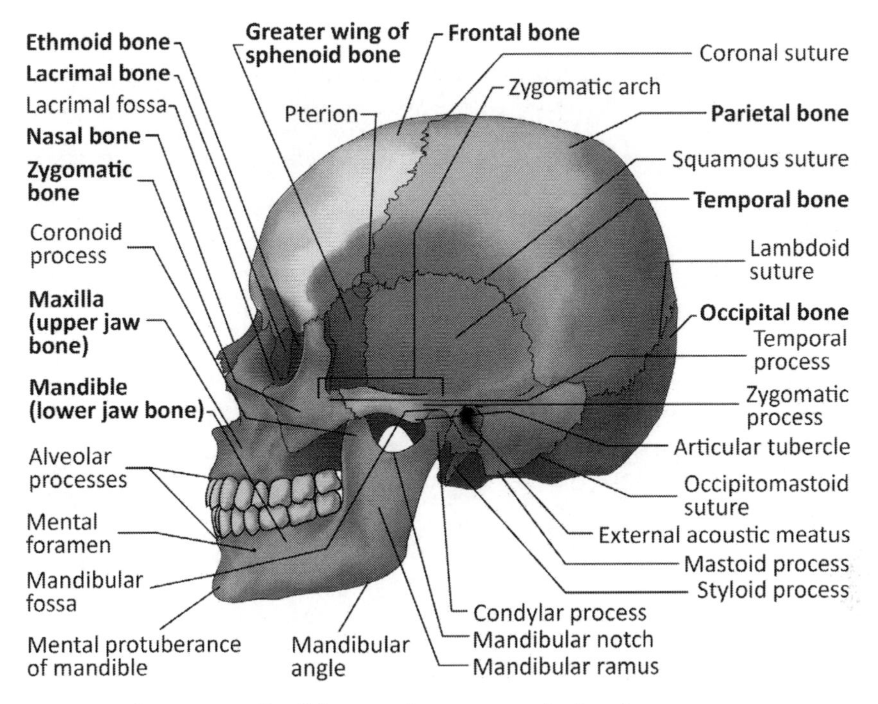

Figure 67: Skull bones important in brain surgery.

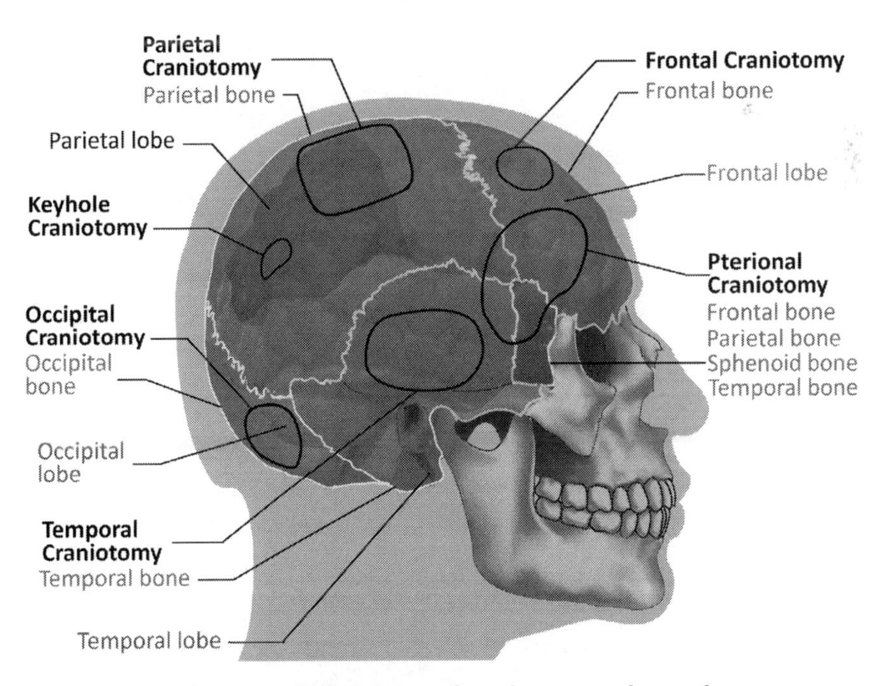

Figure 68: Subdivisions of various craniotomies.

Before surgery, your skull will be shaved near the surgical site and you will be given general anesthesia. Your surgeon will make an incision in your scalp and peel it back to expose your skull, where one or more small holes will be drilled and connected with cuts via a fine saw to remove a piece of your skull—referred to as a bone flap. An additional incision will be made in your dura mater, which will also get peeled back.

The dura mater is the outermost and thickest layer of the meninges—the three membrane layers between the skull and brain—with the other two being the arachnoid mater and the pia mater. These two thinner layers must also be breached either as part of the dura mater incision or subsequent techniques. After the tumor is removed, the meninges are folded back and secured with sutures. Your bone flap is reattached using metal plates and screws while your scalp is folded back and secured with staples.

If there is significant brain swelling, your surgeon may choose to leave the skull flap unattached to relieve the pressure. This variation in a craniotomy is called a craniectomy. At a later date your surgeon will reattach the skull flap and secure with staples in a procedure called a cranioplasty. Some specific craniotomy procedures are discussed below.[66]

Extended Bifrontal Craniotomy

A bifrontal craniotomy is used to remove tumors at the base of the brain, near the brainstem. An incision is made from ear to ear at the top of the head and the front of the scalp, where forehead layers of skin are peeled back. Similar steps as discussed above are made to remove the frontal portion of your skull and the brain is angled upward to allow access beneath it. This surgery is most commonly used to treat tumors at the base of your skull beneath your brain, called skull base tumors. An example is a pituitary tumor. Additionally this surgery may be used to treat esthesioneuroblastoma, a rare nasal cavity cancer, as well as meningiomas, tumors that form around the meninges. The meninges are connective tissue around the brain and spinal cord, consisting of the dura mater, arachnoid mater and pia mater, shown in Figure 69.

66 "Keyhole Brain Surgery (Minimally-Invasive Retro-Sigmoid Craniotomy," *Johns Hopkins Medicine,* http://www.hopkinsmedicine.org/neurology_neurosurgery/centers_clinics/brain_tumor/treatment/surgery/key-hole-retro-sigmoid-craniotomy.html.

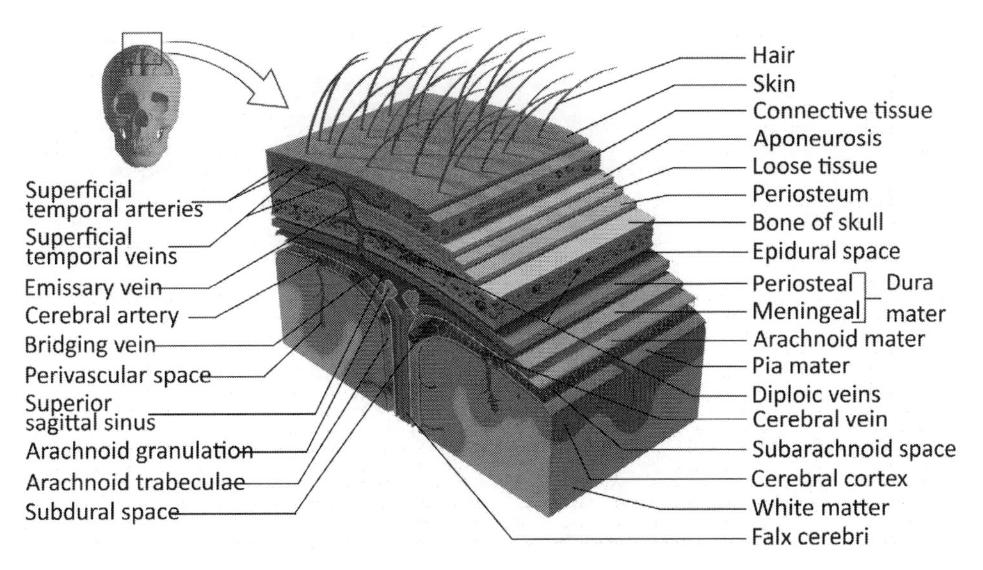

Figure 69: Skin layers around the brain.

SUPRA-ORBITAL "EYEBROW" CRANIOTOMY

This procedure is used to remove tumors toward the front of the skull or near the pituitary gland. A small incision is made through the eyebrow as skin and membranes are peeled back, with a piece of skull then removed. Some specific tumors treated using this procedure include skull base tumors, pituitary tumors and Rathke cleft cysts, which are benign tumors that form between the pituitary gland and skull base.

RETRO-SIGMOID "KEYHOLE" CRANIOTOMY

Similar to the eyebrow and bifrontal craniotomies, the retro-sigmoid craniotomy is used to treat tumors at the base of the skull, most commonly near an ear. An incision is made through the base of the ear, allowing access to the brainstem and cerebellum. Tumors treated include acoustic neuromas (vestibular schwannomas, i.e. tumors that impact hearing and balance), meningiomas, metastatic brain tumors (cancers that have metastasized and spread into the brain) and skull base tumors.

ORBITOZYGOMATIC CRANIOTOMY

An orbitozygomatic craniotomy is used to access hard-to-reach tumors by partially removing your orbital and cheekbones. Holes are drilled and a piece of skull connecting these bones is removed. Tumors accessed include pituitary tumors such as benign craniopharyngiomas and meningiomas.

TRANSLABYRINTHINE CRANIOTOMY

A translabyrinthine craniotomy is used for tumors near the ear. With a similar location as a retro-sigmoid craniotomy, this surgery has a larger piece of bone removed compared with its minimally invasive counterpart, which can also be used for tumors away from the ear. A piece of skull, including parts of the mastoid and inner ear bones, is removed to gain access to the tumor. The main type of tumor targeted with this surgery is acoustic neuroma.[67]

TRANSSPHENOIDAL SURGERY

This type of brain surgery involves access through the nose and sphenoid bone, which is a bone towards the front of the skull near your eyes and nose. The technique is primarily used to remove tumors on the pituitary gland. There are various subtypes of transsphenoidal surgery.

ENDONASAL ENDOSCOPY

Using an endoscope to go through the nose and sinuses, this surgery operates on the front of your brain or top of the spine. When this option is available, it can prevent the surgeon from having to remove a section of your skull with a craniotomy. Generally this technique is used to remove tumors, although it is also used for non-cancerous surgeries such as improving breathing through your sinuses.[68]

NEUROENDOSCOPY

A neuroendoscopy is like an endonasal endoscopy, but an opening is made in the skull for an endoscope rather than going in through the nose.

TRADITIONAL MICROSCOPIC

For traditional microscopic surgery, an incision is made under the top lip and a large region of the nasal septum is removed. This allows access to tumors located near the pituitary gland, which can impact

---

67 "Types of Brain Tumor Surgery," *Johns Hopkins Medicine,* https://www.hopkinsmedicine.org/neurology_neurosurgery/centers_clinics/brain_tumor/treatment/surgery;
"Craniotomy," *MedIndia,* https://www.medindia.net/surgicalprocedures/craniotomy.htm.

68 "Endoscopic Endonasal Approach," *University of Pittsburgh,* https://www.neurosurgery.pitt.edu/centers-excellence/cranial-base-center/endoscopic-endonasal-approach.

your hormones. An endonasal endoscopy is considered safer and more effective compared to a traditional microscopic surgery. Being less invasive, the former also reduces your recovery time to days instead of weeks or months.[69]

BREAST

There are various forms of breast surgery that are segregated based on the amount of breast tissue removed.

LUMPECTOMY (PARTIAL MASTECTOMY)

A lumpectomy is performed for patients with early stages of breast cancer where the tumor can be isolated and most of the breast preserved. During a lumpectomy, also referred to as a partial mastectomy or breast-conserving surgery (BCS), a small incision no more than a few inches at most is made and the cancerous tissue (as well as some healthy tissue and nearby lymph nodes) is removed. Often radiation, chemotherapy or both are required when a lumpectomy is chosen over a mastectomy (entire breast removal) due to the increased chance of cancer cells remaining after surgery. Speak with your oncology team prior to surgery in order to determine the pros and cons of each option and which is best for you. Depending on the stage of your cancer, a lumpectomy may not be an option. Also speak with your team regarding reconstructive surgical options and what your breast(s) will look like post-surgery.

MASTECTOMY

A mastectomy removes the entire breast or both breasts for more advanced stages of cancer. There are various subcategories of mastectomies.

SIMPLE (TOTAL)

The most common form of mastectomy is where a single breast is removed, including the nipple but sparing muscle tissue underneath the breast. This option is chosen when the cancer is localized to one breast but either has spread too much for a lumpectomy or the patient and oncology team decide a mastectomy is preferred rather than additional radiation or

69 Yang Gao, Chunlong Zhong, Yu Wang et al., "Endoscopic Versus Microscopic Transsphenoidal Pituitary Adenoma Surgery: A Meta-Analysis," *World Journal of Surgical Oncology* 12 (Apr. 2014): 94, https://www.ncbi.nlm.nih.gov/pmc/articles/PMC3991865.

chemotherapy. Discuss with your oncology team the specifics of your diagnosis when deciding which surgical treatment is your best option.

Double

A double mastectomy is the same procedure as a simple mastectomy but with both breasts rather than one. This is chosen if the cancer occurs in both breasts or if the likelihood of cancer occurring in the other breast is high. Speak with your oncology team about reconstructive options if this surgery is chosen.

Skin-Sparing

This is a more recent mastectomy option where the same amount of tissue is removed when compared to a simple mastectomy; however the skin (minus the nipple and areola) is left intact. Other parts of the body or synthetic implants are used to reconstruct the breast. This option may be preferred for some women since it leaves less scar tissue and gives a more natural breast appearance compared with a standard mastectomy. If the tumor is too large or near the surface of the skin, a skin-sparing mastectomy may not be an option.

Nipple-Sparing

A form of skin-sparing mastectomy, a nipple-sparing mastectomy spares the nipple and areola in addition to the skin. During surgery, the surgeon will examine the tissue underneath the nipple and areola to see if cancer cells exist. Often the surgeon will radiate the nipple separately during surgery to reduce the chance of cancer forming post-surgery.

While this surgery may be preferred in order to preserve the nipple and areola, it is not yet considered standard and has some drawbacks. For instance if the nerves and blood vessels are significantly cut, this will lead to the nipple not getting the proper blood supply, causing it to wilt. Additionally, with severed nerves there will be little to no sensation depending on the severity of the cutting. Women with larger breasts are especially susceptible to these side effects and therefore surgeons generally recommend this technique for small to medium-sized breasts.

RADICAL

A radical mastectomy includes removal not only of the entire breast but also the nearby underarm lymph nodes and pectoral muscle underneath the breast. This was more common in the past but has given way to modified radical mastectomies since the latter has fewer side effects and is just as effective. The exception is if cancer has spread to the pectoral muscles and underarm lymph nodes, which will then require a radical mastectomy.

MODIFIED RADICAL

The modified radical mastectomy includes the entire breast and underarm lymph nodes but spares the pectoral muscle. This is preferred to preserve the patient's mobility and strength if cancer has not spread to the pectoral muscle.

CHEST (THORACOTOMY)

A thoracotomy involves opening the chest to allow access to internal organs such as the lungs and heart. As heart cancer is extremely rare, a thoracotomy is primarily used to treat other chest cancers, with lung cancer being the most common open surgery. Esophageal cancers can also be treated through a thoracotomy, generally for the removal of the esophagus—or esophagectomy. For lower stage cancers, minimally invasive techniques are typically used with a small incision, deemed a mini-thoracotomy.

LUNG SURGERY

The right lung has three lobes while the left lung has two lobes (Figure 70). The upper lobe is segregated into the superior and lingular divisions, with the latter being analogous to the middle lobe of the right lung, albeit smaller due to the heart taking up space. Lungs are further divided into segments, with the right lung having ten and the left lung having eight to nine, depending on how the original segments are fused together. This distinction is delineated in Figure 70 by 's^{1}' for segment 1 and so on for the original segment numbers, with fused segments grouped together, such as 's^{1+2}'.

The lung shapes differ due to the location of the heart impinging on the left lung, with the left lung being longer and thinner than the right. The combined volume of the right lung remains larger than the left one, however, with a higher oxygen capacity.

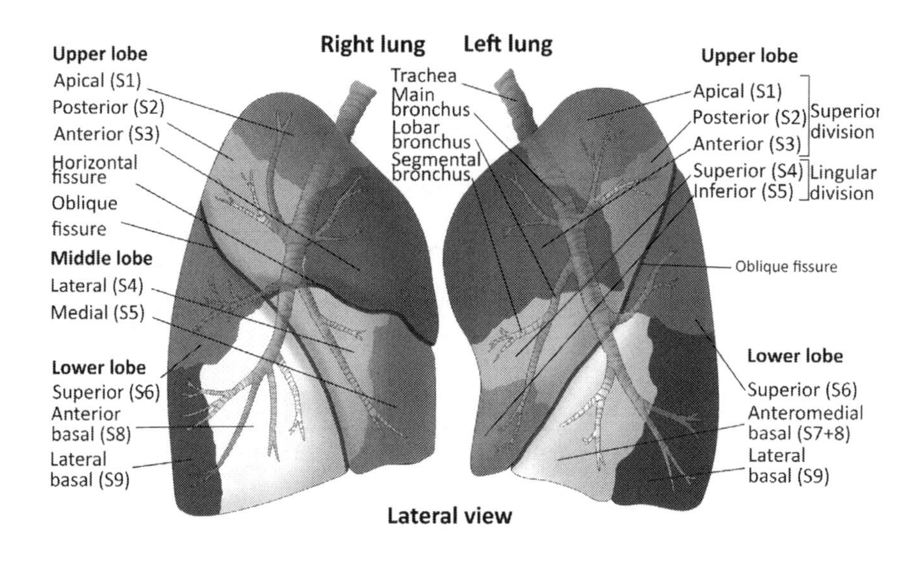

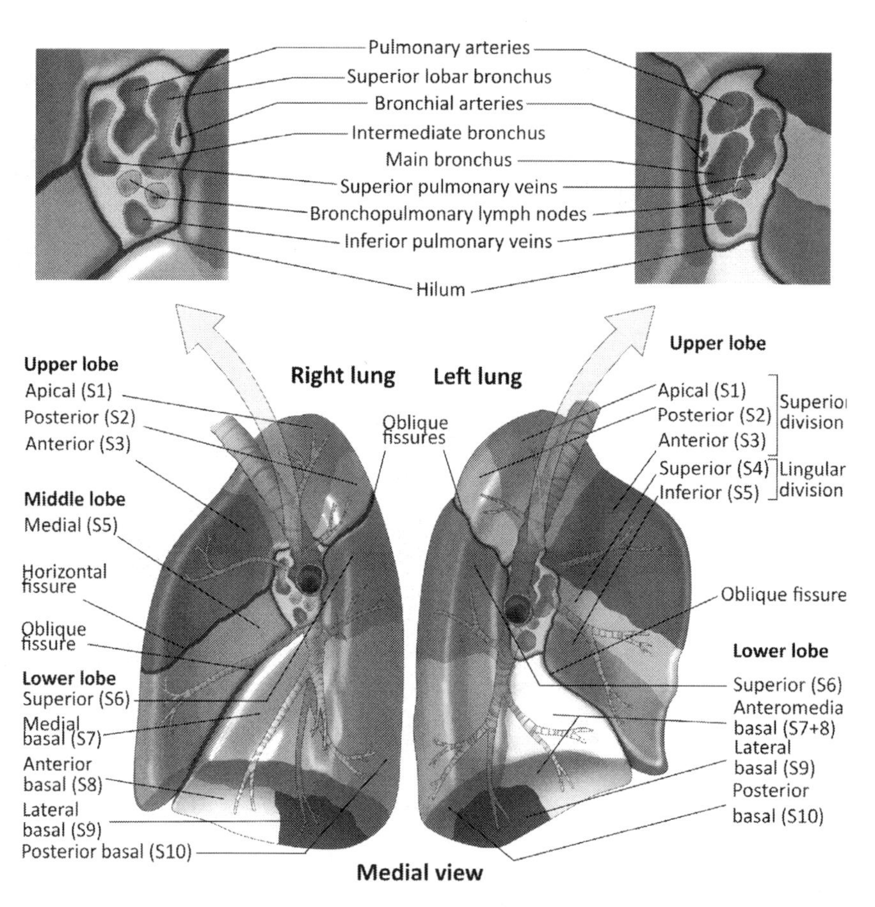

Figure 70: Breakdown of the lungs.

Air passes through your larynx and trachea before entering your right and left main bronchus, as depicted in Figure 70. These bronchi subdivide further into secondary bronchi (also known as lobar bronchi), which are broken down into three categories: superior, inferior and middle. These branch out into tertiary bronchi (also known as segmental bronchi), which further branch out into terminal bronchioles, which then divide into respiratory bronchioles and finally, these are divided into alveolar ducts (Figure 71). Alveolar ducts contain five to six alveolar sacs, which are where blood pumped from the pulmonary vein carries carbon dioxide rich blood to be released while absorbing oxygen via diffusion. Resulting oxygen-rich blood is carried off by the pulmonary artery to be distributed throughout the body.

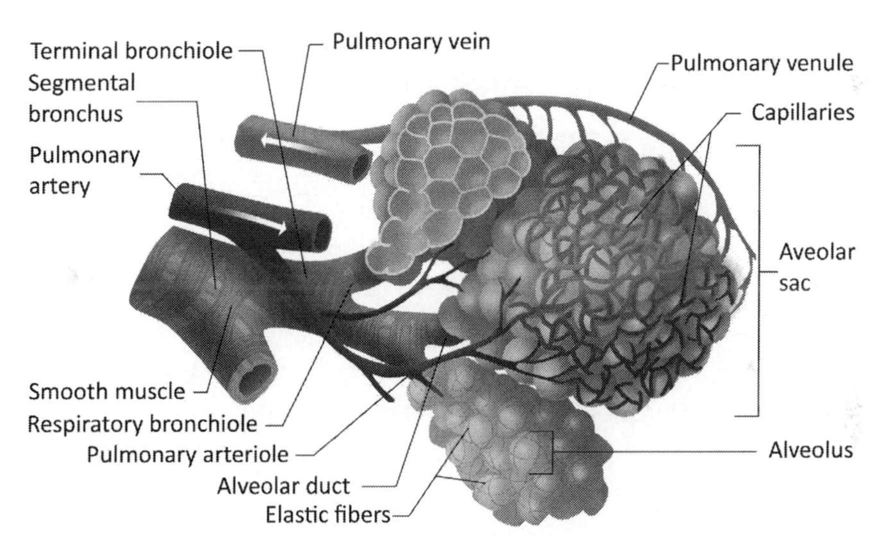

Figure 71: Air passages to, from and within the lungs.

Lung cancer surgeries can be divided into the following categories.

SEGMENTECTOMY

More localized tumors can rely on a segmentectomy, where a segment containing the tumor is removed, sparing the majority of the lung.

LOBECTOMY

When the cancer is isolated to one lobe, the entire lobe is removed from a lung in a lobectomy.

BILOBECTOMY

In a bilobectomy, two right lung lobes are removed: the middle and either the upper or lower.

PNEUMONECTOMY

For advanced stages of lung cancer a pneumonectomy may be required, where an entire lung is removed.

SLEEVE RESECTION

A portion of your bronchus is removed and reattached, along with remaining lobes, to remove cancerous cells while preserving parts of the lung.

WEDGE RESECTION

The tumor is removed along with some surrounding healthy tissue to ensure all cancer cells are eradicated.

MINIMALLY INVASIVE SURGERY

Minimally invasive surgery aims to limit the amount of tissue damage a patient receives during surgery. This reduces the recovery time and scaring the patient receives. There are multiple types of minimally invasive surgery.

ENDOSCOPIC SURGERY

Endoscopic surgery utilizes scopes to go in through natural orifices in the body or small incisions cut by your surgeon. There are various subcategories based on surgical location and body part. As it relates to cancer, endoscopic surgery can be divided into three regions of the body, namely the head, chest and abdomen.

ENDONASAL ENDOSCOPY

An endoscope is inserted through the nose in order to image and remove a tumor, as discussed in the *Transsphenoidal Surgery* section.

NEUROENDOSCOPY

Also discussed in the *Transsphenoidal Surgery* section, a neuroendoscopy is like an endonasal endoscopy but an opening is made in the skull for an endoscope rather than going in through the nose.

LAPAROSCOPIC SURGERY

As opposed to open surgery where incisions of approximately 20 cm or more are made, laparoscopic surgery utilizes one or more small cuts (generally three to five 0.5 to 1.0 cm incisions or one 1.5 to 2.0 cm incision) to minimize the impact to the patient while reducing post-operation pain, risk of infection and speeding up the recovery time. A trocar is used to create an opening for instrumentation to pass through, including a camera to show the image on a screen and cutting tools such as forceps or a scalpel. The trocar consists of a cannula (hollow tube for the tools to pass through), an obturator (sharp tip at the end for penetrating the skin) and a seal at the top of the cannula to prevent air from escaping the surgical cavity while instruments are passed through. Although there are various forms of endoscopic surgery, laparoscopic surgery refers to surgeries in the abdominal region.

Endoscopic Mucosal Resection

Endoscopic mucosal resection (EMR) utilizes an endoscope to remove tumors or cancerous lesions—damaged tissue—in the digestive tract. This includes cancers such as esophageal cancer and squamous-cell carcinoma.

Thoracoscopic Surgery

Another endoscopic surgery relevant to cancer is thoracoscopic surgery, which refers to endoscopic surgery in the chest cavity (i.e. thoracic cavity, the chamber of the body protected by the rib cage). Traditionally chest surgeries required either a thoracotomy or sternotomy incision, which are large incisions that allow access to the heart, lungs and esophagus. Thoracoscopic surgery is often referred to as video-assisted thoracoscopic surgery (VATS) and like laparoscopic surgery, uses small incisions to allow a camera linked to a 5–10 mm fiber optic scope and trocars to pass through. Additional cutting and grabbing tools are used to operate on the patient via the trocar(s). VATS became widespread in the 1990s to reduce the amount of cutting necessary, minimizing pain, risk of infection and recovery time.

Robotic Surgery

Robotic surgery involves a surgeon using a machine autonomously to perform surgery, which can allow more precise cutting and in some cases perform surgery from another building or even across the globe (although typically it is performed with a surgeon sitting in the same operating room at a console). While robotic surgery can be used for traditional open surgery, minimally invasive techniques are generally used. Either a telemanipulator or computer console is primarily utilized for minimally invasive robotic surgery. The former is less common and involves robotic arms that are manipulated via human arms using either mechanical, hydraulic or electronic means. The more common computer console assisted technique relies on robotic arms controlled through a surgeon's hand movements at a computer console. Since the robotic arms are typically connected to the console wirelessly via Wi-Fi or Bluetooth, this type of surgery is referred to as telerobotic surgery.

Various robotic systems have been employed since the mid-1980s, starting with the PUMA 560 in 1985, followed by the PROBOT, developed in Imperial College London; the ROBODOC, created by Integrated Surgical Systems in 1992; and the AESOP/ZEUS Robotic Surgical System by Computer Motion based in Goleta, California in the 1990s. In parallel with the ZEUS Robotic Surgical System, initial research that would result in the Da Vinci Surgical System was being performed at SRI International, a research institute based in Menlo Park, California. SRI International was initially formed as part of Stanford University and called the Stanford Research Institute (SRI) in 1946 before breaking off in 1970 with its new name.

After some burgeoning interest from the Defense Advanced Research Projects Agency (DARPA) to use the device for wounded soldiers, the SRI System (as the robotic device was referred to at the time) had its intellectual rights sold to the newly formed Intuitive Surgical Devices, Inc., founded by Dr. Frederic Moll, John Freund and Robert Yonge in 1995. The SRI System was refined with various spin-offs of Leonardo da Vinci's namesake—Lenny, Leonardo and Mona—before the Da Vinci Surgical System version was coined. Initially marketed in Europe, the Da Vinci was approved by the United States Food and Drug Administration (FDA) in 2000.

Computer Motion's ZEUS Robotic Surgical System was Da Vinci's main rival and while it was developed earlier and approved in Europe, the FDA approved Da Vinci before ZEUS, which led to Computer Motion suing Intuitive Surgical Devices for intellectual property infringement. Since the litigation impacted both company's bottom line, the two decided to merge in 2003, with Intuitive buying out Computer Motion and retaining its name. After the merger, the Da Vinci won out over ZEUS and is the company's mainstay used in the majority of today's robotic surgeries.

The Da Vinci has three arms with pincers to grab tissue or hold surgical tools such as a scalpel, scissors, needles or Bovies, cauterizing devices developed by biophysicist William Bovie.

The robotic "wrists" have more dexterity and degrees of freedom than a human hand (540-degree vs. 270-degree rotation), allowing the surgeon to perform maneuvers that would be more difficult or impossible with traditional laparoscopic surgery. The control system of the arms enables the surgeon to dial back the severity of hand movements for increased control as well as dampen shaking motions.

The fourth arm has cameras that provide a 3-D view of the surgical site that the surgeon views through an ergonomic console. This console also contains the fingertip controls for the surgeon to use during surgery, with the console generally situated next to the patient along with a surgical assistant, nurse and anesthesiologist. Additional controls (such as for focusing the camera) are manipulated by the surgeon via foot pedals, with the entire console designed so the surgeon can comfortably sit rather than stand over the patient during surgery. The viewing screen is directly in front of the surgeon and does not have to be manipulated by an assistant during the procedure. There is also a touch screen located on the vision tower, which sits near the patient and allows for additional viewing beyond the surgeon's console.

One important advantage of robotic surgery over laparoscopic surgery is that the wrist near the pincers moves, which means the long arm near the surgical incision needs to move less to manipulate the tissue or instruments. This is similar to the surgeon having his or her hands right next to the point of interest rather than manipulating tools from the outside. Less motion means smaller required incisions and less bleeding, pain and patient recovery time.

Robotic surgery is used for all types of non-cancerous surgical procedures. The main cancer types being treated are adrenal, bladder, kidney, prostate, testicular and uterine cancer. Partial or full removal of these organs is done via an adrenalectomy (adrenal glands), cystectomy (bladder), nephrectomy (kidneys), prostatectomy (prostate), retroperitoneal lymph node dissection (RPLND; abdominal lymph nodes) and hysterectomy (uterus).

A good source for updates on the latest technology is AVRA Medical Robotics, Inc.[70]

LASER SURGERY

Lasers are used for surgery in place of traditional tools such as a scalpel when very precise cuts are required (on the order of a human hair) or cancer cells are in hard-to-reach locations. In addition to removing small regions of cancer cells, lasers can be used to heat up a tumor and surrounding tissue in order to kill the cancer cells directly or deprive them of nutrients from nearby healthy cells. Light can be bent via tubes, making lasers accessible to remote regions of the body. Lasers can also be used in conjunction with drugs that are activated by laser light to kill cancer cells, a procedure called photodynamic therapy (PDT).

There are many advantages to using lasers including reduced recovery time and the increased temperatures acting as a natural sterilizer to limit the chance of infection. Scarring is also less prevalent with lasers as less healthy tissue must be cut during surgery. A downside of laser surgery is that it may take multiple treatments versus other more standard surgical techniques, depending on the location and size of the tumor. Discuss with your oncologist the various surgical options and the pros and cons of each.

Some specific cancers that often use a laser for surgical treatment include cervical, esophagus, lung, skin, vaginal, vocal cord and vulvar cancer.

Various types of lasers can be used for treatments:

ARGON LASERS

These penetrate skin and the outer layers of organs and are used to treat eye and skin disorders as well as cancerous applications; argon lasers are often used during a colonoscopy to remove polyps. This is also the laser used in PDT, which uses a photosensitizing agent to enter the bloodstream and get absorbed by healthy and cancerous cells. As the agent stays in cancerous cells longer, timing is critical. When the argon laser is incident on the PDT absorbing cells, a chemical reaction is activated that kills the

70 "Allaboutroboticsurgery," *AVRA Medical Robotics, Inc.*, https://allaboutrobotic surgery.com.

cells. Since argon lasers only penetrate slightly, PDT is used for cancers near skin tissue or outer layers of an organ. Examples include cancer of the esophagus and lung cancers. Other cancers that are being researched to use PDT include brain and prostate cancer.

Carbon Dioxide Lasers

Like the argon laser, CO_2 lasers only penetrate the skin slightly. Therefore, these are commonly used to treat skin cancers where only the outer skin layers are desired to be removed while retaining the deeper tissue. This laser can also be used to remove polyps in other regions of the body before they form cancerous tumors. Very little bleeding occurs using CO_2 lasers, speeding the recovery process.

Laser-Induced Interstitial Thermotherapy (LITT)

This laser technique is relatively new (trials were ongoing as of 2011 for liver cancer patients) and relies on a high-energy laser to heat up body tissues near organs with tumors rather than the tumors themselves. Heat from the heated body tissue destroys nearby cancer cells in addition to surrounding healthy tissue.

Neodymium:Yttrium-Aluminum-Garnet (Nd:YAG) Lasers

The Nd:YAG laser is preferred for hard-to-reach cancers such as esophageal and colon cancer. Where other surgical techniques are difficult, the laser can be guided through a flexible endoscope and reach the cancerous cells for a direct strike. These lasers are higher powered when compared to the argon and CO_2 lasers, allowing them to penetrate deeper into tissue. Additionally this laser allows blood to clot quickly, causing less impact and a shorter recovery time for the patient.

MRI-Guided Laser Ablation

MRI-guided laser ablation is a technique primarily used for the treatment of brain tumors. As the name implies, lasers are used in conjunction with real-time MRI imaging to ablate, or burn, tumors. A small hole is made in the patient's skull to gain access while the laser is guided into place. Since the hole is smaller than a typical craniotomy, a patient's pain and recovery time are reduced. Furthermore, this procedure may be required when a tumor is not near the surface and difficult to access when compared with a traditional craniotomy.

CRYOSURGERY

Also referred to as cryotherapy, cryosurgery uses cold gasses to freeze and kill cancerous tissue. Nitrogen is often used for external treatment and nitrogen, argon or carbon dioxide for internal treatment. For external skin cancers, this is done similarly to how treatments are performed for benign skin conditions such as wart removal: liquid nitrogen is expelled in its gaseous form directly onto the outer tissue via the Joule-Thomson effect, where liquid is pushed through a valve and cooled while expanding into a gas. Liquid nitrogen can also be applied directly with a Q-tip or swab.

Internally, a cryoprobe or needle is inserted either through a natural opening such as the anus for prostate cancer, or if the cancer is less accessible, an incision is made. Cold liquid (typically nitrogen) or gas (generally argon or carbon dioxide) is introduced into the needle, cooling it to between -160 and -180°C (temperature range will vary depending on liquid or gas being used). Tissue directly touching or nearby the needles reaches a temperature of around -20 to -40°C (again may vary depending on liquid or gas), which is sufficient to freeze and kill the tissue cells. Depending on the size of the tumor and the liquid or gas used, this takes an average of 8 minutes.

One or multiple needles may be used during the procedure, which also will impact the cooling time. The tumor is allowed to thaw and the process is repeated a second time. A patient will be asked to come back for checkups to monitor the tumor, generally every three months until the tumor is confirmed to be eradicated. It takes roughly one to two months for the tumor to be excreted by the body as waste.

An imaging technique will be used during the procedure to monitor the needle placement and freezing process. Generally a CT scan will be used for this monitoring, although an ultrasound or an MRI is also an option depending on the facility. An example of a cryosurgery imaging technique for prostate cancer is the transrectal ultrasound (TRUS), which is an ultrasound through the rectum.

Oncologists will most likely choose this therapy to supplement chemotherapy, radiation and traditional surgery, although cryosurgery can be a stand-alone treatment option depending on your prognosis and discussions of the pros and cons with your oncology team. Bone, cervical, kidney, lung, prostate and spinal cancers are some of the more common cancers that currently utilize internal cryosurgery. Retinoblastoma, a childhood eye cancer, also utilizes cryosurgery. Additional cancers that are being researched for cryotherapy include breast and colon cancer.

Cryosurgery is a relatively quick procedure and can be done under general or local anesthesia. Recovery time is reduced compared with traditional surgery and since this is a minimally invasive procedure, pain is also lessened due to the fewer cuts. Similar to laser therapy in terms of pros, cryosurgery shares the same con that the cancer is less likely to be completely removed. However, due to its relatively minimal risk and recovery time, this treatment can be used to clean up and reduce the cancer from returning.

Conversely, depending on the stage of your cancer, this may be a judicious choice to try and eradicate the cancer early before more severe treatments such as chemotherapy, radiation or other surgical options are performed. These all have more severe side effects so using cryo or laser surgery can be a smart option. This is especially true if the tumor is rather small or various forms of pre-cancer cells are detected.

More mature cancers likely will not be able to use cryotherapy or laser surgery alone, as the cancer will be too far spread and chemotherapy is likely required. However, cryosurgery and laser therapy can help reduce pain and prolong a patient's life in these instances. Discuss with your oncology team the best course based on your staging, risks and side effects with the various treatment options.[71]

RADIOFREQUENCY ABLATION

Radiofrequency ablation (RFA) is a minimally-invasive approach used for tumors inside the body, most commonly for bone, liver and lung cancers. A needle is inserted into the body and directed by either a CT scan, ultrasound or an MRI until contact is made with the tumor. Alternating current (AC) is used at a frequency of between 350 and 500 Hz to heat up and kill the

71 "Cryosurgery in Cancer Treatment," *National Cancer Institute,* https://www.cancer.gov/about-cancer/treatment/types/surgery/cryosurgery-fact-sheet; "Cryosurgery," *Canadian Cancer Society;* https://www.https://www.cancer.ca/en/cancer-information/diagnosis-and-treatment/tests-and-procedures/cryosurgery; "Cryotherapy: Destroying Cancer With Ice," *MEDICATradeFair,* https://www.youtube.com/watch?v=W2gd7hC5I60.

tumor. Microwaves are also used and are even more effective due to their higher energy, which allows larger tumors to be treated. Both techniques can be administered using a local as opposed to a general anesthetic.

This surgery is effective when multiple tumors are present and you want to preserve as much of the organ as possible since the needle can be moved around to pinpoint the cancer; the downside is that cancer cells are more likely to be missed. Traditional open surgery may also be risky depending on your medical condition, making a minimally invasive approach more prudent. Difficult-to-reach tumors are also good candidates for this treatment as the needle can penetrate remote locations. Chemotherapy or radiation can also be used in conjunction with RFA.

Mohs Micrographic Surgery (Microscopically-Controlled Surgery)

Mohs surgery was developed in 1936 by Frederic Mohs while he was still in medical school at the University of Wisconsin, where he earned his medical degree and practiced throughout his career. The surgery is used to treat skin cancers and although practitioners were skeptical at first, the technique proved to be the most effective at treating skin cancers and today has a more than 98% success rate for new patients. Additionally, 96% of patients that had previous treatment for skin cancer and developed a recurrence were able to be cancer-free after undergoing Mohs surgery.

The original technique utilized zinc chloride to preserve the skin cancer, which Mohs found allowed him to remove skin samples without bleeding or changing the cell structure. Doing so allowed the sample to be examined under a microscope to determine if the skin sample was cancerous. Furthermore, Mohs devised a mapping scheme where he cut the sample into sections, carefully labeling the orientation with respect to the removal site. This enabled Mohs to pinpoint the location of the cancer under a microscope and more precisely remove subsequent samples while minimizing the amount of healthy tissue removed.

Oftentimes skin cancer extends past the outer layer of the skin in an irregular pattern, with offshoots referred to as roots. Standard surgical techniques would remove a large swath of skin to ensure all cancer was removed. Unfortunately this led to large amounts of healthy tissue being removed, which increased pain, recovery time and scarring. Furthermore, sections of cancer were still being missed using this rather crude technique.

Mohs continued practicing and slowly other physicians, primarily dermatologists, began adopting his technique. In December 1965, a newly formed chemotherapy society convened in Chicago at the American Academy of Dermatology (AAD) with twenty participants. That faction would grow into the American College of Mohs Surgery (ACMS), which provides post-residency training and as of 2020 had roughly 900 members.

There have been some important modifications to Mohs's original technique. In 1953, he treated a patient with basal cell carcinoma (BCC). The zinc chloride used in his original treatment (referred to as chemosurgery due to the chemicals) caused skin irritation. To avoid this irritation around the sensitive eye, Dr. Mohs forwent the chemical treatment and took a sample directly using local anesthesia rather than his chemical paste. Other physicians learned of this alteration and began practicing it on other locations of the body, including Dr. Theodore Tromovitch in 1963 and later Dr. Sam Stegman, who presented their results at the 1970 annual Chemosurgery Conference.

Perry Robins, presently professor emeritus of dermatology at New York University Medical Center, also used the fresh tissue technique (as it was dubbed, i.e. using local anesthetics rather than the chemical paste) throughout the 1970s and presented a paper of his results in the 1980s, where 2,900 patients were given treatment. Results showed that 98.2% of newly diagnosed patients with BCC and 96.6% with recurring BCC were cured. Similar results were found for squamous cell carcinoma, which is the second most common type of skin cancer behind BCC. Actinic keratoses (AK) is also a form of skin cancer but is considered precancerous and easily treatable. The last major form of skin cancer, and the deadliest, is melanoma, which was not part of the study.

Although fresh tissue techniques have today replaced the chemical technique (also referred to as fixed tissue technique), the surgery is still referred to as Mohs surgery. More specifically, the term was coined Mohs micrographic surgery by Daniel Jones in 1985 at the annual meeting of the American College of Chemosurgery, based on the technique's usage of a microscope and skin grafts. The use of a microscope makes this procedure a form of microsurgery, which is a technique used in many non-cancerous applications including gynecological surgery, neurosurgery, ocular surgery, oral and maxillofacial surgery, orthopedic surgery, otolaryngology[72] surgery, plastic surgery, pediatric surgery and podiatric surgery.

72 Otolaryngologists are also known as ear, nose and throat doctors.

Electrosurgery

Electrosurgery is a form of surgery for treating skin cancer (primarily BCC and SCC). This technique is generally not as accurate and is more invasive than Mohs surgery. Electric current is used to either burn the skin or heat up an instrument to burn the skin. There are two main categories of electrosurgery: high-frequency electrosurgery and electrocautery.

High Frequency Electrosurgery

High frequency electrosurgery includes four subcategories: electrocoagulation, electrodesiccation, electrofulguration and electrosection. Diodes are placed on the skin and alternating current allowed to pass through the tissue, burning and killing the tissue. Electrodesiccation is the technique generally used to treat skin cancer (often combined with curettage, i.e. the removal of the skin using a curette, an instrument designed for scraping and scooping). After application of a local anesthetic, the curette is first used to remove the cancerous skin by scraping in three directions. Once the scrapping is completed, the tissue is electrodessicated in short, rapid bursts, also in three directions. This process is repeated a second and often third time in succession to maximize the likelihood of removing all the cancerous cells.

Electrocautery

Unlike high-frequency electrosurgery, electrocautery does not use current to directly burn skin. Instead, current is used to heat up an instrument that then touches the tissue to kill skin cells. This technique can be used to treat both basal cell and squamous cancers.[73]

73 William Holmes, "Electrosurgery Technique," *Medscape*, https://emedicine.medscape.com/article/1997619-technique#showall;
"Specialized Types of Cancer Surgery," *University of Rochester Medical Center*, https://www.urmc.rochester.edu/encyclopedia/content.aspx?contenttypeid=134&contentid=115;
"Curettage and Cautery for Skin Cancer," *Cancer Council NSW*, https://www.cancercouncil.com.au/skin-cancer/treatment/curettage-and-cautery.

Chapter 13

Side Effects

MANAGING SIDE EFFECTS is a very important part of your cancer treatment process. In addition to the treatment you will receive to destroy your cancer, you will also be taking pills to manage your side effects. Expected side effects such as hair loss and nausea will be discussed ahead of time with your oncologist, but there may also be some unexpected side effects during the course of your treatment, which is what happened with me. The goal of this section is to go through potential side effects you will experience (covered in detail in the *Radiation* and *Chemotherapy and Pills* chapters of this book) and how to best manage them. I will also discuss my own trials and tribulations to help others prepare for what to expect. Although each cancer type and experience is unique, there are many commonalities that can be shared and learned from.

Nausea

Many chemotherapy agents and radiation treatments cause nausea and this side effect was particularly challenging for me. I have a predisposition for nausea and have always struggled with motion sickness and thrown up almost anywhere you can imagine, including cars, boats, airplanes, underwater, on an amusement ride—you get the picture. I actually studied motion sickness in astronauts in graduate school since I was particularly interested in what caused motion sickness and how to deal with it.

Motion sickness is caused by a loss of balance and the central nervous system (brain and spinal cord) receiving mixed signals from their sensors (i.e. your eyes, ears, pressure receptors in the lower extremities of the spine and muscle/joint sensory receptors throughout the body). While nausea from your treatment is caused by the chemicals or radiation, it is important to understand what causes nausea naturally so you can reduce your symptoms. Your oncologist will prescribe certain types of drugs to counter your nausea, with various ones being sampled if the initial pills are not having much of an effect. In addition to your pills, you can best help your nausea by limiting your body movement—particularly your eye focus—after treatment. You will always want someone to drive you home and I recommend closing your eyes so as not to focus on anything while riding home from treatment. Lie down and rest with your eyes closed and avoid reading or watching television right after.

Looking at my journal in the *Mental Health* chapter you will see that I experienced nausea quite often, which included a bad taste in my mouth. During and after treatment you can taste the chemical poison in your body, which turns you off from eating. This can curb your appetite and should be countered with frequent, small meals rather than large meals spaced out.

I drank a lot of juice during chemotherapy, which helped offset nausea and purge the bad taste and chemicals from my body. One drawback is that certain types of juices or foods you take during chemotherapy may have a negative connotation for you after your treatment. To this day I still get nauseous thinking about the particular type of mango and guava juices I drank during treatment. I stopped buying those flavors after treatment and never went back, so you might not want to choose your favorite foods and beverages during treatment.

Throwing up can be an unfortunate result if your nausea worsens. Fortunately, this only happened a few times for me and I was able to increase or try new nausea pills to counter it. As you would expect, right after treatment is when you are most likely to throw up. Practice the techniques above and make sure to take your nausea pills as prescribed, which for me was right after my chemotherapy session. I continued taking pills for nausea after chemotherapy ended and radiation therapy began. While there can be some nausea during radiation, the effects are not as severe as during chemotherapy.

JOINT AND MUSCLE PAIN

Throughout chemotherapy and radiation, you are likely to feel joint and muscle pain. This is due to the chemicals attacking cells throughout your body, including bone cells. Radiation has a similar effect but is more localized. Unfortunately there is no specific medication for this other than painkillers, which I chose not to take because of how many other pills I was already on. Please discuss with your oncologist if painkillers are an option, which they will need to evaluate along with all your other medications.

I noticed pain in my bones after my second treatment. It was during my first attempt at working out, which made the pain more pronounced. Refer to the *Exercise* chapter on recommended workouts to help reduce muscle loss while not overstressing your body. Bone pain occurred primarily during my first cycle (month) of chemotherapy and then was not as conspicuous. This may have been because of my steady routine of working out or simply my body adjusting to the chemical agents. Muscle pain was not as severe as the bone pain, which enabled me to work out.

Working out with muscle pain is not advised, as you can potentially tear a muscle and your body is already in a weakened state.

FATIGUE

Fatigue is a ubiquitous side effect of the poison and radiation you receive during treatment. Your body uses a lot of energy fighting the chemicals in your body, which leads to fatigue. Unfortunately, healthy cells are being attacked in addition to the cancer cells. Radiation damages cells that your body tries to repair, also depleting energy. Plan rest around treatments and realize you will be much less active than before your diagnosis. Listen to your body and rest frequently to promote healing. Push yourself to exercise when you are able, but when your body is at its worst, especially after a chemotherapy treatment, rest!

I recall coming home after chemotherapy treatments and going straight to my bedroom, turning off the lights and staying in bed until I felt well enough to get up. My energy level fluctuated with treatments during chemotherapy, but since my radiation treatment was every weekday, my energy remained at a steady, fatigued level. This made exercising regularly easier since I could better predict my energy. I was able to do a few more activities such as visiting the city with relatives in town or seeing a tennis match. If you are feeling well enough to do an activity, make sure you plan around your weakened state accordingly. When visiting San Francisco, I was only able to walk around for an hour or so before my energy was depleted. Exercise and activities are important both physically and mentally, just make sure you understand your limits and plan activities around them.

Headaches

Headaches can occur during chemotherapy and radiation treatments for many reasons, but simply put, your body is exhausted. Similar to when you have headaches before treatment, rest and water is the best medicine. Pain relievers such as paracetamols (e.g. Tylenol) or ibuprofen (e.g. Advil) may be an option, but make sure you discuss with your oncologist first, especially if you are on a blood thinner.

I experienced headaches immediately after my first chemotherapy treatment and again during the second cycle. These were intermittent and subsided for the most part during radiation treatment. If your headaches persist, discuss with your oncologist.

Discolored Urine

Various chemotherapy drugs will cause changes in your urine. There is nothing to be alarmed about; discoloration is normal as your body removes toxins. You often can smell the chemicals as they excrete from your body, reminding you of how potent these drugs are. Read up on your specific chemotherapy agent to see if discoloration is expected. If it is not, make sure you notify your oncologist to rule out any unexpected or potentially harmful drug side effects.

In addition to discoloration, you will also urinate frequently due to the amount of fluids entering your body during and after an IV. Some of my treatments were long enough and the fluid ingested large enough where I had to wheel my IV stand to the restroom and urinate during a treatment. I must say, moving around like an old man at just 28 was sobering. Drinking plenty of fluids after treatment is highly recommended to flush the chemicals out of your system. Water or healthy not from concentrate juices are your best choice and will also help fight many side effects such as nausea and headaches.

Appetite Loss

Your appetite will decrease during treatment, which needs to be countered to give your body proper nutrition for fighting cancer. Nutrition is discussed in detail in the *Nutrition* section of this book, providing healthy foods, drinks and supplements during and after treatments. I recommend experimenting with different foods to find what you like and can tolerate during treatment. You may find that one type of food or drink will be palatable during one phase of your treatment, while another will be easier to take in during another portion. Variety is recommended because your taste buds will not be functioning normally due to the toxins in your body, so mixing up what type of food you consume will increase your ability to eat. I drank a lot of juices during treatment since they provided an abundance of vitamin C and had strong flavors to help combat the toxic taste in my mouth.

After my cancer treatment ended, Moy and I started making smoothies, which I wish we had tried during my treatment. These are ideal in that you can create a strong flavor with ingredients that are healthy and more palatable, mixing them up as needed. Drinks and smoothies are easier to take in than solid food during treatment. You can also mix in fruits and vegetables to your smoothie to make sure you are getting enough vitamins and nutrients. One of my favorite smoothie recipes contains frozen strawberries and blueberries, apple juice, chocolate and vanilla protein, peanut butter and your choice of nut. Another one is the Caribbean Passion recipe from Jamba Juice, which contains frozen peaches and strawberries, passion fruit mango juice and orange sherbet. Similar to a smoothie is an acai bowl,[74] which I was lucky enough to discover and fall in love with in Hawaii. This would also be an ideal snack for cancer, as it contains an abundance of fruits and vitamin C, allowing you to tailor it with your favorite fruits. Ask your caretaker to make these healthy treats for you. Otherwise stop by a local Jamba Juice or smoothie store of your choice. (My mom is jealous that they do not have Jamba Juice in Michigan!)

I ended up losing around 30 lbm during my chemotherapy treatment but was able to gain some of that back during the radiation portion. The Stanford V radiation treatment I was on included prednisone, which helps increase your appetite. Without this, my lack of appetite would no doubt have been even worse. Ask your oncologist if prednisone or another corticosteroid is an option if your appetite becomes too poor and you are experiencing severe weight loss. There may be additional options your oncologist will recommend as well to combat your weight loss.

74 An acai bowl is like a parfait but with blended acai berries rather than yogurt as the main ingredient. Complete with granola, fruit toppings and, sometimes, additional ingredients such as peanut butter, it makes for a delightful treat!

Blood Clots

An unexpected and potentially dangerous side effect can be a blood clot. I would have liked to have been warned about this, since both cancer and chemotherapy can cause thickening of the blood, leading to clots. Not only can blood clots be painful, but if they break loose and travel to your brain, lungs or heart, they can kill you.

I began experiencing pain in my right biceps around the middle of my second chemotherapy cycle on January 18. The pain began on a Friday after my treatment and worsened over the weekend. Monday was a holiday and my oncologist and normal doctors were not in, so I drove to the hospital. By then my forearm and biceps were very painful, especially when extending my arm. There was redness and swelling and I began to think it could be a blood clot, although I was uncertain since I had never had one and did not realize that cancer patients were at risk.

I had trouble finding an oncologist, so I walked past the front desk and found one working in his office. After describing the pain and symptoms he ordered an ultrasound that day, which revealed I had a blood clot near my elbow. This is another example of being proactive with your health. Had I waited until later in the week when my oncologist was available, I could have done further damage to my arm or worse, the blood clot could have loosened and traveled to my brain, heart or lungs.

I was immediately put on a blood thinner (warfarin, specifically the brand name Coumadin) and the pain decreased over the next few days. My forearm still experienced numbness later in the week but the sharp pain improved throughout the week. The numbness continued throughout my treatment.

I attended a blood thinning class that explained the medication and dosage as well as answered questions. Being the only person there under 60 (at only 28 no less) was a bit humorous and depressing! Monitoring your blood thickness is important so a nurse calls you daily initially, then less frequently to discuss your dosage. This was one of the few instances where my pill dosage was a moving target, as they tried to find the optimal quantity to thin my blood while preventing excessive bleeding (such as when getting an IV). They provided me with a device to cut my pill in half since that is the level of fidelity required for the blood thinning dosage (why they do not decrease the per-pill dosage to prevent having to cut them is beyond me).

The biggest lesson I would like other patients and caretakers to take away from my experience is to ask your oncologist *before* treatment about the risk of blood clots and whether you should be on a blood thinner. This is a serious and potentially life-threatening side effect that your oncologist should discuss with you and take preventative action for so you do not develop a blood clot. Unfortunately, this was not discussed with me nor covered in the abundance of literature and pamphlet handouts I read. Blood clotting is an important risk and side effect that needs more attention in the cancer community.

INSOMNIA

Getting enough sleep is extremely important during your cancer treatment (and for non-cancer patients as well). One theory I have about what contributed to my cancer forming was my frequent bouts of insomnia, since a lack of sleep reduces your immune system's ability to kill cancer cells developing daily in your body.

During treatment with toxic chemicals and radiation, your body needs as much strength as it can muster. Sleep is imperative to gather your energy and allow your body to most effectively fight your cancer and treatment side effects. But the many unpleasant side effects—nausea, bone and muscle pain, headaches—can make sleeping difficult. I took a sleeping class and read books on various techniques to help me sleep. One book I particularly recommend is *Say Good Night to Insomnia,* by Gregg Jacobs.[75]

SLEEP TIPS

All of these tips can be practiced during and after your cancer treatment.

Go to Bed Around the Same Time Every Night: Your body has a natural circadian rhythm that when disrupted, decreases the amount of sleep you get each night. Going to bed around the same time every night helps manage your circadian rhythm and promotes restful sleep.

75 Gregg D. Jacobs, *Say Good Night to Insomnia: The Six-Week, Drug-Free Program Developed at Harvard Medical School* (New York: Henry Holt and Company, 1999).

Read, Watch Television or Have a Stress-Free Activity in the Hours Before Bed: Days are filled with stress related to work, finances, family drama and personal issues. Your brain needs an escape from these stresses to reduce your anxiety level and allow you to rest. Having a routine of activities you enjoy before bed helps your mind decompress and relax for sleep.

Note that there are mixed opinions on watching television or using a tablet before bed. Some studies have shown that the artificial light from your screen can reduce melatonin and negate sleep. However, the relaxation of watching television or using a tablet may have a net calming benefit, especially if you take a melatonin supplement before bed as I do. I leave it up to readers to experiment with different activities and determine which ones work best for them.

Have a Calming Ritual That You Repeat Every Night: I sit in a massage chair and listen to Hawaiian music, which also serves to relax my back muscles and prevent me from waking up with back pain. This is a shorter activity to be done after your relaxing routine. Meditation is an additional example.

Sex Has Natural Physiological Calming Effects: Chemicals released during sex include dopamine, norepinephrine, testosterone, oxytocin and serotonin. While all are healthy, the latter two in particular aid in sleep. Oxytocin is released after climax; it is a natural tranquilizer that lowers blood pressure, reduces pain, decreases stress and induces sleep. Serotonin is also released after an orgasm and creates a calming effect. Anti-depressants use elevated levels of serotonin, so having an orgasm is a natural anti-depressant![76]

Take Melatonin Chewable Tablets Right Before Bed: I take one 10 mg tablet. Melatonin is a hormone produced naturally in your body's pineal gland and helps regulate your body's circadian rhythm. Levels are highest between midnight and 8:00 a.m. when low light triggers melatonin production, so increasing your body's amount during those times will aid your sleep.

76 Krishna G. Seshadri, "The Neuroendocrinology of Love,'" *Indian Journal of Endocrinology and Metabolism* 71, no. 4 (June 2016): 558–563 , http://www.ijem.in/text.asp?2016/20/4/558/183479.

Do Not Drink Caffeine in the Evening: Caffeine has a half-life between 3 and 7 hours depending on an individual's liver enzyme functionality, pregnancy and what other drugs are being taken. For most people that average half-life is 6 hours. So if you drink caffeine at 5:00 p.m., half of it will still be in your system at 11:00 p.m. Backing up even further, a quarter of the caffeine you consume at noon will still be in your system around midnight.

Caffeine consumption in general should be curtailed and is not recommended daily but rather intermittently, such as on weekends. If you are a coffee or tea drinker, doing so in the morning is advised due to the amount of time it takes to leave your body. Drinking too much caffeine can also cause headaches when you withdraw from the drug, which I experienced once as a teenager working in a fast-food restaurant. Drinking multiple Cokes and root beers[77] throughout the day caused me to have painful headaches on days I would not work. Once I figured out what was causing the pain, I backed off on the amount of caffeine I consumed.

Do Not Eat a Lot of Sugar Before Bed: Sugar intake in general is not healthy in abundant quantities. As pointed out in the *Nutrition* chapter, sugar gets converted to stored energy in the form of fat. Consuming sugar intermittently as an occasional indulgence is fine or if you are low in energy such as before a sporting event or during an afternoon lull at work. However, in excess it can not only lead to obesity, but if taken too close to bedtime your energy level will increase and you will find it more difficult to go to bed.

Do Not Use Your Bedroom as Your Office: Your mind will associate your bedroom with stress rather than relaxation, making it more difficult to fall asleep. When I first moved to Silicon Valley after graduate school, I lived in a one-bedroom apartment, so my office and bedroom were one in the same. This likely was one of many factors that contributed to my difficulty sleeping, which I was able to remedy when I bought a larger townhome with a separate office downstairs. Now I can work in my office and leave my bedroom as a place for rest. You want your brain to associate the bedroom with sleep, calming and sexual activities, not stress.

When you cannot sleep, do not stay in bed brooding, stressing, tossing and turning. Get up and do a calming activity, eat or drink a beverage that helps you sleep, then return to your bedroom.

77 I drank Barq's root beer, which is one of the few caffeinated root beer brands.

Have a Snack Before Bed: I have read that you should not eat before bed because your metabolism decreases and you will build up excess fat. Although I have also found articles discussing why eating before bed is healthy, so there is debate in the nutrition community about the efficacy of not eating before bed. While you should not have big meals right before bed (this may cause an upset stomach), going to bed on an empty stomach will make it harder to sleep. Certain foods help induce sleep as well. Before bed or when I wake up and cannot get back to sleep, I have found a glass of milk with a snack like Pop-Tarts, a protein bar or cereal to be helpful.

Protein-rich foods such as milk contain tryptophan, which aids with sleep; however, protein can also limit the ability of tryptophan to enter the brain. Foods rich in carbohydrates, on the other hand, produce insulin, which helps tryptophan cross the blood-brain barrier. Therefore, combining milk with a carbohydrate snack is a great combination before bed or if you wake up and have difficulties going back to sleep.

Make Your Bedroom as Quiet as Possible: There are factors (such as a newborn baby) that can make this difficult, but as much as possible, create a noise-free environment. If your spouse or significant other snores or is restless in bed, consider earplugs, which are a great way to maintain sleep. A separate bed or bedroom can also be considered although that may not be an option or desired. With all the effort you put into getting to bed, the last thing you want is to wake up, as it can be very difficult to return to sleep.

Do Not Keep Your Phone in Your Bedroom: Unless you are the president of the United States (and I imagine even he would be woken up by staff in person), there is no need to keep your phone in your bedroom and potentially disrupt your sleep. Your sleep and health should be your number one priority. Not picking up the phone at all hours of the night and morning will make you more successful at work and in life.

Do Not Set Your Alarm or Have Your Clock Facing You: You may have a job that prevents you from not setting the alarm, but there is nothing preventing you from not having a clock-face in your sightline. Preconditioning your brain that it has to wake up at a certain time creates anxiety. You subconsciously or consciously may worry about making sure you get up in time and what the consequences will be if you do not. This can keep you tense and prevent you from falling or staying asleep. If your job allows a flexible work schedule—push for this if it is an option—let yourself wake up naturally without an alarm. Once you follow these other strategies and start sleeping at the same time every night and not waking up frequently, you will find your body naturally waking up at the same time every morning. This is what I do, and I only set the alarm if I have an important presentation, plane flight or another event where I need to make certain I arrive on time.

With regard to facing your clock away from you, staring at the clock when you cannot sleep only induces additional anxiety. You worry about not getting enough sleep and what the consequence will be the next day if you do not. The harder you try to force yourself to sleep, the more difficult it will be to do so. This simple tip of turning the clock face away from you will reduce some anxiety.

If you need to know what time it is when you wake up in the middle of the night or wake up in the morning, you can simply turn the clock toward you and then back away again. I have an obstacle in front of my clock, so I do not have to physically move it, but rather must sit up in order to see it.

Make sure your bedroom has proper blinds and curtains to keep out the majority of natural light. Your body wakes up when seeing light due to your circadian rhythm. There are special shades for windows that will keep out light if your blinds or curtains are insufficient. Additionally, adding curtains or blinds in areas with sunlight is recommended. In my home, the master bathroom connects to the master bedroom via a hallway without a door. Light from the sun would penetrate my bedroom early each morning, so I added curtains. Wearing an eye mask while you sleep can be an extra layer of light blockage, which I also started using. While this reminds me of the mom in the movie *Overboard*, it does help! Incidentally, wearing a mask and earplugs on airplanes is also recommended when trying to rest.

Invest in a Quality Mattress to Promote Better Sleep: I did not realize how important a bed could be until I upgraded. Going through college you get used to sleeping on cheap mattresses furnished by your dorm or apartment building. When I first moved to California, I brought my childhood, full-size bed. Moy complained about the springs and joked that I didn't have insomnia; I just needed a mattress that did not stab me with its coils! I upgraded to a California king Beautyrest, which definitely provided a better night's rest. Recently I replaced my mattress again and am a proponent of getting a new one every ten years or so. You can find a great mattress at an affordable price if you do your homework.

All these tips can be practiced during and after your cancer treatment. This is the number one tip I can give to cancer patients to prevent a relapse. During the treatment itself, the days during and immediately after chemotherapy is administered will be more uncomfortable and thus harder to sleep. This can be combated by sleeping more frequently and taking more naps, rather than just one long sleep at night. During my treatment, I found it hard to sleep near the beginning due to the pain; however, I was able to take more frequent naps as the treatment progressed to improve my overall sleep.

Shaking and Twitching

Many of the chemotherapy agents you take have the unfortunate side effect of impacting your nervous system. Mustargen in particular is a nitrogen gas derivative developed in secrecy during World War II, where it was spun off for clinical uses (with the first patient being treated in 1942). Incidentally, while chemical warfare has been prevalent for centuries, nitrogen gas was first used extensively during World War I with symptoms of internal and external burns and bleeding.

You will likely experience some tingling, numbing, shaking or twitching after treatments, depending on the chemicals used. There was nothing provided to help me with this, but you should discuss with your oncologist if the symptoms worsen.

To this day I still have moments where my body twitches, usually induced by drowsiness or stress. I have had a couple of surgeries since my treatment and both times when I woke up I was twitching uncontrollably and the nurses and doctors did not know what was going on. This has not been a health risk but is something to be aware of during and after chemotherapy treatment. Do not confuse these side effects with having a seizure or epilepsy.

DENTAL

Chemotherapy and radiation both can impact your teeth. This can lead to cavities, dry mouth, thickened saliva, canker sores, difficulty swallowing, change in your taste, difficulty chewing, bone disease in the jaw, teeth discoloration and gum inflammation. Confer with your dentist before and during treatment to ensure you minimize these potential impacts.

Everyone should brush their teeth and floss daily regardless of whether or not you have cancer, but during and after your treatment this is even more important due to your weakened state. I never had a cavity until after my treatment, which I was quite proud of! However other than the persistent bad taste in my mouth, the cavity and some staining, I avoided most of the side effects listed above. Since I was on a blood thinner, I had to be a bit more careful when flossing to avoid excessive bleeding, but it did not become a major issue.

CHEST PAINS

You may experience chest pains during treatment, either due to chemotherapy, radiation or both. Discuss any symptoms with your oncologist to ensure they are aware and can assess the severity. It is natural to stress out about pains during treatment and wonder what a normal side effect versus a major concern is. I experienced frequent chest pains and since my tumor was wrapped around my heart and lungs, I had good cause to be concerned, as they received unwanted, stray radiation.

I felt chest pain mostly during the radiation portion of my treatment, for instance while doing cardio at the gym. Monitor your pain and scale back your exercising and if the chest pain persists, discuss it with your oncologist. My radiation oncologist explained that some pain was not abnormal. After my treatment ended, the pains lessened and my goal was and still is, to strengthen my heart through exercise to help combat the damage done during radiation treatment.

VEIN PAIN, SKIN DISCOLORATION AND BRUISING

Receiving IVs during chemotherapy can be uncomfortable, both in having your veins poked and the chemicals burning and discoloring your veins. My veins are quite accessible in my forearms and some nurses were competent at administering an IV on the first try while others struggled. One had to try five or six times with pokes on my arm before getting the needle in! I recall seeing the needle push my vein but not puncture it over and over. If you are having trouble, do not feel uncomfortable asking for another nurse. This one specifically had just started and was obviously still learning.

Many of the drugs themselves will discolor your veins and cause unpleasant burning sensations. This happened during many of my chemotherapy treatments, with some chemicals being worse than others (see the *Chemotherapy and Pills* chapter for additional side effect details). Furthermore, you may have bruising around your IV sites due to the needle and potentially leaked chemicals. As the latter is serious, ensure your nurse adequately tests your IV with saline prior to injecting the chemicals. Bruises due to the needles themselves are unpleasant but nothing serious to worry about. Moving your injection site from one arm to another and different locations on the arm is useful. Coupled with the multiple blood tests you will be getting during your treatment, having all these IV piercings can leave your arm bruised. If your bruises become a concern, discuss with your oncologist.

Note that although this option was not discussed with me, some cancer patients use a port-a-cath (also known simply as a port) as a semi-permanent means to administer IVs. This has the advantage of reducing the amount of needle pokes and potential bruising or discoloration due to vesicants. The downside is that you leave the port in, which may cause discomfort or prove annoying when it comes to activities such as showering. Discuss the option of using a port with your oncologist, along with the potential pros and cons.

Numbing in Extremities

Many of the chemical agents you receive impact your nerves and can cause numbing in your extremities. I experienced this throughout my treatment, especially in my fingertips where they would have a tingling sensation and be less sensitive to the touch. This went on throughout my treatment and even a while after, although my fingers months later would be back to normal. There were no pills given to me to combat this side effect and although uncomfortable and inconvenient, there is no major health risk.

Patients receiving vincristine or vinblastine (both of which I received) are particularly impacted and can develop Raynaud's phenomenon during or after cancer treatment, wherein blood vessels in the hands or feet constrict, reducing blood flow. Symptoms can be heightened with cold temperatures or emotional stress. If symptoms are observed, discuss with your oncologist to ensure they are understood and there is not a more serious underlying issue.

HAIR LOSS

Almost all chemotherapy and radiation treatments cause hair loss. This seems to be the adopted symbol (perhaps since it is the most noticeable symptom) that someone is dealing with cancer. Although there is nothing physically harmful about losing your hair, psychologically this can be difficult. I began losing my hair during my first cycle in about week three. I noticed clumps coming off in the shower, which although depressing, I was mentally prepared for. I waited until the beginning of the second cycle to shave my head, which I decided to do both because of the inconvenience of cleaning up after a shower and to prevent an unseemly lopsided head! My brother and dad were visiting so my brother cut mine and I reciprocated. When and if to shave your head is a personal choice, although I recommend doing it earlier rather than later since you have enough to focus on physically and mentally without the added distraction.

You will also notice hair loss in other parts of your body. I began to lose some chest hair as well as eyebrow hair and some eyelashes, although thankfully I did not lose them all! By the time I reached the radiation portion of my treatment, my head was completely bald (including my face) and most of my leg, arm and chest hair was gone, although the radiation treatment completed the process. My hair on my head began to grow back toward the end of my radiation treatment and the rest of my body followed. Interestingly enough, my hair came back blonder and curlier than before treatment. After about six months of growing out my hair and cutting it again, my hair returned to its pre-cancer treatment color and texture.

Being bald was not a big deal for me, as I have shaved my head in the past (e.g. when playing football). For others, especially women, this can be an issue. Many women wear hats or wigs, while others prefer to be bald. Choose whichever option is the most comfortable and try not to stress out about it, as your hair will come back!

Chapter 14

Mental Health

Going through cancer treatment is an extremely emotionally trying affair to both the patient and support network. Depression, anxiety and insomnia are common effects caused by this experience. Managing your mental state is just as important as controlling your physical state. I have long had an interest in mental health and the importance of training yourself mentally for life's trials and tribulations. One benefit of getting older is the wisdom you gain and emotional strength of past experience you can build off when undergoing new life challenges.

Going through cancer can quite literally be the fight of your life. Realize that battling the mental aspects of cancer must be managed just like your medicine, nutrition and exercise. Rely on family and friends as well as self-help books, professional therapy or both. Most treatment centers offer counseling and support groups with others going through similar challenges as you. Find out what you need to help you. Maybe it is as simple as spending time with your family or doing hobbies between treatments. I found myself working on puzzles as I did with my dad growing up, reading and playing video/computer games. Friends and family visited often, which was very important and uplifting for me.

I am a strong proponent of self-help books, including those related to mental health and specifically depression, insomnia, anxiety and relationships. The references at the end of this book provide the ones I recommend based on my research. In particular, David Burns's *Feeling Good* is an outstanding mental health book that is the first on the list I recommend to anyone going through trying times.[78] The premise of his teaching is

78 David D. Burns, *Feeling Good: The New Mood Therapy* (New York: HarperCollins Publishers, 2008).

that we are in control of our thoughts. This sounds simple, but when I read this book years ago it was profound. When you really start to separate the act of thinking with the thought itself, you realize that most negative thoughts in your head are self-imposed. Sure, there are realities of life that are tragic, depressing, cause you to feel anxious, etc. However, how you react to these events and challenges really is up to you and the thoughts you produce in your brain.

As an example that relates to cancer, it is natural to have a *Why did this happen to me?* mentality upon being diagnosed. If you are religious, perhaps, *Why is God punishing me?* Then follow-up negative thinking occurs: *I must be a bad human being for God to want to punish me,* or *Nobody can have this much bad luck; I must have brought this on myself and it is my fault.*

Examples of cognitive training to reverse these negative thoughts are: *It is not logical that God would be punishing me. Millions of good people get diagnosed with cancer every year and I am no different.* Following the second example: *Everyone has bad things happen to them and challenges. Perhaps mine have been greater, but it is not self-induced or me being a lousy human being. Cancer happens to loving and accomplished human beings and I am no different.*

David Burns also has workbook versions of his books as well as relationship books. Although not necessarily related to cancer but relevant in the sense of relating to loved ones before, during and after cancer, *The Five Love Languages* by Gary Chapman is the best relationship book I have ever read.[79] The author breaks down people's love needs into five categories and you take a quiz to find out the results. It makes a lot of sense to me and can help you and your significant other enhance your love for one another by identifying what you need to feel loved and what they need to feel loved, which will have some overlapping needs but many separate needs as well.

Returning to *Feeling Good*, cognitive therapy is very strong and an alternative (or supplement should you choose) to medication. The latter is a personal choice and I have read many articles and books on this as well regarding the neurological connections in the brain and how

---

79 Gary Chapman, *The Five Love Languages: The Secret to Love That Lasts* (Chicago: Northfield Publishing, 2014).

drugs try and bridge neurological signals that are deteriorated due to depression. There are many debates as to whether these neurological signals, or lack thereof, are hereditary or due to experiences. My personal opinion is drugs can be useful for short-term help in overcoming depression, anxiety or insomnia (which often come together and feed off one another) but one should strive to heal themselves mentally for long-term health. Again, this is an opinion and I am not a mental health professional. However, the important takeaway is there are many tools out there to fight cognitive challenges, which can be extremely important during cancer treatment.

LOGBOOK

Throughout my cancer treatment I kept a log of my mood along with notes on what went on that day to influence my mood score. I did this as a way of tracking my mental health and how treatment, family/friend interactions and exercise influenced my moods (long ago I found exercise to be the most important tool for me to maintain a healthy mental state). Hopefully this will prove useful to others in seeing what type of challenges you may experience during your treatment and that there is light at the end of the tunnel. Despite all the hardship you are going through, you have the ability to be strong and make it through!

Looking at my journal entries Tables 5 through 9, I provided a scale of 1 to 10 with 1 being as miserable as you can imagine and 10 being on cloud nine without a care in the world (results are also plotted in Figure 72. Now of course every individual will have a different baseline (e.g. some may be around a 5 before treatment, an 8, a 3 or so forth). Rather than focus on the quantitative value, of more importance is the impact of my treatment to my overall mental state throughout my treatment.

You will notice a general upward trend with a somewhat cyclical variation throughout. At the beginning of my therapy I was down in the dumps, as can be expected when finding out you have a football-sized tumor in your chest and stage IV cancer at 28 years old. The first point on the graph corresponds to the first day of my treatment on Thursday, December 3. My mood remained low throughout the first couple of weeks as I dealt with the physical tolls on my body, as well as the lingering psychological challenges of going through cancer treatment.

Continuing to the middle of December my mood started an upward shift towards the 4 to 6 range, a substantial increase from the 2 to 3 range at the beginning of my treatment. I also included *exercise, nausea* and *description* columns along with my *mood* column in Table 5, since the former three directly contributed to the latter. Specifically, you see that my mood shifted upward right around the time I started exercising. With exercise always being extremely important to my overall mental and physical well-being, during cancer treatment this can be a powerful combatant to the doldrums you are experiencing.

Through the end of December my mood remained around the 5 range. Being around the holidays was depressing; the last thing you want to be doing on Christmas Eve is getting cancer treatment. However, having family visit helped offset the overall dour mood. My oncologist came by to check on my treatment and wish me a merry Christmas, which was a very nice gesture.

The beginning of January saw my mood take a dip down to 2 again, which was mostly attributed to the strong impact of the medication, extreme nausea and lack of sleep and energy. I have found over my years of studying mental health that sleep is so vital to your overall mood and when your body is getting poisoned, sleep unfortunately suffers. Despite this setback I continued with my exercise regime and continued having family support, which was not easy since my family lived in Michigan and Missouri. Early on my brother had suggested I come out to Michigan to receive treatment to be closer to family members; however with my insurance and the overall increased stress of traveling and living away from my home, girlfriend and cats, staying in California made the most sense.

January 7 was a bit of a milestone in that I had lost enough hair where I decided to shave it all off before it got worse. Having my brother shave his head was a cool show of love and support! It brought me back to our time playing football in high school where most teammates shaved their heads at the end of our preseason camp (Camp Greilick). Having my dad and brother as well as my mom, sister, cousins, aunt and uncle spend time with me also really made a big difference. A strong cancer support team is vital to anyone's struggle; you can see that my mood generally increased on days when family and friends visited. Conversely, treatment days and days after treatment saw my mood sour, as can be expected.

The remainder of January and my cycle two saw similar ups and downs with a mood score high of 7 and low of 3. My overall mood score from cycle two was 4.4, remarkably similar to a cycle one score of 4.5. While the cumulative treatment and side effects took their toll during cycle two, my increased exercise and family and friend visits helped to offset this. Getting out of the house was helpful, including trips to San Francisco and a few dinners with friends when my energy level was high enough toward the end of a treatment week right before the next treatment. These outings had to be managed, as nausea and poor energy limits the amount of walking and exercise you can do.

Cycle three over the month of February had a low of 3 and high of 6 for an overall score of 4.8. This small increase from cycle two is in the noise, with my mood generally fluctuating with treatment times, side effects and friend interaction.

I played a lot of competitive tennis leading up to my cancer treatment (and after), so during cycle three I did try some light tennis a couple of times. My leg strength was not near what it should have been, so it was mostly just to get some friendly interaction and manage my strength and nausea. I recall the first time I tried to play a competitive practice after my treatment was over; I charged the net like I normally do and my brain was too far in front of my legs. I stumbled and wiped out badly, doing a roll of sorts. The blood drained from my partner Andy's face and he told me to take it easy!

It does take time after treatment to get back to your pretreatment body and routine, which I eventually did. It took me months to get back to my full strength in the gym, although I made strides quickly and was back to playing tennis even sooner. While as mentioned previously I lost around 30 lbm overall during treatment, the fact that I exercised made a big difference in my ability to resume my active lifestyle soon after treatment.

My radiation treatment began soon after my chemotherapy treatment ended. Here my mood made a noticeable increase up to 5.6 from cycle three's 4.8, the biggest change in the roughly four and a half months of combined chemotherapy and radiation treatment. This mood increase was likely due to seeing light at the end of the tunnel and an increase in what I could do physically.

While the radiation treatment was every day during the week and therefore a more constant mental drain, the peaks and valleys were not as high and low as during chemotherapy (which saw a more pronounced day-to-day change in my side effects). We even managed to fit in a weekend trip to Los Angeles to visit my uncle and watch a bit of tennis, which although physically taxing, was mentally uplifting.

My final day of treatment was on April 9, yet I continued to keep a journal for the remainder of April. As expected, my mood began to increase as I attempted to resume my normal activities as soon as possible. The tennis incident I mentioned happened four days after my last treatment, so it took a while for me to get back into shape. However from a mental standpoint, knowing my cancer treatment was over and the cancer was in remission was very uplifting.

Fighting for your life is a tremendous struggle and journey; the relief you feel at the end is palpable! I continued exercising and playing tennis throughout the month—my post-treatment mood score in April had an average of 6.9 with a peak of 8. I transitioned to heavier weights and more intense exercises such as deadlifts a few weeks after treatment, along with increasing the protein intake in my diet.

I returned to work a few weeks after my last treatment, giving my body and mind additional time to recover. As discussed in the *Treatment Logistics* chapter, when I first began treatment I contemplated working while going through cancer. By the end of my treatment I was happy with my decision not to do this. From a physical standpoint you can take better care of your body and manage side effects at home.

Furthermore, work for most people is quite stressful and takes a certain amount of mental fortitude. My specific job as an aerospace engineer requires mental focus both in solving problems and dealing with a plethora of personalities at work, some of which are quite challenging. Having this added stress on top of your already-trying cancer treatment can contribute to insomnia, anxiety and depression. I highly advise working out a situation where you can take time off work to focus on your health, which must be your number one priority.

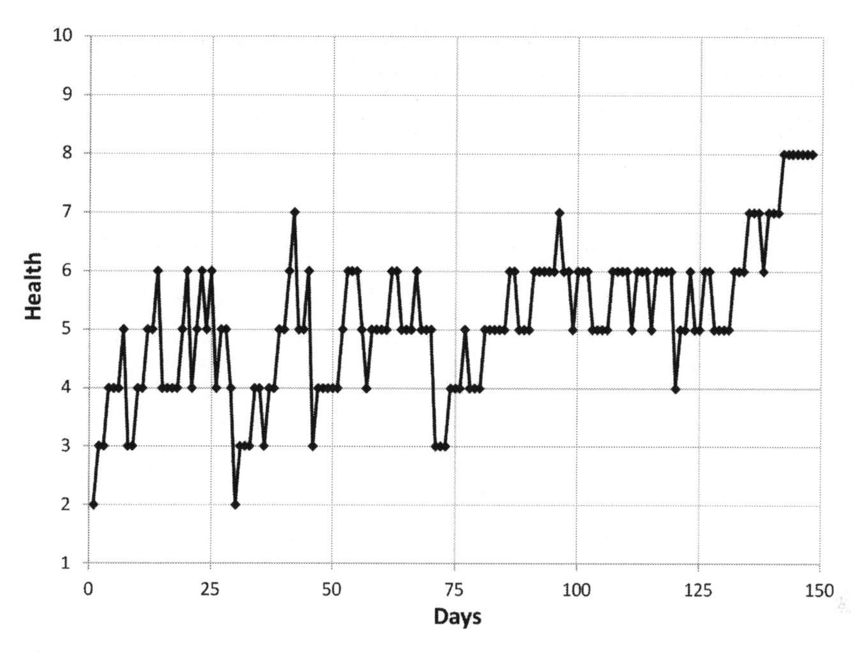

Figure 72: Mood scores over the course of my cancer experience.

Table 5: Mood journal, chemotherapy cycle one (Dec. 3–30, 2008).

Day	Date	Mood	Event	Exercise	Nausea Pills	Description
Thurs.	3	2	Treat-ment	N/A	N/A	Headache, fatigue, bad taste in mouth, some pink urine.
Fri.	4	3	N/A	N/A	N/A	N/A
Sat.	5	N/A	N/A	N/A	N/A	N/A
Sun.	6	N/A	N/A	N/A	N/A	N/A
Mon.	7	N/A	N/A	N/A	N/A	N/A
Tues.	8	N/A	Blood work	N/A	N/A	N/A
Wed.	9	N/A	N/A	N/A	N/A	N/A
Thurs.	10	3	Treat-ment	N/A	N/A	Shaking, fatigue, veins hurt in arm, teeth stain, bad taste.
Fri.	11	N/A	N/A	N/A	N/A	N/A
Sat.	12	N/A	N/A	N/A	N/A	N/A
Sun.	13	N/A	N/A	N/A	N/A	N/A
Mon.	14	N/A	N/A	N/A	N/A	N/A
Tues.	15	5	Blood work	Legs and cardio	N/A	Leg bones painful.

Day	Date	Mood	Event	Exercise	Nausea Pills	Description
Wed	16	6	N/A	Chest/ back/cardio	N/A	N/A
Thurs.	17	4	Treatment, blood work	N/A	N/A	Bad taste, some pink urine.
Fri.	18	4	Treatment	N/A	Neupogen	Bad taste.
Sat.	19	4	N/A	Cardio	Neupogen	Bad taste.
Sun.	20	4	N/A	Legs and cardio	Neupogen	Bad taste.
Mon.	21	5	N/A	Arms/cardio	Neupogen	Bad taste. Arm bones hurt while lifting. Hard to sleep.
Tues.	22	6	N/A	Chest/ back/cardio	Neupogen	Slept during the day. Taste better. Shoulder bones painful while lifting.
Wed.	23	4	Blood work	Legs and cardio	N/A	Sore shoulders, nausea. Started losing hair in shower.
Thurs.	24	5	Treatment	N/A	N/A	Some pain in left arm where injected. Overall not as fatigued as Week 2 when I got same treatment.
Fri.	25	6	N/A	N/A	Neupogen	N/A
Sat.	26	5	N/A	N/A	Neupogen	N/A
Sun.	27	6	N/A	Cardio (16 minutes)	Neupogen	Didn't do much; stayed at home for most of the day. Leg and shoulder bones a little sore after elliptical, which was cut short due to closing.
Mon.	28	4	N/A	Arms/cardio	Neupogen	Nauseous working out after dinner.
Tues.	29	5	N/A	Chest/ back/cardio	Neupogen	Spread meals out more, seemed to help with nausea. Bones sore (legs, shoulder, back, etc.).

Day	Date	Mood	Event	Exercise	Nausea Pills	Description
Wed.	30	5	N/A	N/A	N/A	Went to San Francisco with Mom/uncle/Moy, etc. Felt OK during, tired afterwards.

Table 6: Mood journal, chemotherapy cycle two (Dec. 31–Jan. 27, 2009).

Day	Date	Mood	Event	Exercise	Nausea Pills	Description
Thurs.	31	4	Treatment	Legs and cardio	N/A	Bad taste in mouth, sore throat, nausea.
Fri.	1	2	N/A	N/A	Neupogen	Threw up, sore throat.
Sat.	2	3	N/A	Wii	Neupogen	Sore throat, congestion, fatigue, night sweats.
Sun.	3	3	N/A	Wii	Neupogen	Sore throat, congestion, fatigue.
Mon.	4	3	N/A	Chest/back/cardio	N/A	N/A
Tues.	5	4	N/A	Arms/cardio	N/A	Shaved head, still a little stuffed up.
Wed.	6	4	Blood work	N/A	N/A	N/A
Thurs.	7	3	Treatment	Legs and cardio	New nausea pills (NNP)	Fatigue, bad taste, less nausea. Ate at Red Robin.
Fri.	8	4	N/A	Cardio	Neupogen, NNP	Nausea better than last week.
Sat.	9	4	N/A	N/A	Neupogen, NNP	N/A
Sun.	10	5	N/A	Cardio	Neupogen	N/A
Mon.	11	5	N/A	Chest/back/cardio	N/A	N/A
Tues.	12	6	N/A	Arms	N/A	Dinner with friends at Thai restaurant.
Wed.	13	7	Blood work	N/A	N/A	Felt pretty good. Al in town. Still congested but getting better. House-hunting.

Day	Date	Mood	Event	Exercise	Nausea Pills	Description
Thurs.	14	5	Treat-ment	Legs and cardio	NNP	N/A
Fri.	15	5	Treat-ment	N/A	Neupo-gen, NNP	Veins in arm hurt. Ben and Minal visited.
Sat.	16	6	N/A	Lots of city walking	Neupo-gen, NNP	Went to city with Al and Moy; lots of shop-ping. Veins still hurt. Co-worker visited with card and Wii gifts from work.
Sun.	17	3	N/A	N/A	Neupo-gen	Exhausted after drop-ping Al off in the morn-ing. Stayed up late the night before. Headache and arm pain worsen-ing.
Mon.	18	4	N/A	Legs and cardio	Blood thin-ning (BT)	Exhaustion and head-ache better, but still arm pain. Went to doc-tor's during holiday and got ultrasound. Blood clot in right arm near the elbow. Redness and some swelling from middle of forearm to middle of biceps. Hurts to move arm, especial-ly extend it. Ate lots of seafood, soup and ice cream!
Tues.	19	4	N/A	Cardio	BT	N/A
Wed.	20	4	Blood work	N/A	CT scan, BT	Blood class to monitor my blood density for clot.
Thurs.	21	4	Treat-ment	Legs and cardio	N/A	CT scan results good, approximately 25% of tumor left (10x5 cm, from a football to an or-ange). Will do PET scan at end of treatment.
Fri.	22	4	Blood work	Chest/back/ cardio	Neupo-gen	Numbing in forearm like frostbite, even though biceps region near elbow where clot is feels better.
Sat.	23	5	Blood work	Arms/car-dio	N/A	Still have forearm num -bing.

Day	Date	Mood	Event	Exercise	Nausea Pills	Description
Sun.	24	6	N/A	Legs and cardio	Neupo-gen	Still have forearm num-bing.
Mon.	25	6	Blood work	Chest/back/ cardio	Neupo-gen	Still have forearm num-bing. Week of Moy fighting, real estate looking, loan uncertain-ty, work paperwork.
Tues.	26	6		Arms/car-dio	N/A	Still have forearm num-bing.
Wed.	27	5	Blood work	N/A	N/A	Still have forearm num-bing. Slept a lot, tired.

Table 7: Mood journal, chemotherapy cycle three (Jan. 28–Feb. 28, 2009).

Day	Date	Mood	Event	Exercise	Nausea Pills	Description
Thurs.	28	4	Treat-ment	N/A	N/A	N/A
Fri.	29	5	Blood work	N/A	N/A	Dermo visit, moles sliced, bump on face.
Sat.	30	5	N/A	N/A	Neupo-gen	Tired.
Sun.	31	5	N/A	N/A	Neupo-gen	N/A
Mon.	1	5	N/A	Legs and cardio	Neupo-gen	Slept a lot.
Tues.	2	6	N/A	Chest/back/ cardio	N/A	Arm still tingles. Blood clot is getting better but still some pain in the forearm.
Wed.	3	6	Blood work	Arms/car-dio	N/A	N/A
Thurs.	4	5	Treat-ment	Legs	N/A	N/A
Fri.	5	5	N/A	N/A	Neupo-gen	Tired, slept a lot. Arm feeling better; still have tingling in fingertips.
Sat.	6	5	N/A	Chest/back/ cardio	Neupo-gen	N/A
Sun.	7	6	N/A	Tennis with Matt	Neupo-gen	N/A
Mon.	8	5	N/A	Arms/car-dio	N/A	Sore from tennis.

Day	Date	Mood	Event	Exercise	Nausea Pills	Description
Tues.	9	5	N/A	Legs and cardio	N/A	Leg press hard as usual. Slept a lot. House hunting past few weeks.
Wed.	10	5	Blood work	Chest/back/ cardio	N/A	Tired, drained and a little nauseous after workout. Arm still a little numb but much better. Fingertips have been numb on both hands for the past three weeks or so. Slept a lot. Moy stressed.
Thurs.	11	3	Treatment	N/A	N/A	Urinating a lot, some nausea, tired, chemo sucks!
Fri.	12	3	Treatment	N/A	N/A	Tired, peeing...
Sat.	13	3	N/A	Arms/cardio	N/A	N/A
Sun.	14	4	N/A	Tennis with Elisban/ Toby	Neupogen	Couldn't serve, tired, noticed large green bruise around infusion site on right forearm after playing tennis. Doesn't hurt though.
Mon.	15	4	N/A	Arms/cardio	Neupogen	Slept a lot, fight with Moy.
Tues.	16	4	N/A	Chest/back/ cardio	Neupogen	Slept a lot.
Wed.	17	5	Blood work	Legs and cardio	N/A	N/A
Thurs.	18	4	Treatment	Cardio (17 minutes)	N/A	N/A
Fri.	19	4	N/A	Cardio (32 minutes)/push-ups/sit-ups	N/A	Slept in since couldn't get to bed. Numbness in fingers.
Sat.	20	4	N/A	N/A	N/A	N/A
Sun.	21	5	N/A	Arms/cardio	N/A	Trouble sleeping past week or so. Went to bed around 3 AM, slept in until 11:30 AM. Played World of Goo.

Day	Date	Mood	Event	Exercise	Nausea Pills	Description
Mon.	22	5	N/A	Chest/back/ cardio	N/A	Still have numbness in fingertips, bruises on forearms from IVs and blood test. Getting massage today. Haven't talked to Moy since Thursday.
Tues.	23	5	N/A	Legs and cardio	N/A	N/A
Wed.	24	5	Blood work	Arms/cardio	N/A	N/A
Thurs.	25	5	N/A	N/A	N/A	N/A
Fri.	26	6	N/A	Chest/back/ cardio	N/A	Felt like I had a little more energy than previous workouts, but still not normal.
Sat.	27	6	N/A	Legs and cardio	N/A	Felt like I had a little more energy than previous workouts, but still not normal. Stuck in traffic for 4 hours going to Warriors game...
Sun.	28	5	N/A	Tennis/bike	N/A	Legs sore from yesterday, a little congested, some stomach problems (bad granola bar?), did the bike hard for 10 minutes, then tennis, only lasted through half of the ball machine, then nice massage from Jin.

Table 8: Mood journal, radiation (March 1–April 9, 2009).

Day	Date	Mood	Event	Exercise	Description
Mon.	1	5	PET scan/ mask/ tattoos	N/A	Butt sore, must have pulled a muscle working out or playing tennis. Legs feel week, lack of energy from PET scan.
Tues.	2	5	N/A	Chest/back/ cardio	Still sore in legs, a bit more energy than yesterday. Results show cancer is mostly gone, although scar tissue still present and tumor is about the size of an orange (4x4x2 inches).

Day	Date	Mood	Event	Exercise	Description
Wed.	3	6	Blood work	N/A	Still sore in butt/legs, but getting better. More energy. Mom is visiting.
Thurs.	4	6	N/A	Legs and cardio	Pretty good energy, still some leg/butt soreness but a little better.
Fri.	5	6	N/A	Arms/cardio	N/A
Sat.	6	6	N/A	Wii tennis	Spent day in Tiburon, felt pretty good physically, not too great emotionally after how Moy acted. Fingers still tingle, legs still feel a bit weak, but some progress.
Sun.	7	6	N/A	N/A	Went to Cirque du Soleil. Mom left, fingers numb, a bit stuffed up, slept in after taking her to airport.
Mon.	8	7	N/A	Chest/back/cardio	Decent energy at the gym, not normal yet but better. Still have numb fingers, toes a bit cold, still a bit congested. Gained a little more weight, now up to 200 lbm. Was down to 180 lbm at my lowest (about a month after chemotherapy), then was at 185 lbm for a while, then 190, 195 and 200 lbm.
Tues.	9	6	Treatment	Legs and cardio	1st treatment today, went fine. They wrote a lot on my chest and shoulders to help with the aiming (wanted more tattoos, I said no). Didn't feel much during treatment. Five angles, about -135, -45, 0, 45 and 135°, with two bursts each of about 1–20 seconds in duration. Maybe 1–2 minutes between firings, then a few minutes to align the next angle. Whole firing was about 15 minutes, and they spent about 15–20 minutes before marking me and taking some X-rays. Felt pretty good at the gym, even though I didn't get as much sleep (6 hours) the night before. Carpet got cleaned.

Day	Date	Mood	Event	Exercise	Description
Wed.	10	6	Treatment	Arms/cardio	A bit more tired, stayed up late plus radiation.
Thurs.	11	5	Treatment (BW)	Chest/back/cardio	A bit more tired, stayed up late plus radiation. Still numbness in legs, less energy than yesterday.
Fri.	12	6	Treatment	Legs and cardio	Indian Wells.
Sat.	13	6	N/A	N/A	Indian Wells.
Sun.	14	6	N/A	N/A	Indian Wells. Fun tournament, overall felt pretty good.
Mon.	15	5	Treatment	N/A	Tired from traveling, slept a lot.
Tues.	16	5	Treatment	Arms/cardio	Tired, slept in. Fingers still numb, energy OK, better after working out.
Wed.	17	5	Treatment (BW)	Chest/back/cardio	Slept in, got tired after working out and napped for a while. Overall not a lot of energy, still numbness, a bit congested.
Thurs.	18	5	Treatment	Legs and cardio	N/A
Fri.	19	6	Treatment	Arms/cardio	Still numbness, not too bad with the fatigue.
Sat.	20	6	N/A	N/A	N/A
Sun.	21	6	N/A	N/A	N/A
Mon.	22	6	Treatment	Chest/back/cardio	N/A
Tues.	23	5	Treatment	Legs and cardio	Still numbness, congested (due to radiation near throat?). Been going to bed late, waking up late. Met with work to discuss paperwork. Congestion worse. Throat a bit irritated.
Wed.	24	6	Treatment (BW)	Arms/cardio	Took QVAR,[80] congestion seems a bit better.
Thurs.	25	6	Treatment	Chest/back/cardio	N/A

80 QVAR is a corticosteroid prescription inhaler used for asthma symptoms such as wheezing and shortness of breath. I was given it to combat chest congestion.

Day	Date	Mood	Event	Exercise	Description
Fri.	26	6	Treatment	Legs and cardio	Numb fingers, decent energy.
Sat.	27	5	N/A	N/A	A bit tired, still numbness.
Sun.	28	6	N/A	Arms/cardio	A bit tired, still numbness.
Mon.	29	6	Treatment	Chest/back/cardio	N/A
Tues.	30	6	Treatment	Legs and cardio	N/A
Wed.	31	6	Treatment (BW)	Arms/cardio	Had some chest pains (heart) at the gym. Slept in as usual, still going to bed late. Still numbness. Hair is starting to grow back on head and face, but not full.
Thurs.	1	4	Treatment	Chest/back/cardio	Slept a lot, very tired.
Fri.	2	5	Treatment	Legs and cardio	N/A
Sat.	3	5	N/A	N/A	N/A
Sun.	4	6	N/A	N/A	Got a massage and it felt good. Helped blood circulation in fingers/toes.
Mon.	5	5	Treatment	Chest/back/cardio	N/A
Tues.	6	5	Treatment	Arms/cardio	N/A
Wed.	7	6	Treatment (BW)	Legs and cardio	Blond facial hair, still numbness but a little better (massage seemed to help). Work appointment on May 17 to discuss paperwork. Planning on going back to work April 19. Doing a little more weight in the gym, but still well below previous levels. Doing 135 lbm on squats (did one set of 185 lbm last Friday, very sore) and started doing bench with 185 lbm. Lat pull-downs up from 140 to 150 lbm, then 160 lbm.
Thurs.	8	6	Treatment	Chest/back/cardio	N/A
Fri.	9	5	Treatment	Arms/cardio	N/A

Table 9: Mood journal: post-treatment (April 10–April 29, 2009).

Day	Date	Mood	Event	Exercise	Description
Sat.	10	5	N/A	N/A	N/A
Sun.	11	5	N/A	N/A	N/A
Mon.	12	5	N/A	N/A	Slept a lot, still numbness but a little better.
Tues.	13	6	N/A	Tennis, four sets of doubles	Slept a lot, almost 12 hours. Fell on 1st point of tennis, but started to play better. Legs felt like Jell-O, will take a while to get back into playing shape.
Wed.	14	6	Blood Work	Legs and cardio	Legs very sore from tennis yesterday. Took some Monster Milk.
Thurs.	15	6	N/A	Chest/back/ cardio	N/A
Fri.	16	7	N/A	Arms/cardio, hour of tennis with Matt	N/A
Sat.	17	7	N/A	Legs and cardio	N/A
Sun.	18	7	N/A	Chest/ back/15 minutes of tennis	Getting stronger; did 180/170/160 lbm lat pulldown sets, but still less on bench (195 lbm, up from 185 lbm) and other exercises. More energy, but still not 100%. Hit tennis for a bit.
Mon.	19	6	N/A	Tennis at Kona Kai with Oz, Tim and CJ	First week of work. Felt tired and a little dizzy on 1st day. Only got 7 hours of sleep.
Tues.	20	7	N/A	Arms/cardio	Slept in a bit more, felt better at work. Stayed late Tuesday through Thursday.
Wed.	21	7	Blood Work	Legs and cardio	N/A
Thurs.	22	7	Ultrasound	Chest/back/ cardio	Hair seems to be lighter than before; people tell me I look pale.
Fri.	23	8	N/A	Tennis with Rouz and Matt	Results of ultrasound show no blood clot, last day of Coumadin! Legs felt better during tennis, forehand good, backhand needs work. Conditioning getting better, but still need work.

Day	Date	Mood	Event	Exercise	Description
Sat.	24	8	N/A	Arms/cardio	N/A
Sun.	25	8	N/A	Tennis with Ferdi	Energy OK, still numbness, tennis skills need sharpening.
Mon.	26	8	N/A	Tennis at Kona Kai	N/A
Tues.	27	8	N/A	Chest/back/cardio	Benched 205 lbm eight times, getting stronger...
Wed.	28	8	N/A	N/A	N/A
Thur	29	8	N/A	Legs and cardio	Increased weight on deadlift to 175 lbm for 3rd set, still doing 185 lbm for squats. Still sleeping 8–9 hours, but energy at work feels pretty good. Getting to work late and staying late.

Chapter 15

Nutrition

Eating a healthy diet is important no matter what your health, yet especially important when fighting for your life! There are many books and support material on proposed diets for cancers. Reading through the litany of offerings is a daunting task, especially while dealing with the physical and physiological effects of cancer. I will summarize my findings on nutrition based on the material I read as well as what worked for me. It was a bit of trial and error and each diet can be tailored based on the patient's individual needs and attributes (age, gender, size, metabolism, exercise routine, etc.).[81]

Your food and beverage consumption can be broken down into three categories: macronutrients, micronutrients and water. All three are vital for life, yet the subcategories and quantities of the former two are complex and important when maintaining a healthy diet. Each category will be discussed in detail to educate patients and caretakers on what entails a healthy diet. How these nutrients can be tailored specifically for cancer patients will also be discussed.

Macronutrients

Humans require three macronutrients: proteins, carbohydrates and fats. These are opposed to micronutrients, such as vitamins and minerals. The three macronutrients cannot be produced internally by the body and therefore must come from outside sources. Large quantities are required, which is why they are referred to as *macro* rather than *micro*, the latter of which are consumed or produced internally in much smaller quantities.

81 "Nutrition For the Person With Cancer During Treatment," *American Cancer Society,* https://www.cancer.org/treatment/survivorship-during-and-after-treatment/staying-active/nutrition/nutrition-during-treatment.html.

PROTEINS

Protein is essential for cell and tissue formation as it is made up of amino acid chains, which get broken down by the body so that new proteins can be formed. An analogy is a LEGO set, where a protein is an assembled structure made of individual pieces (amino acids) that have unique shapes. These amino acids can be broken up like LEGO bricks in a process called catabolism, as well as reassembled, called anabolism, to form a new protein (analogous to a new LEGO structure).

There are 21 amino acids that are required for protein synthesis in the human body (and eukaryotes in general), although there are many additional amino acids that are used for processes other than protein formation. The former is referred to as proteinogenic amino acids (PAAs) while the latter are non-proteinogenic amino acids (NPAAs). In this section we will focus on the proteinogenic amino acids required to form proteins in the body.

Although all PAAs are essential to life, they are referred to as either essential or non-essential, which can be confusing. These labels refer to nutrition, i.e. is the amino acid essential in food we eat or not essential? Since the body is able to synthesize 12 of the 21 PAAs, these 12 are referred to as non-essential as we do not need to get them through food intake. The remaining nine are referred to as essential since we require them as part of our diet to live. Some sources further subdivide the non-essential since when our bodies are sick, not all the non-essential amino acids may be able to form. Therefore a third category is used: conditionally-essential. Using this terminology we can further segregate the 12 non-essential into non-essential or conditionally-essential, resulting in the following breakdown of the 21 PAAs:

Essential (Nine): Histidine, isoleucine, leucine, lysine, methionine, phenylalanine, threonine, tryptophan and valine.

Conditionally-Essential (Seven): Arginine, cysteine, glutamine, glycine, proline, serine and tyrosine.

Non-Essential (Five): Alanine, asparagine, aspartic acid, glutamic acid and selenocysteine.

These amino acids all have a central carbon atom (referred to as the alpha carbon) with four attached primary bonds: a hydrogen bond, carboxyl group (carbon bonded to an oxygen atom and oxygen/hydrogen pair, or COOH), an amino group (NH_2) and an R-group. The latter is also referred to as the side chain and while the other three bonds are identical for PAAs, this last group is what makes each amino acid unique. These side chains are combinations of some or all nitrogen, carbon, oxygen and hydrogen atoms.

When two amino acids come together, the OH from one carboxyl group will combine with a hydrogen atom from the other PAA amino group, discharging a water molecule while the two PAAs are linked in what is referred to as a peptide bond. This is called dehydration synthesis, which is a chemical reaction that gives off a water molecule. Dehydration synthesis is a subset of a condensation reaction, where two molecules come together and give off a smaller molecule (such as water, acetic acid, hydrogen chloride or methanol) in the process.

Two amino acids combined via a peptide bond are referred to as peptides. Multiple peptides come together to form proteins. Aside from the chemistry, proteins are complex molecules that have unique shapes and their structure can be broken down and described as follows:

Primary Structure: This is the chemical structure of the molecule, i.e. the sequence of atoms, molecules and bonds that make up the chemical compound.

Secondary Structure: This is the pattern of hydrogen bonds in the biopolymer (an organic polymer), which give the 3-D shape of local segments (peptides). The most common structural elements are alpha helices and beta sheets.

Tertiary Structure: This refers to the overall shape of the protein, which is made up of the individual peptides that retain their secondary structure. As an example, think of the secondary structure as being pieces of fusilli pasta or alpha helices combined with curved chicken wire (beta sheets), giving an overall 3-D shape unique to the protein.

Quaternary Structure: Tertiary structures can be grouped into various folded protein subunits, which are multiple protein chains combined. Further combinations of protein subunits create a protein complex and how this happens uniquely for each protein defines the quaternary structure.

One might not think protein is important for cancer treatment as it is used to build muscle, e.g. when athletes train to break down and rebuild muscles with a high protein diet and often supplements. However remember that during cancer treatment your cells are being attacked—both healthy and unhealthy cells. Increased protein will help repair your healthy cells (i.e. RBCs for oxygen circulation and WBCs to fight your cancer) and also slow down muscle atrophy when combined with exercise.

When it comes to what foods are best to meet your protein needs, it should be noted that some foods have all the essential amino acids (complete protein foods) while others have only some of them (incomplete protein foods). Protein sources derived from animals such as meat, milk, cheese and eggs are generally complete proteins. Plant-derived proteins are generally incomplete proteins (except for soybeans); however, even soybeans have small amounts of methionine and therefore large quantities are required to equal the required essential PAAs found in animal sources such as eggs.

Although there are some additional plant-derived sources that are considered complete, they have much fewer quantities of essential proteins than animal-derived foods. Examples include potatoes, rice and some forms of nuts and grains. Therefore while vegetarian diets can have all the essential amino acids when various foods are combined, it is much more difficult to get all the protein you need without protein supplements such as whey protein.

Protein gets broken down starting in the stomach via digestive juices that include enzymes. This breakdown continues in the small intestines, where digestive juices also containing enzymes needed to break down protein are delivered from the liver and pancreas by the gallbladder, which stores the digestive juices until needed. Once the protein is broken down into its amino acid constituents, it is delivered to the bloodstream to be re-assembled into new forms of protein required for cell and tissue growth. Note that unlike fat and cholesterol, protein is not stored long-term in the body and therefore excess is burned away as fuel. This is one reason why carbohydrates and fat are an important component of nutrients in your diet, so your body doesn't burn protein for fuel that is needed for cell and tissue growth.

Not all foods are created equal when it comes to the body's ability to absorb protein. Table 10 provides a list of foods and their associated net protein utilization (NPU).[82] NPU is the number of amino acids converted to proteins divided by the number of amino acids supplied.

Table 10: Protein efficacy for various foods (1 of 2).

Food	NPU [%]
Eggs	88
Fish	78
Dairy Products	76
Meat	68
Soybeans	48
Natural Brown Rice	40
Red Beans	39
Coconut	38
Nuts	35
White Beans	33
Maize	25
Whole Wheat Bread	21
White Bread	20

As demonstrated by the table above, eggs are the most efficient source of protein you can eat, followed by fish, dairy products and meat—both red meat and poultry. Plant sources fall below with soybeans the highest at 48%, although as noted above, animal sources have a better balance of the essential PAAs; therefore eating twice as many soybeans will not equal the same nutritional protein as eggs.

Table 11 provides an additional source of NPU, as well as a different methodology for measuring protein intake: the protein digestibility corrected amino acid score (PDCAAS). NPU values are like Table 10 for milk, eggs and beef, with all values being biased slightly higher (4 to 5%). Soy protein is significantly higher at 61% versus the previous 48%, although still well below animal products.

PDCAAS values are provided for reference, although there is a litany of literature on which technique is better and why. PDCAAS takes into

82 Franco Columbu, *The Bodybuilder's Nutrition Book* (New York: McGraw-Hill Education, 1985), 35.

account human excrement and claims to be more adjusted to the needs of the human body's protein intake. However values above 1.0 were rounded down, which makes it more difficult to ascertain the efficacy between these sources. The main difference in this data is that soy protein (plant source) is on par with animal sources (whey and casein from animal milk, along with milk itself and eggs). Additional research is required to explore these differences. Regardless, due to the lower amount of methionine in soy protein, I recommend eggs, whey, fish and chicken as excellent protein sources to choose from, which include vegetarian and vegan options.[83]

Table 11: Protein efficacy for various foods (2 of 2).

Protein Source	NPU [%]	PDCAAS [%]
Eggs	94	100
Whey Protein	92	100
Milk	82	100
Casein Protein	76	100
Beef	73	92
Wheat Gluten	67	25
Soy Protein	61	100
Black Beans	Not included	75
Peanuts	Not included	52

Finding high-quality protein sources is essential. Organic eggs and meats that avoid chemical pesticides and hormones should be purchased. Fresh fish from natural oceans versus farm-raised are healthier due to both the protein/fat ratio (fish in the wild swim much farther) and lack of man-made chemicals. Mercury content is higher in fish than in land animals, so I recommend having a healthy balance of fish, poultry and other meats (beef, pork, lamb, etc.), leaning more toward fish and poultry due to their lower fat content. Fast food restaurant meat has much higher fat content than meat you buy from the store, while store meats also vary in their fat content and overall quality. Buy only from sources with free-range, as it is both healthier for you and more humane for the animals.

83 Jay Hoffman and Michael Falvo. "Protein—Which is Best?," *Journal of Sports Science Medicine* 3, no. 3 (Sept. 2004): 118–130, https://www.ncbi.nlm.nih.gov/pmc/articles/PMC3905294.

Protein shakes are also a good way to supplement your diet, especially if you are a vegetarian or vegan.[84] Whey and soy proteins are the most common, with a multitude of brands varying in price and quality. Search for brands with the highest protein content per serving. Additional types of protein drinks include casein protein (which takes 5 to 7 hours to digest so is good for sleeping and as a meal replacement) and egg albumin—egg whites generally found in liquid form. Protein powders such as whey, soy and casein can come in concentrate, isolate or hydrolysate forms. The former is a concentration of around 80% protein, 10% carbohydrates, 5% fats and the remaining 5% for additional ingredients (percentages vary by brand). Isolate contains a higher concentration of protein, generally around 92% and is therefore a higher quality (and more expensive) choice. Hydrolysate protein is exposed to heat and enzymes as part of the manufacturing process, therefore already having some of the protein broken down into amino acids; this allows the protein to absorb and digest quicker. However concentrate and isolate proteins digest rather quickly (1 to 3 hours) as-is, so there is a negligible benefit for the higher price.

You may also see branch chain amino acids (BCAAs) available, which as opposed to the protein powders mentioned that have all nine essential amino acids in protein form, have only three PAAs (leucine, isoleucine and valine) in amino acid form. While some fitness experts and companies advertise this as a way to increase muscle mass, if you are already eating large amounts of protein in your diet you likely do not need additional BCAAs. Yet taking this drink during a workout can help replenish some of the BCAAs that are used up, which will help counter fatigue.

When it comes to how much protein is enough, the recommended daily allowance (RDA) is 0.8 g of protein for every 1 kg of body mass (i.e. 2.2 lbm).[85] However, this is a minimum quantity to maintain your health and many nutritional consultants recommend a value closer to 1 g of protein for every 1 kg of body mass. For athletes or individuals

84 Note that maintaining weight is often difficult during cancer treatment, particularly with chemotherapy. This may be especially challenging for vegetarians or vegans, where protein shakes can be particularly beneficial.

85 Institute of Medicine, *Dietary Reference Intakes for Energy, Carbohydrate, Fiber, Fat, Fatty Acids, Cholesterol, Protein, and Amino Acids* (Washington D.C.: The National Academies Press, 2005).

working out regularly, this can be increased further to between 1 and 2 g of protein per every 1 kg of body mass.[86] With cancer patients, while you require increased protein due to treatment in order to repair cells, you will be exercising less strenuously than before treatment. Therefore with these effects countering one another, aiming for the 1 g of protein per 1 kg of body mass is recommended. I work out regularly with weights, play tennis and train Muay Thai, taking around 60 g of protein at night after exercise and getting the remaining 40 g or so from natural sources. During chemotherapy and radiation treatment I maintained a similar routine with one protein shake a day and mixed in other forms of protein with meals. However due to nausea induced by the treatment, my appetite was not always high. Exercising and taking protein helped reduce muscle atrophy while helping with my cell and tissue recovery.[87]

CARBOHYDRATES

Carbohydrates are the body's main source of energy and consist of carbon, hydrogen and oxygen atoms bonded together (hydrates of carbon). The majority of carbohydrate molecules come in a ratio of two hydrogens to one oxygen with a form of $C_x(H_2O)_y$, although there are exceptions (such as deoxyribose, a sugar constituent of DNA with ten hydrogen and four oxygen molecules). Carbohydrates (also referred to as saccharides) include sugars, starch and cellulose. They can be broken down into four subgroups: monosaccharides, disaccharides, oligosaccharides and polysaccharides.

MONOSACCHARIDES

These are the simplest form of carbohydrate and are also referred to as simple sugars. They have the chemical formula $C_x(H_2O)_y$ and are colorless, sweet tasting crystalline solids that are soluble in water.

Monosaccharides can be subdivided by how many carbon atoms they have, as shown below. They are further subdivided with the prefix *aldo*, referred to as aldose, or *keto*, referred to as ketose, with molecules in each subgroup containing the same position in the chain of their functional group (grouping of atoms and bonds within a chain responsible for chemical reactions). The further division within an aldo or keto group is the

86 Columbu, *The Bodybuilder's Nutrition Book*, 45-47.

87 Katrina VB. Claghorn, "Protein Needs During Cancer Treatment," *OncoLink*, https://www.oncolink.org/support/nutrition-and-cancer/during-and-after-treatment/protein-needs-during-cancer-treatment.

orientation of the atoms and bonds, not including the functional group or groups, which are referred to as stereoisomers.

For example, you may have an O-OH-H sequence versus an H-O-OH sequence. Finally, the D and L prefixes of a molecule represent a chain that is symmetric, with the non-functional group atoms being flipped with respect to the functional group. This is a brief summary of the molecular organic chemical makeup of these molecules and readers are encouraged to explore further references if interested.

Below is an organized listing of monosaccharide molecules broken down by their subgroups.

TRIOSE (THREE CARBONS)

Aldotriose: D-glyceraldehyde, L-glyceraldehyde
Ketotriose: Dihydroxyacetone

TETROSE (FOUR CARBONS)

Aldotetrose: D-erythrose, D-threose (L forms of both do not occur naturally)
Ketotetrose: D-erythrulose (L form does not occur naturally)

PENTOSE (FIVE CARBONS)

Aldopentose: D-arabinose, D-lyxose, D-ribose, D-xylose, L-arabinose, L-lyxose, L-ribose, L-xylose
Ketopentose: D-ribulose, D-xylulose, L-ribulose, L-xylulose deoxyribose (used in DNA)

HEXOSE (SIX CARBONS)

Aldohexose: D-allose, D-altrose, **D-glucose,**[88] D-mannose, D-glucose, D-idose, **D-galactose**, D-talose (D-altrose and L forms of all do not occur naturally)
Ketohexose: D-psicose, **D-fructose**, D-tagatose, L-sorbose (corresponding L and D forms do not occur naturally)

HEPTOSE (SEVEN CARBONS)

Aldoheptose: L-glycero-D-manno-heptose, D-mannoheptulose
Ketoheptose: Sedoheptulose (also known as D-altro-heptulose)

---

88 Monosaccharides in bold are particularly important nutritionally.

Although monosaccharides come in many variants, the most common forms and those that are most relevant to you nutritionally are glucose, fructose and galactose. All three of these have the same chemical formula ($C_6H_{12}O_6$) but are hexoses with different chemical structures as noted above.

Glucose is the primary metabolic fuel used by most animals and circulates throughout the bloodstream, impacting your blood sugar levels. Humans use glucose as an energy source for aerobic respiration. Broken down into water and carbon dioxide, glucose is vital as an energy source in the ATP process for cellular metabolism and muscular work (discussed in more depth under the *Biotin (B7)* section). Insulin is produced to absorb glucose and convert it into either glycogen (stored energy for later use) or fat (long-term energy storage) via liver, fat or muscle cells.

Glucose is made in plants via photosynthesis and consumed by humans either directly or through the conversion of starch to glucose or glycogen (to be turned into glucose later when needed). Processed or refined sugars are also a form of glucose but less healthy, as they do not contain the additional vitamins and minerals that complex carbohydrates do.

Fructose is found naturally in fruit but in much smaller quantities than high-fructose corn syrups, which are, for example, added to soft drinks. Fructose does not impact blood sugar levels like glucose does but can lead to obesity. While glucose can be metabolized by cells throughout the body, fructose is primarily metabolized in the liver. Once your body's glycogen limits are reached, excess amounts of fructose or glucose will get turned into fat rather than converted to glycogen in the liver.

Galactose is sweeter than glucose and fructose and when combined with glucose in a condensation reaction, creates the disaccharide lactose. Within the body, galactose gets converted to glucose used for energy. Galactose is found in dairy products, avocados, sugar beets, natural gums (found in woody elements of plants or seed coatings, used as thickening, gelling and emulsifying agents or stabilizers in cooking as well as adhesives for non-cooking applications) and mucilage (a thick, gluey substance found in plants and some microorganisms). This sugar is also synthesized in the body.

DISACCHARIDES

When two monosaccharides bond together by means of a glycosidic linkage[89] they form a disaccharide. These sugars are soluble in water and the most common forms are sucrose (table sugar), lactose (milk sugar) and maltose (malt sugar). Other common forms include lactulose, trehalose and cellobiose. All six of these organic polymers have the chemical formula $C_{12}H_{22}O_{11}$ and are therefore isomers, meaning they have the same chemical formula but are distinct based on the order of their bonds. There are multitudes of less common disaccharides that are not as important nutritionally, with the majority having the same chemical formula as the common disaccharides.

Sucrose is glucose bonded to fructose and comes from sugar cane and sugar beets. This is also what people are familiar with as table sugar. Often sucrose is refined, where it is stripped of its vitamins and minerals, leaving an odorless crystalline powder with a sweet taste. Sucrose is a primary ingredient in foods throughout the world. While sugar cane and sugar beets are the primary sources, sucrose exists in many other plants, particularly in their roots, fruits and nectars. For instance, bees accumulate sucrose to produce honey.

Lactose is sugar found in milk and is a combination of glucose and galactose. All mammalian infants rely on their mother's milk, which among other nutrients and minerals contains lactose for energy and cell growth. Some ethnicities tend to be less tolerant (i.e. lactose intolerant) than others based on their ancestry and habit of drinking milk into adulthood. For instance, more than 70% of western Europeans can drink milk without issue while fewer than 30% of Africans and eastern/southern Asians are able to.

Maltose is sugar derived from malt and contains two glucose molecules bonded together. Malt is created from raw grains such as barley, wheat, rye and sorghum through a malting process, where the enzyme amylase catalyzes the hydrolysis (accelerates the breakdown) of malt starch into maltose. This malting and subsequent mashing process is used in the distillation of various alcohols such as whiskey and beer.

89 A glycosidic linkage is a specific type of covalent bond between a carbohydrate molecule and another molecule, either a carbohydrate or non-carbohydrate.

OLIGOSACCHARIDES

This carbohydrate contains between three and ten monosaccharides joined together. These organic polymers are not digestible and their main role is to feed bacteria (known as probiotics, bacteria introduced into the body that are helpful) primarily located in the large intestines and secondarily in the colon. Oligosaccharides fall into the category of prebiotics (foods that promote the growth of probiotics). Raffinose ($C_{18}H_{32}O_{16}$), stachyose ($C_{24}H_{42}O_{21}$) and verbascose ($C_{30}H_{52}O_{26}$) are three common oligosaccharides containing three (trisaccharide), four (tetrasaccharide) and five (pentasaccharide) monosaccharides, respectively. Raffinose is composed of one galactose, one glucose and one fructose molecule, while stachyose has two galactose molecules and one each of glucose and fructose molecules. Verbascose consist of three galactose molecules, one glucose molecule and one fructose molecule.

All three are found primarily in legumes such as beans, with raffinose also found abundantly in asparagus, sugar beet molasses, cabbage, broccoli, Brussels sprouts, sweet potatoes and whole grains, while stachyose is additionally found copiously in peas.

Oligosaccharides can link to compatible amino acid side-chains, creating glycoproteins and glycolipids. The former is important in processes such as cell binding while the latter enables cell recognition via cell receptors involved in the immune response. (See *WBCs* section.)

POLYSACCHARIDES

These are the most complex carbohydrates and are made up of more than 10 monosaccharides bonded together by glycosidic linkages. They can be categorized by two types of functions, either to store energy or for cell structural properties. The most common forms of polysaccharides are broken down by their function discussed below.

Energy Storage: Starch and glycogen.

Starch $(C_6H_{10}O_5)_n$ is found in plants as their main means to store energy. Glycogen is the equivalent energy storage system for animals and can be broken down more quickly, which is valuable for mobile animals who require energy quicker than plants. Starch is the largest percentage of all food sources in the human diet. Cereals and grains are high in starch and all fruits and vegetables have some starch in them, although many of them have only trace amounts that are negligible.

Pure starch is a white powder that is odorless and tasteless while being insoluble in water. Starch can undergo malting by drying cereal grains soaked in water, developing enzymes required to break the starch down into more basic sugars. Fermentation—converting sugars to acids, gases or alcohol—can then follow to make consumables such as beer and whiskey. Starch is also used in processed foods (for instance with corn starch) where it serves roles such as improving structure (e.g. in bread) or increasing viscosity (e.g. in salad dressing).

Glycogen ($C_{24}H_{42}O_{21}$) is a secondary energy storage source in animals (behind fat as the primary source) to be converted into glucose when needed for quick energy (as opposed to fat, which is more for long-term energy use). Stored primarily in the liver (up to 10% by weight) and muscles (1–2% by weight), due to the amount of muscle in the body the total amount of glycogen in muscle compared to the liver is approximately 2:1. Like starch, glycogen is a white powder soluble in water—tasteless and odorless. Glycogen storage is impacted by eating habits, exercise and basal metabolic rate (BMR), as more muscles allow for more glycogen storage.

BMR is a reflection of your metabolism and is a measure of how much energy it takes for your body to function in a resting state (i.e. while sleeping) to perform basic functions like breathing and blood circulation. Individuals with more muscle and less fat have a higher BMR since muscle burns 3–5 times more calories than fat. BMR slows down with age (which often leads to obesity in individuals who age without increasing their exercise) and is approximately 5–10% lower in women.

Starch is also used in non-nutritional applications including as a stiffening, gluing or thickening agent. For instance, garments may receive starch treatment to stiffen them prior to ironing.

Cell Structure: Cellulose and chitin.

The two other forms of polysaccharides are cellulose and chitin, which are fibers used for cell structure. Cellulose has the same chemical formula as starch, $(C_6H_{10}O_5)_n$, but with a different rotational orientation: every other repeat unit is rotated 180° about the backbone. Cellulose is found in plant cell walls as well as in the cell walls of oomycetes (fungi) and is the most ubiquitous organic polymer on Earth.

Cellulose is a non-soluble fiber that is indigestible. Note that there are soluble and insoluble types of nutritional fiber, with cellulose being one form of insoluble fiber. Although cellulose passes through your digestive system, it likewise aids other food through the digestive track and prevents constipation. Cellulose can be used as a filler in wood pulp form, where it is found in processed foods such as frozen breakfasts (e.g. waffles), ice cream bars, pancakes and syrup.

Chitin $(C_8H_{13}O_5)_n$ is found in the cell walls of fungi, the exoskeletons of arthropod crustaceans (such as crabs, lobsters and shrimp) and anthropoid insects, radula (structure used for feeding) of mollusks (e.g. clams, mussels, oysters and scallops), cephalopod beaks (e.g. beaks of octopuses and squids) and the scales of fish and lissamphibia (group of tetrapods that includes amphibians, such as frogs).

Like cellulose, chitin is an indigestible fiber and is the second most abundant natural carbohydrate on the planet behind cellulose. In addition to its benefits as a fiber, chitin also is a prebiotic helpful for probiotic bacteria in your digestive tract that aid in breaking down your food.

While western diets do not typically include insects, many eastern diets do. Sources include edible crickets and roasted mealworms as well as cricket powder, cricket flour and insect candy! Chitin is also found abundantly in the shells of crustaceans such as shrimp, crab and lobster. However for those less adventurous souls, chitin is also found in mushrooms, which accounts for the meaty texture.

Dietary Sources of Carbohydrates

When it comes to your diet, since fibers primarily pass through your digestive tract, starch should be the main focus with regard to fat buildup and energy use. Fibers are still important for digestion and as a prebiotic, but in terms of your glucose and energy levels, starch is going to turn into glucose directly, glycogen for near-term storage or fat for long-term storage. Glycogen storage is limited to around 350 g in the body (varying based on many individual factors such as muscle mass). As mentioned previously, around twice as much glycogen can be stored in the muscles throughout the body as in the liver. However, glycogen in the liver differs from glycogen in muscles since the latter lacks glucose-6-phosphate enzymes, which are required to pass glucose into the bloodstream. Therefore, liver glycogen will make glucose available to organs throughout the body while muscle glycogen is available only to local muscles.

When used by the muscles, glycogen gets converted to lactic acid, which then travels through the bloodstream to be converted by the liver into glucose or glycogen for storage. Glycogen typically remains available for 2–15 hours depending on how strenuous the activity (athletes will go through their stores very quickly), although it can remain for up to 24 hours. This is a primary reason why athletes need time to recover after a competition, to replenish their glycogen reserves before the next event. Glycogen levels generally take 24 hours to replenish, although this can be up to 48 hours for extremely strenuous workouts.

Glucose is the first energy storage that your body will look to burn, followed by fat. If both are depleted, the body will break down your muscles into amino acids, which get converted to glucose in the liver. You never want to be in this state as muscle atrophy is unhealthy, although most humans will not reach this point as they have high carbohydrate and fat diets. Any excess starch is therefore generally converted into fat. This is why a high carbohydrate diet is beneficial for athletes and active individuals to fuel their muscles yet can lead to obesity, as starchy carbohydrates get converted to fat for inactive individuals.

Also of interest is the speed at which carbohydrates, proteins and fats digest. Carbohydrates begin digesting in the mouth and are processed the quickest, continuing into the stomach and absorbing in the small intestines. Protein is digested primarily in the stomach and takes longer than carbohydrates but is quicker than fats. Fats take the longest to digest, which primarily occurs in the stomach. This is another reason why carbohydrates prior to exercise are desirable—they will get into your bloodstream the quickest.

When it comes to your dietary choices, you should be aware of which foods contain large amounts of starch and which foods have low amounts. You should eat starchy foods early to provide fuel throughout the day and fruit (apples and grapes are great)[90] is recommended 30 to 60 minutes before exercise to provide additional sugar for energy while minimizing your fat intake. Proteins are needed after exercise to repair muscle, supple-

90 Fruit can be in the form of solid or dried. If you are watching your weight, however, fresh fruit is recommended as you get similar nutritional value but are less likely to consume large quantities of sugar (e.g. 1 cup of sliced apple is roughly equivalent to 1/4 cup of dried apple in terms of calories, sugars, vitamins and nutrients). Also note that while dried fruit is slightly higher in certain nutrients, it is substantially lower in vitamin C due to heating while it is processed.

mented by smaller amounts of carbohydrates. I recommend less starchy foods before bed since you need less energy and excess starch is more likely to be converted into fat. This is when fruit and vegetables are particularly helpful, as they are low in starch but high in vitamins and nutrients. Milk is excellent before bedtime as mentioned in the *Proteins* section due to the protein content taking 5 to 7 hours to digest, reducing the likelihood of you waking up hungry.

Below is a list of foods divided by their starch content, with the amount of starch listed per 100 g of serving. Note that while all vegetables contain starch, some have only trace amounts and are considered non-starchy while others have large amounts and are deemed starchy.

Foods High in Starch:

- ***>50 g:*** Cereal (brand dependent), Chex Mix, chips, cornmeal, crackers, flour, pancakes, popcorn, pretzels and ramen noodles.
- ***>25 g:*** Bagels, cereal (brand dependent), French fries, kidney beans, navy beans, pasta, pinto beans, rice, wheat bread and white bread.
- ***>5 g:*** Bananas, corn, potatoes, refried beans and sweet potato.
- ***>1 g:*** Carrots, chilis, edamame, milk and peas.
- ***<1 g:*** Asparagus, broccoli, cabbage, celery, cucumber, kale, lettuce, mushrooms, onions, peppers, spinach, summer squash, tomatoes, okra, snap beans and zucchini.[91]

Foods High in Fiber:

Fiber is also an important carbohydrate to include in your diet to help with digestion as previously mentioned.

Good sources of cellulose fiber include: Beans, bran, brown rice, fruit skins, nuts, seeds, vegetable skins, whole grains and whole wheat.

FAT

Fats are large carbon chains that store more energy than carbohydrates, which cause them to build up in the body if not properly balanced with a healthy diet and exercise. A fat molecule consists of at least one fatty acid molecule bonded with at least one glycerol molecule. Fatty acids are

91 Jessica Bruso, "Starchy Fruits and Vegetables," *Livestrong,* https://www.livestrong.com/article/316069-starchy-fruits-vegetables;
Ryan Rayman. "19 Foods That Are High in Starch," *Healthline,* https://www.healthline.com/nutrition/high-starch-foods#section9.

complex organic molecules containing carbon, hydrogen and oxygen. For instance one subgroup of fatty acids, saturated fatty acids, contains multiple fatty acid molecules that are all variations of the chemical formula $CH_3(CH_2)_nCOOH$. Glycerol is also an organic molecule consisting of carbon, oxygen and hydrogen, having a chemical formula of $C_3H_8O_3$.

One gram of fat has nine calories, while a gram of either carbohydrates or protein has four calories. There are four types of fats, with two being good for you and two that should be avoided.

Saturated Fat: Saturated fat is a solid at room temperature (e.g. butter or margarine) and is considered unhealthy for you. It causes an increase in low-density lipoprotein (LDL) cholesterol, which can clog arteries (as opposed to healthy high-density lipoprotein (HDL) cholesterol that can actually reduce heart disease by carrying away portions of LDL cholesterol to the liver for waste processing). Animal foods such as meat, milk and cheese contain primarily saturated fat (e.g. the gristle on a steak). This fat is also found in tropical oils such as coconut and palm oil as well as cocoa butter.

Trans Fat: This is the worst kind of fat you can eat. Also a solid at room temperature, trans fat is found in processed food such as crispy crackers or chips and is even harder than saturated fat. Trans fat increases your LDL cholesterol and should be avoided.

Monounsaturated Fat: A liquid at room temperature, this fat is considered healthier for you and may lower your risk of heart disease. Eating a diet high in monounsaturated fats will increase your ratio of healthy HDL cholesterol when compared with LDL cholesterol. Foods high in mono-fats include avocados, nuts, vegetable oil, canola oil, olive oil and peanut oil. Omega-7 fatty acids are included in this group.

Polyunsaturated Fat: Also liquid at room temperature, these oils are healthy and consist of omega-3, omega-6 and omega-9 fatty acids. Vegetable oil, corn oil, sunflower seeds, sesame seeds and soybeans have fat included in this category.

Known as essential fatty acids (EFAs), omega-3 and omega-6 fatty acids are required for human biological processes (unlike the saturated fats used for fuel). These fats are also considered essential, as they are not synthesized by the human body; alternatively, omega-9 is not categorized as an EFA as it is produced within the body.[92]

92 "Essential Fatty Acids," *Oregon State University,* https://lpi.oregonstate.edu/mic/other-nutrients/essential-fatty-acids.

The EFAs are straight-chain hydrocarbons with a carboxyl (COOH) located at the end while the omega-3 versus omega-6 comes from the length of chains before reaching a double-carbon bond. While there are many forms of these hydrocarbons chains, two specific EFA forms are required for humans: alpha-linolenic acid (ALA, an omega-3 fatty acid found in plant oils and seeds) and linoleic acid (LA, an omega-6 fatty acid found in vegetable oils and nuts). Two additional omega-3 fatty acids are considered "conditionally-essential" since they are synthesized by the human body in small amounts, but supplementation is recommended. These two fatty acids are eicosatetraenoic acid (EPA) and docosahexaenoic acid (DHA), both found in marine oils. Each is obtained from various fish oils, seaweeds and human breast milk.

Omega-3 plays an important role in brain function and helps prevent heart disease. Additionally, research has shown omega-3s to reduce inflammation, helping to prevent cancer and arthritis. Omega-6 is important for the stimulation of skin and hair growth as well as cell metabolism and a healthy reproductive system.

Representations of saturated versus unsaturated and omega-3 versus omega-6 carbon chains are shown in Figure 73.

Monitoring body fat is an important aspect of maintaining your overall health. The best way to do this is measuring your percent body fat (PBF), which is the percentage of fat in your body by weight compared with the remaining body constituents (primarily water and lean body mass including bone, organs, muscle and other tissues). Body mass index (BMI) is often incorrectly used to track obesity but PBF is a much more accurate indicator. For instance, I have a high BMI that would suggest I am obese if you did not look at my PBF, which is in the athlete range, with added muscle causing the high BMI. Table 12 summarizes where your PBF range puts you in terms of your health.

Inflammation

EFAs are important factors in the body's level of inflammation. There are two types of inflammation: acute and chronic. Acute is a natural and necessary process the body uses to deal with injury and tissue damage. This should last for a 24 to 48-hour time period and then subside. Chronic inflammation, on the other hand, is not healthy and is a constant inflammatory state in which your fat expands and produces excess hormones, which leads to additional inflammation.

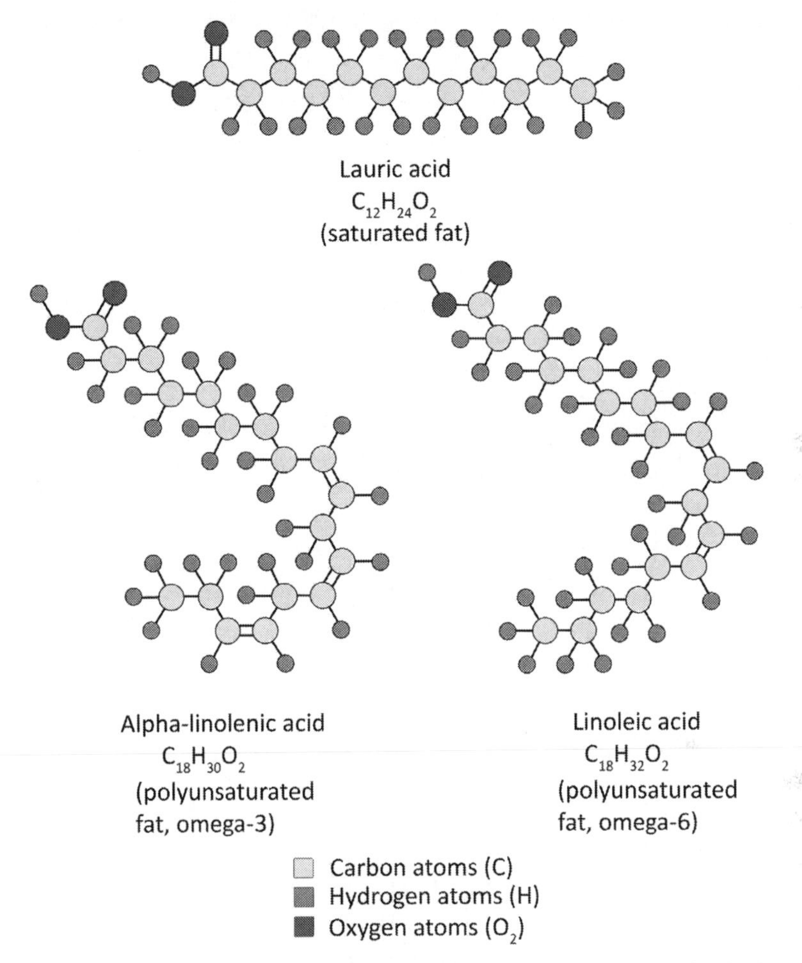

Figure 73: Example of fatty acid (type of carboxylic acid) chains: bottom left is an omega-3 and bottom right is an omega-6.

Table 12: Recommended PBF.

Description	Men [%]	Women [%]
Essential Fat	3–5	10–13
Athletes	6–13	14–20
Fitness	14–17	21–24
Average	18–24	25–31
Obese	25+	32+

To understand how inflammation works, one needs to understand the biochemical response to an injury. Signaling proteins (particularly cytokines) play an important role in conjunction with your WBCs to fight an infection or injury. There are four signaling proteins specifically important in the role of inflammation.

Tumor Necrosis Factor Alpha (TNFa): This cytokine acts as a messenger to call WBCs to the scene. These are the first responders.

Nuclear Factor Kappa B (NFkB): A protein that controls the transcription of DNA (DNA copied to RNA), cytokine production and cell survival. This protein enters the nucleus of WBCs and turns off their apoptosis (programmed cell death) function, allowing WBCs to multiply and accumulate more rapidly around an infection.

Interleukin-6 (IL-6): Also a cytokine, IL-6 signals monocytes to arrive at the infected site. Monocytes create macrophages, which help destroy pathogens via phagocytosis.

C-Reactive Protein (CRP): This protein increases in levels due to the presence of IL-6. It binds to dead or dying cells in order to activate the complement system, which is the part of the immune system that enhances the ability of antibodies and phagocytic cells.

While inflammation is an important process that destroys invading pathogens, too much inflammation leads to unwanted health concerns as WBCs accumulate in excess throughout the body and create dilatory, incessant swelling.[93]

Having a healthy balance between omega-3 and omega-6 fatty acids in your diet is important, as the latter produces inflammation while the former acts as an anti-inflammatory. Excess inflammation leads to inflammatory diseases and conditions including cardiovascular disease, type-2 diabetes, obesity, metabolic syndrome, irritable bowel syndrome, inflammatory bowel disease, macular degeneration, rheumatoid arthritis, asthma, cancer, psychiatric disorders, Alzheimer's disease and autoimmune diseases.

A typical American diet is biased toward omega-6s, as much as 15 to 25 times more than omega-3s, which is in opposition to the ideal 1:1

93 "Types of Cytokines," *Sino Biological.* https://www.sinobiological.com/Types-of-cytokines.html;
"What Causes Inflammation: A Comprehensive Look at the Causes and Effects of Inflammation," *PeerTrainer,* https://www.peertrainer.com/health/inflammation.aspx.

ratio. Mediterranean diets high in fish, olive oil, garlic and fresh fruits and vegetables are much healthier and able to obtain the proper EFA balance the body needs.

As seen in Table 13, olive oil has a higher ratio of omega-6 to omega-3 fatty acids when compared with many other oils such as canola oil; however, there remain fewer omega-6 fatty acids per serving.

Table 13: Fatty acid breakdown for various oils, per 100 g (3.5 oz).[94]

Plant Fats and Oils*	Omega-6 [g]	Omega-3 ALA [g]
Palm Kernel Oil	1.6	0
Coconut Oil	1.8	0
Macademia Nut Oil	2.4	0
Cocoa Butter	2.8	0.1
Sunflower, High Oleic 70% & Over	3.6	0.192
Shea Nut Oil	4.9	0.3
Palm Oil	9.1	0.2
Olive Oil	9.8	0.761
Hazelnut Oil	10.1	0
Avocado Oil	12.5	0.947
Flaxseed Oil	12.7	53.3
Canola Oil, Natreon High Oleic	14.5	9.137
Safflower Oil, High Oleic	14.3	0
Mustard Oil	15.3	5.9
Almond Oil	17.4	0
Peanut Oil (Not Recommended)	31.7	0
Rice Bran Oil	33.4	1.6
Sesame Oil	41.3	0.3
Soybean Oil	50.2	7.033
Cottonseed Oil	51.5	0.2
Walnut Oil	52.9	10.401
Corn Oil	53.5	1.161
Wheat Germ Oil (Not Recommended)	54.8	6.901
Sunflower(Linoleic)	65.7	0
Grape Seed Oil	69.6	0.1
Safflower (Linoleic)	74.6	0

*Sorted lowest to highest by omega-6 quantity.

94 "Omega 6 and 3 in Nuts, Oils, Meat and Fish. Tools to Get it Right," *Paleozone Nutrition,* https://paleozonenutrition.com/2011/05/10/omega-6-and-3-in-nuts-oils-meat-and-fish-tools-to-get-it-right.

Other oils, such as coconut oil, have low omega-6 fatty acids but little to no omega-3s. Many foods are made with vegetable oils high in omega-6, which contribute to the skewed ratio in American diets. Vegetable oils in general are not recommended unless the ingredients are well understood, as they are a blend of various vegetables (such as olive, canola and sunflower) with an overall average higher in omega-6. Using olive oil and consuming either fish or taking a fish oil supplement is an excellent method of reducing your omega-6 to omega-3 ratio.

WATER

H_2O is the most essential substance for your body. As up to 60% of your body is made up of water, drinking adequate amounts is the most important step you can take to improve your health (coupled with getting enough sleep). There are numerous roles water plays in your body. Your blood plasma has large amounts of water and moves RBCs throughout your body for oxygen transport and WBCs to fend off invaders. Lymph contains ubiquitous amounts of water to transport WBCs throughout your lymphatic system and capture pathogens. Chronic cellular dehydration from a lack of water weakens cellular walls, making you more prone to infections.

Vitamins, minerals, enzymes, carbohydrates and amino acids are among the many nutrients that require water for transportation and cell health. Dehydration causes muscle fatigue as water enables muscles to contract and function properly (a lack of water can also lead to muscle cramping). The electrolytes absorbed in water are not only used for muscle contraction but also for electrical stimulation on the surface of your heart and within your brain. Water keeps organs moist and makes up a large amount of protective spinal cord fluid.

In addition to these vital needs at the micro level, your body regulates your temperature via water in the form of sweat on a macro scale. Additionally, water lubricates your joints and removes waste and toxins from your body (constipation is a sign of dehydration). Water is essential to the acids and saliva used in breaking down food. Your brain is approximately 85% water and reduced amounts can impair your thinking and lead to headaches.

Drinking pure water is the best way to ensure you are getting your adequate daily intake. Many popular beverages contain caffeine or alcohol, which are natural diuretics that expel water in the form of urine. Therefore if caffeinated coffee, soft drinks and alcoholic beverages are your main source of water, you still run the risk of dehydration (especially if you are exercising and active). Natural fruit juices that do not contain large quantities of manufactured sugar can also be a good source of water as well as additional vitamins and minerals. My favorite refreshment is a good smoothie made with fresh or frozen fruits along with not from concentrate apple or orange juice (often adding protein sources like peanut butter or almonds).

In terms of cancer prevention, maintaining an overall healthy body will reduce your chances of cancer based on all the benefits listed above (most notably a healthy immune system). Abundant water consumption has specifically been shown to decrease the chances of bladder cancer, as your urine is flushed more regularly and cancer-causing agents in your urine are diluted.

During cancer treatment, drinking water will help flush out poisonous chemotherapy chemicals and provide your immune system a vital medium for fighting cancer cells. Radiation and surgery will leave your body fatigued and proper water intake will help reduce the fatigue. Fruit juices are particularly helpful with their additional vitamin C.

I recommend having water around with you most of the day. If at work, have a bottle you can refill. At the gym or playing sports, bring a bottle as well or make frequent trips to the water fountain. When at home, there is no excuse not to drink plenty of water, as it is just a few steps away. Remember how important water is to your health!

MICRONUTRIENTS

Micronutrients can be divided into two categories, namely vitamins and minerals. While these nutrients are consumed in much smaller quantities than macronutrients, their functions are essential to human health and life. Vitamins are organic, i.e. they contain carbon atoms, while minerals are inorganic. This leads to complex organic molecules for vitamins versus simple elements from the periodic table for minerals. While vitamins come from plant and animal food sources directly, minerals come from the earth and are thus found in soil and unfiltered water (thereby showing up in plants grown from soil and animals that eat these plants, as well as other animals). Vitamins are susceptible to breaking down and becoming less useful to humans with heat and sunlight exposure while minerals do not have this issue, owing to their less complex chemical makeup.

VITAMINS

Vitamins cannot be synthesized in sufficient quantities by your body and therefore must come from food sources. There are 13 essential vitamins for humans: vitamins A, B1, B2, B3, B5, B6, B7, B9, B12, C, D, E and K. Each of these vitamins play an important role in your health and must be consumed either naturally from your diet or synthetically as vitamin supplements.

VITAMIN A

Vitamin A refers to a group of vitamins coming from both plant and animal sources. These can be broken down into the following categories:

Retinoids: Consist of bioactive forms of retinol, retinal and retinoic acid along with a bio-inactive storage form (retinyl ester). These can be ingested directly from animal meat and are absorbed by fat.

Carotenoids: Plant pigments that give plants color. Over 600 are known and include two main categories, carotenes and xanthophylls. Carotenes consist of alpha, beta, gamma, delta, epsilon and zeta carotenes while xanthophylls contain astaxanthin, beta-cryptoxanthin, canthaxanthin, fucoxanthin, lutein, zeaxanthin, violaxanthin and neoxanthin. Note that only a small percentage of these listed carotenoids can be metabolized into retinoids and thus bioactive forms of vitamin A for the body to use.

Retinol is important in maintaining healthy skin and mucous membranes throughout the body such as the nose, mouth and throat. It is also important for vision. Retinal is likewise vital for vision and deficiencies can cause issues seeing when lighting is dim. In fact, the Egyptians first discovered the benefits of vitamin A when they found eating liver helped heal nyctalopia (night-blindness), a condition that makes it difficult to see in low light and can be caused by retinol deficiencies. Retinoic acid is important for WBC health and fetal growth for pregnant mothers.

Alpha-carotene, beta-carotene, gamma-carotene and beta-cryptoxanthin are all provitamins in that they can be converted into active forms of vitamin A (retinoids) used by the body. However, converting carotenoids to retinoids is only efficient in around a 1:6 ratio, meaning only a small percentage of carotenoids get converted by the body to retinoids. For this reason, a balanced diet of meat and vegetables is recommended for vitamin A intake or a vitamin supplement in addition to vegetables for vegetarians due to the low conversion rate. Beta-carotene is also high in antioxidants, which help protect cell DNA from being damaged by free radicals. The role of antioxidants is discussed further in the *Riboflavin (B2)* section.

Symptoms of low vitamin A include scaly skin, acne, respiratory infections, kidney stones and bladder stones. Alcohol and caffeine, along with deficiencies in vitamin D, destroy vitamin A. When cooking vegetables, 15–20% of vitamin A is lost while 30–35% is lost when cooking non-green vegetables. Boiling is the biggest offender, as vitamins and minerals are secreted from the vegetables and carried away when the water is drained. Steaming is best, as it limits this secretion and vitamin A boiling off. When grilling vegetables, the leftover juices should be retained for the nutritional value.

Specific foods high in vitamin A include:[95]

- ***Animal Products:*** Butter, cheese, egg yolks, halibut, kidney, liver, milk, oysters, salmon, swordfish and tuna.
- ***Colorful Vegetables:*** Carrots, peppers, sweet potatoes, tomatoes and winter squash.
- ***Fruit:*** Apricots, cantaloupe, mangos, papayas, peaches, strawberries and watermelons.
- ***Green Vegetables:*** Broccoli, collard greens, kale, parsley, romaine lettuce, spinach and turnip greens.

95 Multiple sources were used for micronutrient information in this chapter, including Columbu, *The Bodybuilder's Nutrition Book* and "Self Nutrition Data: Know What You Eat," *Nutrition Data,* https://nutritiondata.self.com.

VITAMIN B

Vitamin B is a group of vitamins known as complex vitamins. These are water-soluble and therefore do not stay in the body as fat and must be replenished daily. Generally, where one complex vitamin is found in a food, others are likewise present. This is important as they complement one another and are most effective when taken together.

One of the primary functions of vitamin B is to convert carbohydrates to glucose—food to fuel—which is then used to produce energy. Additionally, these complex vitamins break down fat and utilize the body's protein. Vitamin B also enhances your skin, eyes, hair, lips, liver and nails while improving the overall functionality of the nervous system.

There are eight complex vitamins important for humans to get from food: thiamine (B1), riboflavin (B2), niacin (B3), pantothenic acid (B5), pyridoxine (B6), biotin (B7), folic acid (B9) and cobalamins (B12).

Incidentally for curious readers wondering what happened to B4, B8, B10 and B11, these nutrients were initially classified as vitamins but then reclassified as either conditionally-essential or non-essential nutrients. While vitamins and minerals are essential, conditionally-essential and non-essential nutrients are by definition not. For completeness, these four nutrients are: adenine (B4), inositol (B8), pteroylmonoglutamic acid (B10) and salicylic acid (B11).

Adenine is one of the nucleic acids that make up the DNA molecule. Inositol is a nutrient that can be used to treat psychiatric disorders such as anxiety and depression as well as certain types of ovarian cancer. Pteroylmonoglutamic acid (at one point also known as vitamin R) is referred to today as para-amino benzoic acid (PABA) and has various health benefits such as protection against UV rays. Finally salicylic acid plays an important role in health, but like the other three nutrients, it can be synthesized within the human body and therefore is not considered a vitamin. Additional numbers up through B17 were once considered vitamins but also lost their designations for the same reason.

THIAMINE (B1)

A lack of what would be later known as thiamine was prominent around the turn of the 19th to 20th century. The Dutch East Indies in particular had reported cases of beriberi, which has symptoms including muscle aching and fatigue, weight loss, mental confusion and irregular heartbeats. Christiaan Eijkman was a Dutch physician sent to the East Indies to investigate and found that pigeons being fed leftover military white rice soon were cured of their beriberi symptoms after switching to a diet of brown rice.

Eijkman would later go on to share the 1929 Nobel Prize for Medicine with English biochemist Sir Frederick Hopkins for the discovery of vitamins. The Polish chemist Casimir Funk also deserves credit, as he coined the term vitamines (later changed to vitamins) and proposed various vitamins could be used to cure diseases such as rickets, pellagra, celiac disease and scurvy.

White rice goes through a milling process in which it is stripped of its germ and outer bran layer, both of which are rich in nutrients including thiamine. Brown rice, on the other hand, includes the germ and bran layer, explaining why Eijkman's pigeons recovered when switching their diet. A white rice diet caused beriberi in Asians, prisoners and sailors due to lack of thiamine.

Thiamine also aids in the breakdown of carbohydrates into energy rather than these carbohydrates being stored as fat, promoting a healthy body weight.

Fatigue and muscle weakness are symptoms of inadequate thiamine in your body. Raw seafood destroys vitamin B1, while large amounts of tea, coffee and alcohol block thiamine absorption. When whole grains, rice, fruits, vegetables and legumes are cooked, 15–25% of thiamine is lost. This increases up to 50% when meat is cooked.

Foods rich in thiamine include:

- ***Animal Products:*** Albacore and yellowfin tuna, eggs, lean pork, milk, mussels, salmon and trout.
- ***Fruits and Vegetables:*** Asparagus, cantaloupe, green peas, oranges, seaweed, spinach and squash.
- ***Legumes:*** Beans (black, mung, navy, pink, pinto and soy), lentils, peas and red-skinned peanuts.
- ***Nuts and Seeds:*** Macadamia nuts and sunflower seeds.
- ***Whole Grains and Rice:*** Brown rice, oatmeal and whole-wheat bread.

Riboflavin (B2)

Vitamin B2 aids in converting carbohydrates (via glucose), fats (via ketosis) and amino acids (via gluconeogenesis) into energy. In addition, riboflavin helps the body's cells breathe and aids in the production of RBCs. This vitamin also helps create healthy skin, hair and nails along with good vision.

Riboflavin is high in antioxidants. Antioxidants help fight free radicals, which are molecules that have missing electrons from their out shell. When a valence electron is missing in an atom's outer shell, it will attempt to steal the missing electron from a nearby atom. For healthy cells within the body, this can be very damaging as a chain reaction is started where one molecule after the next steals nearby electrons. Eventually this series of events can lead to the destruction of the cells and may contribute to the aging process and cancer.

Antioxidants are stable with and without all their valence electrons, which allows them to fight free radicals. By donating their outer electrons and remaining stable, they effectively act as a shield to healthy cells from getting damaged. Beta-carotene, vitamin C, vitamin E and the minerals zinc and selenium are also high in antioxidants.

Symptoms of low riboflavin include fatigue, slowed growth, digestive problems, cracks and sores around the mouth, brittle nails, chapped hands and eye sensitivity to light.

Alcohol and estrogen destroy vitamin B2, which should be stored away from light, as 50% of the riboflavin is lost when under sunlight for 2 hours. Fortunately, cooking does not have much of an impact on riboflavin degradation, as vitamin B2 is relatively heat-resistant.

Foods high in riboflavin include:

- ***Animal Products:*** Eggs, feta cheese, Gjetost cheese, goat cheese, and whole milk.
- ***Fruits, Vegetables and Fungi:*** Asparagus, beet greens, broccoli, collard greens, Italian mushrooms, spinach and white mushrooms.
- ***Meat:*** Beef, lamb, lean pork sirloin, mackerel, smoked salmon, trout, tuna and wild salmon.
- ***Nuts and Seeds:*** Almonds, cashews, pine nuts, pistachios, pumpkin seeds, sesame seeds, sunflower seeds and squash seeds.

Niacin (B3)

Niacin aids the digestive system, promotes healthy skin, reduces cholesterol levels and helps the nervous system function properly. Additionally, niacin works with thiamine and riboflavin to burn starches and sugars for energy. Niacin (also known as nicotinic acid) helps to reduce cholesterol, although this ability goes away when the vitamin is in the form of nicotinamide.[96]

A lack of niacin can cause headaches, fatigue, nausea and lesions on the skin and mouth. Pellagra is a disease caused by a lack of niacin and was rampant in the early twentieth century where more than three million Americans suffered from it. Dr. Joseph Goldberger studied the cause and found that diets restricted to primarily corn—which lacks vitamin B3—caused the ailment (symptoms included sores and scales covering the skin). Conrad Elvehjem and Tom Spies carried on this work and were able to show specifically that niacin could cure pellagra.

Alcohol, caffeine, tobacco and antibiotics all destroy niacin. Niacin stores relatively well compared to other B-complex vitamins with little loss when kept at room temperature. It is also unique in that canned products only lose around 20% of their vitamin B3.

Foods with abundant vitamin B3 include:

- ***Animal Products:*** Egg yolks.
- ***Fruits, Vegetables and Fungi:*** Asparagus, avocados, bell peppers, broccoli and mushrooms.
- ***Meat:*** Beef, chicken, lamb, organ meats, pork, salmon, sardines, shrimp, tuna, turkey and veal.
- ***Nuts, Seeds and Legumes:*** Coffee, kidney beans, peanuts and sunflower seeds.
- ***Whole Grains and Rice:*** Brown rice and cereal.

Pantothenic Acid (B5)

Vitamin B5 plays an important role in the production of RBCs as well as sexual hormones. Pantothenic acid also is important for maintaining the body's digestive tract and absorbing other vitamins (particularly B12). Cholesterol synthesis with vitamin B5 prevents unwanted cholesterol

96 Niacinamide is found in certain foods and dietary supplements.

accumulation in the bloodstream. Vitamin B5 also acts as a coenzyme to release energy from carbohydrates, fats and proteins.

A lack of vitamin B5 can lead to insomnia, depression, fatigue, irritability, stomach pain, vomiting and upper respiratory infections; however, as the name comes from the Greek word for everywhere, *pantothen,* the vitamin is found in almost all foods and therefore deficiencies are rare.

Alcohol, caffeine, tobacco and stress all destroy pantothenic acid. Food processing also reduces the amount of available vitamin B5 and thus canned and frozen foods should be avoided.

Foods high in vitamin B5 include:

- ***Animal Products:*** Egg yolks and milk.
- ***Fruits and Vegetables:*** Avocados, broccoli, corn, kale, sweet potatoes and tomatoes.
- ***Meat:*** Beef, chicken, duck, lobster, salmon, turkey and veal.
- ***Nuts, Seeds and Legumes:*** Peanuts, soybeans, split peas and sunflower seeds.
- ***Whole Grains and Rice:*** Whole grain bread and whole grain cereal.

Pyridoxine (B6)

Vitamin B6 helps with the metabolism of proteins and amino acid buildup. For this reason pyridoxine is a supplement used by bodybuilders, where large amounts of protein are taken for muscle growth. The amount of pyridoxine taken should be proportional to your protein intake. Pyridoxine also aids in the conversion of glycogen (secondary long-term energy storage behind fat) to glucose and usable energy. This vitamin acts as a natural diuretic by reducing the amount of stored water in muscles—another benefit bodybuilders take advantage of when getting ready for a competition.

For disease prevention, vitamin B6 produces antibodies and properdin, a protein found in the blood that is part of the body's natural immune system, destroying viruses and bacteria. Vitamin B6 also helps regulate body fluids and maintain a healthy nervous system. As with all complex B vitamins, pyridoxine should not be taken alone as it will more likely lead to side effects. Specifically, vitamin B6 is required for B12 absorption.

A lack of pyridoxine can lead to low blood sugar, drowsiness, depression and irritability. Alcohol, caffeine and birth control pills all destroy vitamin B6. Cooking destroys approximately 50% of pyridoxine in beef, fruits and vegetables. Canned food also has a reduced amount of B6 due to food processing techniques and should therefore be avoided.

Best sources include:

- ***Fruits and Vegetables:*** Avocados, bananas, bell peppers, Brussels sprouts, cabbage, cauliflower, dried prunes, garlic, onions, spinach and turnip greens.
- ***Meat:*** Chicken, lean beef, lean pork, salmon, tuna and turkey.
- ***Nuts, Seeds and Legumes:*** Lima beans, pinto beans, pistachios and sunflower seeds.
- ***Whole Grains and Rice:*** Brown rice and whole-wheat cereal.

BIOTIN (B7)

Biotin creates enzymes that are needed for digestion. Additionally, vitamin B7 is important for cell growth and promotes healthy skin. Many topical beauty products contain biotin for strengthening of the skin, hair and nails. Like many of the other B-complex vitamins, vitamin B7 helps break down fats, amino acids and carbohydrates into energy. Specifically, biotin aids in the release of ATP, the main source of energy for muscular work. This nucleotide breaks apart within cells when energy is required such as during a workout and gets replenished with additional ATP via one of the following three processes:

1. ***The Phosphagen System:*** Produces ATP at the quickest rate but for only 8-10 seconds at a time, e.g. performing a deadlift. Creatine phosphate is broken apart and phosphate is transferred to adenosine diphosphate (ADP), which then forms ATP.
2. ***The Glycogen-Lactic Acid System:*** Produces ATP at a medium rate for up to 90 seconds at a time, e.g. sprinting. Cell splits glycogen into glucose, which metabolizes into ATP and lactic acid.
3. ***Aerobic Respiration:*** Produces ATP at the slowest rate but lasts for several hours or longer, e.g. running a marathon. Carbohydrates first, followed by fat second (if necessary) and protein third (if necessary) are broken down into amino acids.

A deficiency in biotin can lead to oily hair, dry and scaly skin, cracking at the corners of the mouth, dry eyes, loss of appetite, fatigue, insomnia and depression. Alcohol and caffeine both deplete biotin levels within the body.

The following foods are rich in biotin:

- ***Dairy:*** Cheese, egg yolks[97] and milk.
- ***Fruits, Vegetables and Fungi:*** Avocados, brewer's yeast, cauliflower, mushrooms and raspberries.
- ***Meat:*** Organ meats, pork, salmon and sardines.
- ***Nuts, Seeds and Legumes:*** Almonds, beans, black-eyed peas, nut butters, peanuts, pecans, soybeans and walnuts.
- ***Whole Grains and Rice:*** Whole grain bread and whole grain cereal.

FOLATE (B9)

Folate is important for cell growth and brain function, which is why it is used to treat Alzheimer's disease. Helping to produce DNA and RNA, this vitamin is particularly important for babies, adolescents and pregnant mothers. B9 works in tandem with B12 to aid in the production of RBCs. These two vitamins (along with B6) also help maintain blood levels of the amino acid homocysteine, large amounts of which are linked to heart disease. Folate has been shown to help prevent certain cancers, presumably due to its DNA and RNA benefits that reduce the chances of a mutated cancer cell from surviving and multiplying. This vitamin also helps produce hydrochloric acid, which breaks down proteins within the body.

Vitamin B9 exists in various forms, primarily folate and folic acid. The former is naturally found in plants (primarily green leafy vegetables) and is converted readily within the bloodstream to the biologically active form of vitamin B9 known as levomefolic acid (or 5-methyltetrahydrofolate: 5-MTHF). Folic acid, on the other hand, is a synthetic version of the vitamin found in supplements and injected into foods such as cereals, as required by US law. This version of vitamin B9 must go through the liver in order to get converted into the usable form of 5-MTHF. Unfortunately this takes time (many hours—up to the time of your next daily dosage)

97 Egg whites alone contain a protein called avidin that prevents the absorption of biotin in your body; however, avidin becomes inactive when cooked.

and can lead to a buildup of excess folic acid in the bloodstream, which can be adverse for a couple of reasons:

- ***Cancer Risk:*** High levels of folic acid have been linked to an increased risk of cancer.
- ***B12 Deficiency:*** Folic acid masks a deficiency in vitamin B12, which is especially prevalent in the elderly. Low vitamin B12 can lead to dementia and have an adverse effect on the nervous system.

Therefore, the best source of vitamin B9 is to get it naturally through a diet high in leafy vegetables (this is how folate got its name—from the Latin word for leaf, *folacin*). If supplements are required, a healthier alternative to folic acid is taking 5-MTHF directly, which when bound to calcium is known as methylfolate or levomefolate calcium (brand names are Metafolin and Deplin).

Alcohol, caffeine and tobacco all attack folic acid in the body. Folic acid deficiency is more common than the other B-complex vitamins, particularly due to its susceptibility to breaking down with heat during cooking, losing up to 95% when boiled in water instead of steamed. This insufficiency makes dietary and supplement choices particularly important. Vegetables should be eaten soon after purchase or stored in the refrigerator, as they lose around 75% of their vitamin B9 after a few days when exposed to room temperature air.

Foods high in vitamin B9 include:

- ***Animal Products:*** Chicken, egg yolks, milk, pork, salmon, shellfish and tuna.
- ***Fruits:*** Citrus fruits (grapefruit, oranges, papaya, raspberries and strawberries).
- ***Legumes:*** Beans (black, garbanzo, green, kidney, lima, navy, pink, pinto and soy (edamame)) lentils and peas (split and green).
- ***Nuts and Seeds:*** Almonds, flax seeds, peanuts and sunflower seeds.
- ***Vegetables:*** Asparagus, avocados, beets, broccoli, Brussels sprouts, carrots, cauliflower, celery, corn, greens (collard, mustard and turnip), okra, romaine lettuce, spinach and squash.
- ***Whole Grains, Rice and Fungi:*** Baker's yeast, bread, breakfast cereals and rice.

Cobalamins (B12)

Vitamin B12 is important for cell metabolism, as it helps produce DNA and RNA. Nerve cells are particularly impacted by levels of vitamin B12 in the body. Along with vitamin B9, vitamin B12 aids in the production of RBCs and the usage of iron in the bloodstream.

Vegetarians and the elderly are most likely to be deficient in vitamin B12. The former is due to only small quantities of cobalamins being present in plants (plants cannot produce vitamin B12 but only absorb small amounts from the soil). The elderly have fewer stomach acids and therefore have a harder time converting meat into vitamin B12. As mentioned previously when discussing vitamin B9, folic acid in the bloodstream can mask this deficiency, making detection difficult.

Deficiency symptoms include fatigue, muscle weakness, numbing sensations in the extremities, nervousness and diarrhea.

Alcohol, caffeine, tobacco, sleeping pills, laxatives and estrogen all destroy vitamin B12 or limit its absorption in the body. Sunlight also reduces the vitamin's efficacy. Cooking destroys up to 35% of the vitamin, so meat should be steamed or broiled. While found in canned foods, vitamin B12 is best obtained from fresh sources.

Foods high in vitamin B12 include:

- ***Dairy:*** Eggs, milk, mozzarella cheese, Swiss cheese and yogurt.
- ***Legumes:*** Soybean products (soy burgers and tofu).[98]
- ***Meat:*** Beef, bologna, catfish, clams, cod, crab, frankfurters, halibut, herring, liverwurst, mackerel, organ meats, pork, salmon, sardines, sausage, scallops, shrimp and tuna.
- ***Whole Grains and Rice:*** Whole grain cereal.

As previously mentioned, vitamins B13–B17 were initially thought to be vitamins but are no longer considered vitamins today.

Orotic Acid (Vitamin B13)

This is not considered a vitamin since the body produces it naturally in sufficient quantities, which makes it a non-essential nutrient. Not a lot is known about orotic acid and it is not available as a supplement in the

98 B12 does not occur naturally in tofu but is instead added as a supplement.

United States (although it is in Europe). Research has shown that vitamin B13 helps in the metabolism of folic acid and vitamin B12. Sources are whey and root vegetables such as carrots, potatoes and turnips.

VITAMIN B14

Vitamin B14 was originally isolated from wine but is now thought to be a combination of vitamins B10 and B11. Disputed as a vitamin by biochemist Earl R. Norris, vitamin B14 is no longer considered a vitamin. This nutrient, which has no official name, may contribute to cell metabolism and can be found in organ meats, wine and yeast.

PANGAMIC ACID (VITAMIN B15)

This is a controversial nutrient that is not acknowledged as a vitamin by the FDA, although its sources were shown to be the same as other B-complex vitamins. Discovered in Russia, this substance was banned in the US in the 1970s since it was believed the Russians were using it to gain a competitive advantage; the validity of this claim is debatable to many. For further confusion, Russians refer to the vitamin as calcium pangamate rather than pangamic acid. Pangamic acid can be found in apricot kernels, whole grains and seeds including sunflower, pumpkin, poppy and sesame seeds.

DIMETHYLGLYCINE (DMG, VITAMIN B16)

DMG is not considered a vitamin since it is non-essential to humans. Manufacturers claim this vitamin can improve athletic performance, stimulate your immune system and help fight autism, epilepsy and mitochondrial disease. According to the Memorial Sloan Kettering Cancer Center, DMG enhances the immune response in humans and animals by improving oxygen circulation. These claims and studies have not been fully vetted by the scientific research community. DMG is a byproduct of choline (discussed in a few sections) and can be found in livers, beans, whole grains, rice, brewer's yeast and many seeds.

LAETRILE (AMYGDALIN, VITAMIN B17)

Similar to pangamic acid, this is another nutrient that is controversial. Banned by the FDA in the 1980s, laetrile (amygdalin) is still legal in numerous countries including Italy, Germany and Mexico. Many proponents claim it helps prevent and cure cancer since it attacks mutated, cancerous cells more than regular healthy cells via apoptosis. A study by the Department of Physiology at Kyung Hee University in South Korea showed that amygdalin injected into cancerous human prostate cells significantly induced apoptosis. Other animal studies have shown amygdalin effectively killing cancerous bladder and brain cells.

In contradiction to these findings, studies in the United States showed no impact on stunting cancerous tumor growth; therefore, the jury is still out on how effective this substance can be in fighting cancer. Amygdalin also has been shown to boost immunity, reduce pain and lower blood pressure. There are concerns over its potential toxicity, however, as intestinal bacteria contain enzymes that can activate the release of cyanide found in amygdalin. For this reason, it is safer to inject amygdalin rather than ingest it orally. Amygdalin can be obtained naturally from the pits of apricots, plums, nectarines, cherries and peaches. If taken orally, no more than one gram should be ingested at a time (do not do this without consulting a medical professional first).

CHOLINE

Choline is not a B-complex vitamin but is more like them than other vitamins or minerals and is therefore listed here. Found in many multi-vitamin supplements and foods, choline impacts your liver, brain, muscles, nervous system and metabolism. The FDA requires infant milk not made from cow's milk to be supplemented with choline, specifically for its brain development attributes. Choline can also be used as a supplement to improve or maintain cognitive functions such as memory, especially important for elderly individuals. A lack of choline can lead to liver disease, atherosclerosis (a disease caused by excess plaque narrowing the arteries) and neurological disorders.[99]

VITAMIN C

Vitamin C produces collagen, a protein vital to the production of tissue cells in your body including skin, ligaments and tendons. Collagen is also important to the bones and cartilage. With its role in tissue formation,

99 "Choline," *Oregon State University.* https://lpi.oregonstate.edu/mic/other-nutrients/choline.

vitamin C collagen production also helps heal wounds and maintain strong blood vessel walls.

Like beta-carotene, vitamin B2, vitamin A, vitamin E, flavonoids, manganese, selenium and zinc, vitamin C is an antioxidant that helps protect the body's DNA from free radicals. Vitamin C also helps in the production of the signaling protein interferon, which protects the body from invading viruses by interfering with the viral replication process. For this reason, vitamin C is often taken to help build up the immune system and can contribute to fending off unwanted viruses (including those that can destroy cells and lead to cancer).

Scurvy is a condition caused by a lack of vitamin C and results in weakened bones and muscles, bleeding gums and an increased susceptibility to infection. This condition is believed to have been noticed as early as 1500 BC by the Egyptians and there are various accounts and claims of discoveries for cures throughout history. Generally the Scottish physician James Lind is credited with the discovery that citrus fruits can prevent scurvy, which was rampant aboard ships where sailors' diets were mainly grains and meat without fruits and vegetables. In 1795, the British Navy began prescribing limes to soldiers to help prevent scurvy. But citrus fruits were not generally accepted everywhere as a cure for scurvy and the condition continued to plague sailors, soldiers and expeditions throughout the 19th and early 20th centuries.

Stress, smoking, painkillers, sugar and antibiotics can destroy vitamin C. Sensitive to air, heat and light, vitamin C is best stored in a cool dark location. Like the B-complex vitamins, vitamin C is water-soluble and therefore must be replenished daily.

Foods high in vitamin C include:

- ***Fruits:*** Acerola cherries, apples, bananas, blackberries, blueberries, cantaloupe, guava, kiwis, lemons, limes, oranges, papayas, peaches, pears, raspberries, strawberries and tomatoes.
- ***Vegetables:*** Asparagus, bell peppers, beets, broccoli, Brussels sprouts, cauliflower, collard greens, dandelion greens, kale, lima beans, peas, red cabbage, spinach, sweet potatoes, Swiss chard, turnip greens and white potatoes.

Vitamin D

Vitamin D works in conjunction with calcium to build and maintain healthy bones. In addition to calcium, vitamin D also increases the digestive absorption of iron, magnesium, phosphate and zinc. Additionally, it has been found that vitamin D helps the body's immune system by activating helper white B-cells and killer white T-cells.

There are two main forms of vitamin D: D2 (ergocalciferol) and D3 (cholecalciferol). The latter is produced naturally by the body when sunlight hits the skin while the former is produced by some plants (e.g. mushrooms) with exposure to UV light. Both are available in supplements, although vitamin D3 has been shown to be more effective, as it is naturally produced by the human body. Since vitamin D is lacking in many healthy foods like fruits and vegetables and vitamin D3 is superior to vitamin D2, the best source is natural sunlight exposure. However, those with darker skin or living in overcast climates lacking adequate sun exposure should obtain vitamin D3 from a supplement (note that multivitamins generally use vitamin D3).

Research has also shown that individuals with a high intake of vitamin D are less prone to developing cancer, likely due to vitamin D's impact on the body's immune system. Colon, breast, prostate and pancreatic cancers have the most available data and there are ongoing studies to determine the significance of vitamin D and cancer. This includes determining how much of the reduced cancer rates are due to vitamin D and how much are due to individuals with higher vitamin D having more active and healthy lifestyles.

Food and vitamin supplements should be kept in a dark location and covered as air, light and acids deteriorate vitamin D.

Foods high in vitamin D include:

- ***Animal Products:*** Butter, caviar, cream, eggs, milk and yogurt.
- ***Fruits, Vegetables and Fungi:*** Mushrooms (chanterelle, maitake and portabella).
- ***Meat:*** Cod liver oil, herring, liver, liverwurst, mackerel, pork, salmon, smoked eel, trout and tuna (packed in oil).
- ***Nuts, Seeds, Grains and Legumes:*** Products such as cereal that have vitamin D additives, soy milk and soy yogurt and tofu.

VITAMIN E

Discovered in 1922 by Herbert Evans and isolated in 1935 by Gladys Emerson of the University of California at Berkley, vitamin E comes in two types, either as a tocopherol or tocotrienol. Each type consists of either an alpha, beta, delta or gamma form. Alpha and gamma tocopherols are the most commonly found forms of vitamin E in the human diet (gamma in the US and alpha in Europe). Alpha tocopherol is the biologically active form of vitamin E and therefore the most important nutritionally, which is why it is the form found in multivitamins.

Like beta-carotene, riboflavin, vitamin C, vitamin E, zinc and selenium, vitamin E is an antioxidant that protects the body's cells against harmful free radicals. Additionally, vitamin E plays an important role in cell respiration, the strengthening of cell membranes and the production of RBCs. Vitamin E also enables lower blood pressure since it helps the heart use oxygen more efficiently, thereby causing heart muscles not to work as hard. Some athletes have noted an increase in endurance with high vitamin E intake due to this effect as well. Far exceeding the daily recommended dosages for vitamin E can have an adverse impact on the body's use of vitamin K; the same can be said for vitamin A.

Deep-frying reduces the efficacy of vitamin E by around 80%. Light and oxygen also destroy vitamin E. High vitamin E berries should not be washed when stored but rather when ready for consumption, as they can mold easily.

The best sources of vitamin E include:

- ***Animal Products:*** Butter, crab, lobster, oysters, rainbow trout, salmon, shrimp, swordfish and tuna.
- ***Fruits and Vegetables:*** Apples, asparagus, avocados, blackberries, broccoli, collard greens, kale, kiwis, mamey sapote, mangos, olive oil, peaches, pumpkin, spinach, squash, Swiss chard, turnip greens, and tomatoes.
- ***Nuts, Seeds and Legumes:*** Alfalfa seeds, almonds, beans, Brazil nuts, hazelnuts, sunflower seeds and wheat germ.

Vitamin F

Originally discovered and classified as vitamin F in 1923, the two essential fatty acids (EFAs) were reclassified as fats in 1929. As noted earlier, these EFAs are alpha-linolenic acid (an omega-3 fatty acid) and linoleic acid (an omega-6 fatty acid). Please refer to the *Fat* section for further details on EFAs.

Vitamin K

The primary function of vitamin K is to aid the body in blood clotting. Additionally, vitamin K helps strengthen bones as it controls the binding of calcium in bones as well as other body tissues. In addition to supplement form, vitamin K is produced naturally inside the body's digestive tract, specifically via gut flora (bacteria living in the digestive tract). This microbe composition is impacted by your diet. For example, yogurt stimulates gut flora growth. An added benefit of vitamin K is that it aids liver function. There are proponents that vitamin K supplements can reduce tumor growth rates, although as of this writing, the claim has not been proven in a medical study.

Sunlight destroys vitamin K so any supplements should be stored in a sealed container. Aspirin destroys vitamin K, leading to thinning of our blood.

Good sources of vitamin K include:

- ***Animal Products:*** Beef liver, cheese, chicken, goose liver, kidneys and pork chops.
- ***Fruit and Vegetables:*** Asparagus, avocados, broccoli, Brussels sprouts, collard greens, kale, kiwi, lettuce, mustard greens, prunes, spinach, Swiss chard and turnip greens.
- ***Nuts, Seeds, Grains and Legumes:*** Cashews, green beans, green peas, natto, oatmeal, pine nuts, pumpkin seeds and soybean oil.

Vitamin P (Flavonoids)

Like vitamin F, vitamin P was once deemed a vitamin but is no longer classified as such. Rather this group of substances is now referred to as flavonoids (or bioflavonoids). Flavonoids were first referenced by the scientist Robert Boyle in 1964 and coined vitamin P in 1935, when Dr. Szent-Gvörgyi and his associate Dr. István Rusznyák isolated it from lemon juice.

Flavonoids provide pigmentation, flavor and smell to plants, where up to 6,000 various subtypes have been identified. These substances are essential for the human body as they aid in cell and organ functionality. Specifically they help absorb vitamin C, maintain your bone and teeth health and help with the production of collagen, a protein used in the creation of blood vessels and muscular tissues. Flavonoids are also important antioxidants.

Curcumin, quercetin and rutin are flavonoids that have been shown to have anti-cancer effects. Quercetin with green tea is believed to be an advantageous combination and curcumin is found in turmeric.

Natural sources of flavonoids are:

- ***Fruit and Vegetables:*** Apples, apricots, artichoke, berries, broccoli, cantaloupes, celery, eggplant, grapefruit, green peppers, lemons, onions (green and red), okra, oranges, papaya, peaches, pears, peppers, plums, red wine, tangerines and tomatoes.[100]
- ***Nuts, Seeds, Grains and Legumes:*** Cashews, chocolate, pecans, pistachios, snap beans, soybeans and walnuts.
- ***Spices:*** Dill, parsley, thyme and turmeric.

MINERALS

Minerals are needed to make your body function properly and you will not survive without proper amounts in your diet. They can be broken up into macrominerals and trace minerals. The former must be consumed in relatively large amounts while the latter are consumed in smaller amounts.

MACROMINERALS

There are seven macrominerals: calcium, chloride, magnesium, phosphorus, potassium, sodium and sulfur. Aside from the major elements in the body that include oxygen, carbon, hydrogen and nitrogen (Table 14), the macrominerals make up the majority of the remaining elements.

CALCIUM

Calcium strengthens your teeth and bones, with a few pounds worth making up your skeletal structure. Additional calcium is found in organs such as the intestines, liver and stomach, as well as in your muscles. Calcium helps prevent muscle cramps and has a calming effect, which is

100 Veronique Desaulniers. "What is Vitamin P and How Can it Prevent Cancer?," *The Truth About Cancer,* https://thetruthaboutcancer.com/vitamin-p-cancer.

why high calcium intake before bed or stressful events via natural foods or supplements is advantageous. As cancer treatment degrades your bones, calcium, vitamin D and lactose (milk sugar) intake can help combat these deteriorating effects (the latter two indirectly aid by helping your body absorb calcium). Calcium is also essential to activating enzymes involved in the ATP muscle energy process. Blood clotting prevention and cardiovascular health are other functions of calcium. Heavy sweating causes a loss in calcium so be sure to replenish after workouts. Signs of a calcium deficiency include nervousness and joint pain.

Foods high in calcium include:

- ***Animal Products:*** Animal bones, caviar, cheese, ice cream, milk, sardines with bones, oysters and yogurt.
- ***Fruits:*** Figs, prunes and raisins.
- ***Grains, Nuts and Seeds:*** Almonds, Brazil nuts, hazelnuts and soybean flour.
- ***Vegetables:*** Bok choy, broccoli, collard greens, kale, okra, spinach, turnip greens and watercress.

Table 14: Elemental makeup of the human body.[101]

Element	Symbol	Percentage
Oxygen	O	65
Carbon	C	18.5
Hydrogen	H	9.5
Nitrogen	N	3.2
Calcium	Ca	1.5
Phosphorus	P	1
Potassium	K	0.4
Sulfur	S	0.3
Sodium	Na	0.2
Chlorine	Cl	0.2
Magnesium	Mg	0.1
Trace Elements: Boron, Chromium, Cobalt, Copper, Fluorine, Iodine, Iron, Manganese, Molybdenum, Selenium, Silicon, Tin, Vanadium and Zinc.	B, Cr, Co, Cu, F, I, Fe, Mn, Mo, Se, Si, Sn, V and Zn.	<1

101 "Composition of the Human Body," *Wikipedia,* https://en.wikipedia.org/wiki/Composition_of_the_human_body.

CHLORINE (CHLORIDE)

Chlorine by itself (uncharged) is rarely found in nature, but rather the negatively charged ionic form chloride typically exists, as chlorine is an oxidizing agent. Oxidation (gaining an electron from a reducing agent such as a metal) occurs rather easily for chlorine and chloride is an essential mineral for humans. Chloride is found with various reducing agents in the body, including hydrochloric acid (HCl), potassium chloride (KCl) and sodium chloride (NaCl). These acids are all used in the digestive process of proteins, iron and B vitamins by activating digestive enzymes.

The best source of chloride is found in common table salt, NaCl. Any salty food is therefore also a good source of chloride (rye bread and certain vegetables are also a good source of chloride, including celery, lettuce and tomatoes). Processed foods contain a lot of salt, so the average American gets plenty of chloride in their diet. Vegetarians may need to be mindful of low chloride in their diet due to low salt intake. Multivitamins also generally contain chloride in the form of potassium chloride.

Good sources of chlorine include:

- ***Animal Products:*** Bacon, butter, cheese, cod, crab, eggs, lobster, oysters, prawns, salami, salmon, sardines, sausage and tuna.
- ***Fruits and Vegetables:*** Kelp, olives and seaweed.
- ***Nuts, Seeds, Grains and Legumes:*** Beans (baked and mung), cereal and peanuts.

MAGNESIUM

Magnesium plays many important roles in the biological process. One of the most important purposes for magnesium is its aid in regulating blood sugar levels since it converts blood sugar into energy. Additionally, magnesium is involved in activating enzymes involved in the ATP process to produce energy for muscular work. Magnesium assists in the absorption of other important vitamins and minerals, including vitamins B, C and E as well as calcium, phosphorus, potassium and sodium. Bone strength and a steady heartbeat are also benefits of magnesium.

Conversely, deficiencies in magnesium may lead to a loss of appetite, muscle spasms, confusion, poor memory, irritability and in extreme deficiencies, hallucinations.

Foods high in magnesium include:

- ***Animal Products:*** Crab, mackerel, milk, sardines, scallops and yogurt.
- ***Fruits and Vegetables:*** Apricots, avocados, bananas, chard, chickpeas, figs, prunes and spinach.
- ***Legumes, Nuts and Seeds:*** Almonds, black beans, Brazil nuts, chocolate, cocoa powder, coconut, peanuts, pumpkin seeds, soybeans, squash seeds and walnuts.
- ***Miscellaneous:*** Sea salt.
- ***Whole Grains, Rice and Fungi:*** Baker's yeast, brown rice, wheat bran and wheat bread.

Phosphorus

Phosphorus serves multiple biological functions, including maintaining strong bones, balancing the body's pH levels and maintaining brain cognition. This mineral is also important in the metabolism of fats, proteins and carbohydrates. Furthermore, phosphorus is involved in the production of ATP muscle energy. Phosphorous is found in virtually all cells throughout the body and is involved in cell mitosis, making it a vital nutrient.

A deficiency in phosphorous can lead to a loss of appetite, bone pain, stiff joints, fatigue, numbness, weakness, irregular breathing and anxiety.

Foods high in phosphorus include:

- ***Animal Products:*** Beef, cheese, chicken, cod, crab, eggs, lamb, lobster, milk, muscles, organ meats, oysters, pork, prawns, salmon, sardines, scallops, tuna and veal.
- ***Legumes, Nuts and Seeds:*** Almonds, Brazil nuts, cashews, chocolate, cocoa powder, coconut, hazelnuts, macadamia nuts, peanuts, pistachios, pumpkin seeds, sesame seeds and walnuts.
- ***Whole Grains, Rice and Fungi:*** Baker's yeast, flour, mushrooms and wheat bran.

Potassium

Potassium is an important mineral electrolyte that helps balance out the body's sodium electrolytes for healthy blood pressure—sodium increases blood pressure while potassium decreases blood pressure.

Electrolytes are important because they form ions in the bloodstream that create an electric potential across a cell membrane. This electric potential is important in the conduction of impulses along nerve cells, most notably neurons. Potassium has a much higher concentration inside the cell membrane while sodium is much higher outside, which is created by a sodium-potassium pump that moves positive sodium and potassium ions across the membrane. This pump is powered by ATP energy.

While the ratio of sodium entering to potassium leaving is 3:2, the main reason for a net positive charge outside of the cell membrane is because the cell membrane is more permeable to potassium ions than sodium ions. This creates a net potential of around -80 mV in neurons.[102] In addition to neurons, sodium-potassium pumps are used in cells throughout the body, most notably for the electrical activity of the heart and for muscle contractions.

In addition to lowering blood pressure, potassium promotes stronger and healthier bones via the preservation of bone mineral density, reduces the chance of a stroke, decreases the likelihood of kidney stones and protects against muscle atrophy. Along with other electrolytes, potassium can reduce the chances of muscle cramps (e.g. due to exercise in hot conditions where a lot of sweat is lost).

A lack of potassium can result in weakened muscles, muscle coordination, cramping and irritability.

Good sources of potassium include:

- ***Animal Products:*** Beef, chicken, cod, eggs, flounder, haddock, ham, lobster, milk, organ meats, oysters, prawns, salmon, sardines, scallops, turkey, tuna and veil.
- ***Fruits and Vegetables:*** Apricots, avocado, bamboo, bananas, currants, dates, figs, grapes, parsley, peaches, potatoes, prunes, raisins and spinach.
- ***Legumes, Nuts and Seeds:*** Almonds, Brazil nuts, cashews, chestnuts, cocoa powder, coconuts, lima beans, peanuts, pistachios, sesame seeds, sunflower seeds and walnuts.
- ***Whole Grains, Rice and Fungi:*** Baker's yeast, flour and wheat bran.

102 The unit abbreviation mV stands for millivolt.

SODIUM

Sodium is an important mineral that has many benefits, but in excess there can be adverse effects. As part of the potassium-sodium pump previously mentioned, sodium helps regulate blood pressure and must maintain a healthy balance with the amount of potassium in your body. Having a healthy blood pressure is vital for transporting oxygen throughout your body without overstressing the heart. This pump system is important for cell functioning and for certain cells (such as neurons) to send electrical signals within the nervous system and muscles (such as the heart) for contraction. Sodium is the most abundant extracellular (external to the cell membrane) electrolyte in the body's blood plasma, whereas potassium is the most abundant intercellular electrolyte (inside the cell membrane).

Replacing electrolytes—such as with sports drinks like Gatorade—is very important for strenuous activities as electrolytes (primarily sodium, potassium and chloride) are lost during sweating. On hot days this is especially important to prevent heat stroke (i.e. an increase in body temperature above 104°F; various sources cite different values). Sports drinks such as Gatorade contain electrolytes to replenish those lost from sweating. With their importance in muscle contraction, loss of electrolytes can lead to muscle cramping, which is especially prevalent in the hot summer months.

Additional benefits of sodium include eliminating carbon dioxide from the body via excretion and aiding cognitive functioning (à la healthy neurons). Sodium helps with the absorption of other nutrients in the small intestines such as chloride, amino acids and glucose. Cellular water retention is also enhanced by sodium (and conversely reduced with potassium), which can be both beneficial and detrimental depending on the individual and level of exercise. For instance, during training and athletic performances, increased water retention is important; if you are trying to lose weight and are not very active, excess water retention can lead to unhealthy bloating.

There are many different forms of sodium intake in the human diet. These include:

- ***Baking Powder:*** Used as a leavening agent in cooking, baking powder is a combination of baking soda and an acid.
- ***Baking Soda (Sodium Bicarbonate):*** Baking soda is used as a leavening agent in cooking, among other non-food uses.

- ***Brine:*** Brine is table salt mixed with water.
- ***Disodium Phosphate:*** Used as an emulsifier for cheese (i.e. it keeps fat globules from dissociating onto the surface of the cheese), disodium phosphate is also used to adjust the acidity in soda, chocolate and sauces such as Worcestershire, soy and steak sauce.
- ***Monosodium Glutamate (MSG):*** MSG is a sodium salt of glutamic acid, which is an amino acid. Used to enhance flavor in cooking, MSG is especially prevalent in meat flavoring for Asian foods.
- ***Sea Salt:*** Sea salt is an organic salt with additional natural minerals.
- ***Sodium Alginate:*** This is a sodium salt of alginic acid that is extracted from brown algae. Used as a stabilizer for ice cream, yogurt and cheese, sodium alginate is also used as a thickener for pudding, jam, tomato juice and canned products. Additionally this salt is used as a hydration agent for bread, noodles and frozen products.
- ***Sodium Benzoate:*** Sodium benzoate is used to preserve certain foods such as sauces, salad dressings, relish, jams, sauerkraut, hot sauce and soda.
- ***Sodium Hydroxide:*** Sodium hydroxide is used in the preparation of canned and frozen fruits. Undergoing a blanching process to remove the skin, fruit is dipped briefly in boiling water and treated with sodium hydroxide, softening the skin and allowing it to peel right off.
- ***Sodium Propionate:*** Sodium propionate is used as a preservative for bread and cheese.
- ***Sodium Sulfite:*** Most commonly used in non-food applications such as chemical cleaning agents, sodium sulfite is also used to bleach food before being dyed with food coloring. Additionally, sodium sulfite is used to preserve certain dry fruits including figs and prunes.
- ***Table Salt (Sodium Chloride):*** This salt is standard table salt that is used for seasoning and in processed foods for storage.

In addition to cramping that was mentioned, a lack of salt can lead to dizziness and headaches. Conversely, increased blood pressure from too much salt can lead to strokes, heart failure, osteoporosis, kidney disease and stomach cancer.

Since most humans, especially Americans and those on a Western diet, intake plenty of sodium, this mineral is treated uniquely in that foods both high and low in sodium are presented so you can get the right amount of sodium in your diet.

- ***Foods High in Sodium:*** Bacon, bologna, brine-based food (such as olives), cereals (brand-dependent), corned beef, lunch meats, Parmesan cheese, potato chips, pretzels, processed cheese and sausage.
- ***Foods Low in Sodium:*** Fresh fruits and vegetables. Note that while many vegetables have natural sodium—with some of the highest being artichokes, beets, bell peppers, broccoli, carrots, celery, radishes, spinach and sweet potatoes—they are still considered low in sodium (i.e. they have less than 140 mg of sodium in a 50 g serving).[103]

SULFUR

Sulfur is vital to the body's cellular metabolic process since it helps create enzyme structure and enzymes are required for metabolic functioning. Additionally, sulfur helps proteins keep their shape. Specifically, keratin is a protein found in your hair, skin and nails, which is why you find many beauty ointments containing sulfur as well as products that treat skin conditions such as acne. Sulfur also helps with the production of collagen, which forms cell structure, artery walls and connective tissues throughout the body. Additionally, sulfur helps the liver produce bile, which breaks down fats and fat-soluble vitamins in the small intestine. Cellular respiration also relies on sulfur as part of the oxidation-reduction reactions that allow cells to utilize oxygen.

Sulfur comes in either two forms with the foods and supplements we take: dimethyl sulfoxide (DMSO) or methylsulfonylmethane (MSM). The former is produced synthetically while the latter occurs naturally in fruits, vegetables, grains and milk. While MSM is healthy, you should consult your doctor before taking any DMSO supplements as too much can cause toxic effects.

Foods high in sulfur include:

- ***Animal Products:*** Beef, cheddar cheese, chicken, cod, crab, duck, eggs, haddock, ham, lamb, lobster, milk, mussels, oysters, prawns, salmon, sardines, scallops, turkey, veal and whiting.

103 "Sodium in Fruits and Vegetables," *PBH Have a Plant,* https://fruitsandveggies.org/stories/best-of-sodium.

- ***Fruits and Vegetables:*** Apricots, bok choy, broccoli, Brussels sprouts, cabbage, cauliflower, chives, figs, garlic, kale, kohlrabi, leeks, onions, peaches, spinach and turnips.
- ***Legumes, Nuts and Seeds:*** Almonds, black pepper, Brazil nuts, navy beans, peanuts and walnuts.
- ***Whole Grains, Rice and Fungi:*** Wheat bran, wheat germ and whole grains.

TRACE MINERALS

Required in smaller amounts than macrominerals, trace minerals remain important. Below is a list of trace minerals along with their function and food sources.

CHROMIUM

Chromium is an essential mineral that helps regulate blood sugar levels and is involved in the metabolism of proteins, fats and carbohydrates. Adequate amounts can prevent hypertension and high blood pressure, as well as reduce food cravings and promote weight loss. Studies have also shown chromium in the form of chromium picolinate can aid the brain's memory and help combat Alzheimer's disease.

Chromium can be found in multi-vitamin supplements. Natural sources of chromium include:

- ***Animal Products:*** Beef, butter, cheese, egg yolks and oysters.
- ***Fruits and Vegetables:*** Apple peels, black pepper, chili, potatoes, spinach and wine.
- ***Grains, Nuts and Seeds:*** Rye bread, spaghetti and wheat bread.

COPPER

The essential mineral copper has multiple purposes for humans. Copper is required to synthesize hemoglobin, a complex protein made up of carbon, hydrogen, oxygen, nitrogen, sulfur and iron that is found in RBCs (96% of a RBC's dry mass is hemoglobin and 35% of its total mass when water is included). As discussed in the *Red Blood Cells* section, hemoglobin transfers oxygen throughout the body and therefore a lack of copper will prevent this from happening. Copper is found primarily in the brain, heart, kidneys, liver and skeletal muscles.

Since copper is responsible for the pigmentation of your skin, a lack of copper will lead to a pallor effect. A lack of copper can also decrease neurological functionality, cause an increased risk of infection and lead to osteoporosis.

Good sources of copper include:

- ***Animal Products:*** Beef liver, cheddar cheese, crab, lamb liver, lobster, oysters and prawns.
- ***Fruits and Vegetables:*** Avocados and green leafy vegetables.
- ***Legumes, Nuts and Seeds:*** Almonds, black pepper, cashews, chocolate, lentils and peanut butter.
- ***Whole Grains, Rice and Fungi:*** Brewer's yeast, rye bread, spaghetti, wheat bran and wheat bread.

IODINE

Iodine is an important element that is primarily used by your thyroid gland (70 to 80% of your body's iodine levels are found in the thyroid while the remaining levels are in your blood) to help synthesize hormones. These hormones are important in bone health, metabolism, immune response, health of your central nervous system and heart functionality. Therefore a lack of iodine can impair your mental capacity and motor skills as well as lead to fatigue, depression and weight gain. Furthermore, your thyroid gland will enlarge with insufficient iodine, which can cause a goiter. Lack of iodine in a pregnant woman can also lead to mental retardation in her child, known as iodine deficiency disorder (IDD).

Table salt generally contains small traces of iodine, which has its genesis from the United States' attempt in 1924 to supplement citizens' lack of iodine in their diet (which was adopted after successful implementation by the Swiss). University of Michigan researchers created this mixture for the Morton Salt Company and iodine has been added to table salts ever since (sea salts generally do not contain iodine, although check the label as there are exceptions). A lack of iodine was originally an issue due to some regions of the country (and world) lacking enough iodine in the soil used for food. Presently, however, food sources in the United States come from farther distances and more varied locations, decreasing the likelihood of an iodine deficiency. This does remain a problem for less developed regions of the world.

Iodine has also been shown to reduce the effects of radiation and help patients recover from radiation treatment. Additionally, iodine has been shown to induce apoptosis in cancer cells or mutant cells that may grow into cancer cells.

Good sources of iodine include:

- ***Animal Products:*** Cheese, cod, eggs, milk, shrimp, tuna (canned in oil), turkey and yogurt.
- ***Fruits and Vegetables:*** Cranberries, navy beans, potatoes, prunes, seaweed and strawberries.
- ***Miscellaneous:*** Table salt.
- ***Whole Grains, Rice and Fungi:*** Brewer's yeast, rye bread, spaghetti, wheat bran and wheat bread.

IRON

Similar to copper, iron is an essential mineral required to create hemoglobin. While copper is required in the metabolic process to make protein, iron is one of the elements mentioned above in the hemoglobin protein itself. Therefore without iron, most vertebrates, including humans, could not survive, as hemoglobin carries oxygen throughout the body.

Due to its role in carrying oxygen, iron is vital to the functioning of all your organs, including brain cognition. Iron also helps make up myoglobin, which is a protein like hemoglobin but found primarily in muscle tissue and used for muscle contraction. High levels of hemoglobin also allow mammals to hold their breath longer, which is why aquatic mammals such as whales, seals, beavers and dolphins have such high amounts (elephant seals can hold their breath for 2 hours, sperm whales for 90 minutes, beavers for 15 minutes, dolphins for 10 minutes and humans for 2 minutes). Incidentally sloths can hold their breath for up to 40 minutes, but this has more to do with their ability to slow their heart rate.

A lack of iron can lead to anemia, which causes one to be lethargic and tired due to a reduced oxygen level. Coffee and tea retard the absorption of iron in your body.

Good sources of iron include:

- ***Animal Products:*** Beef, chicken, clams, egg yolks, halibut, ham, mollusks, muscles, organ meats (heart, kidneys and liver), oysters, sardines, turkey and veal.
- ***Fruits and Vegetables:*** Asparagus, figs, lima beans, parsley, peaches, raisins, spinach and spirulina (a type of algae).
- ***Legumes, Nuts and Seeds:*** Cocoa powder.
- ***Whole Grains, Rice and Fungi:*** Baker's yeast, cereals[104] and wheat bran.

MANGANESE

Manganese serves various biological purposes including strengthening bones, blood clotting, aiding in the formation of connective tissues and blood sugar regulation. Additionally, manganese helps to regulate sex hormones and activate enzymes used for the metabolism of carbohydrates, proteins, fats and nucleic acids. With its specific role in helping activate the enzymes arginase, glutamine synthetase and manganese superoxide, manganese serves as an antioxidant that helps reduce inflammation. Healthy cognitive brain functions are also related to manganese due to its existence in the brain's synaptic vesicles, located within neurons where neurotransmitters are stored for cell-to-cell signaling.

Foods high in manganese include:

- ***Animal Products:*** Bass, chicken and mussels.
- ***Fruits and Vegetables:*** Blueberries, cinnamon, pineapple, raspberries, spinach, strawberries and tea.
- ***Legumes, Nuts and Seeds:*** Black pepper, butter beans, cloves, hazelnuts, lima beans, pumpkin seeds, soybeans and tofu.
- ***Whole Grains, Rice and Fungi:*** Brown rice, oats and wheat bread.

SELENIUM

Selenium is an antioxidant that protects against free radicals, thereby preventing cell damage and reducing inflammation within the body. Antioxidants also help with the aging process by their ability to prevent cell damage. Studies have shown that selenium can help protect against

104 Cereals can reduce the absorption of iron if phytic acid is an ingredient.

thyroid disease as well as promote a healthy heart. Selenium, in combination with vitamin E, can help combat the poisonous effects of mercury and cadmium.

Reproductive health in both men and women can be improved from selenium, with studies showing a decreased chance of infertility when adequate selenium levels are maintained. Selenium has shown some promise in terms of cancer prevention, but studies are ongoing and not yet conclusive.

Deficiencies in selenium can cause hair loss, nail sloughing, fatigue, irritability, mood swings and diarrhea.

Foods high in selenium include:

- ***Animal Products:*** Beef, chicken, crab, eggs, halibut, mollusks, organ meats, oysters, pork, sardines, salmon, shrimp, tuna, turkey and veal.
- ***Fruits and Vegetables:*** Broccoli, cabbage and spinach.
- ***Legumes, Nuts and Seeds:*** Brazil nuts, chia seeds, flax seeds, lima beans, pinto beans, sesame seeds, soybeans and sunflower seeds.
- ***Whole Grains, Rice and Fungi:*** Brown rice, mushrooms, oats, rye bread, spaghetti and wheat bread.

ZINC

Zinc is another important trace element that is present in all bodily tissues. One important function is to help activate enzymes, making it imperative in the cell metabolic process. High concentrations are found in your hair, skin and nails, as well as the prostate gland. Zinc also helps the liver, where it is especially useful in breaking down alcohol. Additional roles of zinc include aiding in the activation of WBCs, specifically T-cells, hormone production and digestion.

As an antioxidant that fights free radicals, zinc fosters a healthy appearance and reduces the effects of aging. This role as an antioxidant also helps prevent cancer cells from forming. Signs of zinc deficiency include changes in appetite, taste, smell, and weight, hair loss, digestive issues, fatigue, poor concentration or memory, decrease in ability for wounds to heal, nerve dysfunction, and infertility.

Foods high in zinc include:

- ***Animal Products:*** Beef, crab, chicken, egg yolks, lamb, milk, mussels, organ meats, oysters, pork, sardines, shrimp, Swiss cheese, turkey, veal and yogurt.
- ***Fruits and Vegetables:*** Garlic, lima beans, kidney beans and spinach.
- ***Legumes, Nuts and Seeds:*** Almonds, black pepper, Brazil nuts, cashews, chickpeas, cocoa powder, flax seeds, hazel nuts, kidney beans, peanuts, pumpkin seeds, sesame seeds, squash seeds, walnuts and watermelon seeds.
- ***Whole Grains, Rice and Fungi:*** Cereal, flour, mushrooms, oatmeal, oats, wheat bran and wheat germ.

Miscellaneous

Various sources list additional trace minerals with varying opinions and research on which are essential versus non-essential. While there is agreement on the ones already discussed as being essential, some sources state the following trace minerals are also essential: arsenic, boron, cobalt, molybdenum, nickel, silicon and vanadium.

Additionally, the following trace minerals are either considered non-essential or are up for debate: aluminum, antimony, bromine, cadmium, fluoride, lead, lithium, nickel, rubidium and tin (stannum). I recommend you ensure you are getting the trace minerals listed previously and research the above nutrients further for their importance to your health.[105]

Multivitamin Supplements

There are a host of nutritional supplements to take for vitamins and minerals. An entire book can be written on this subject based on the number of products out there for consumption. Readers are encouraged to do their own research; however, I will provide an overview and recommendations based on my research.

105 "Essential Nutrients," *Nutrients Review,* http://www.nutrientsreview.com/glossary/essential-nutrients;"The Essential Nutrients," *Center for Magnesium Education and Research,* http://www.magnesiumeducation.com/essential-nutrients-for-humans; "21 Essential Minerals and 16 Trace Minerals Your Body Absolutely Needs," *Juicing for Health,* https://juicing-for-health.com/essential-minerals.

Multivitamins are an easy way to supplement your diet to ensure you are getting the proper vitamins and minerals your body needs. This should not be thought of as a replacement for eating healthy, but rather a way to enhance your dietary intake. While there are multiple supplements that can be taken, we will focus on a multivitamin that can be taken once or twice daily depending on your diet.

Recall that vitamin C and the B-complex vitamins are water-soluble and therefore get absorbed quickly and excess are released through your urine. For this reason, eating fruits and vegetables throughout the day is an important means to replenish these vitamins. Taking a multivitamin in the morning and evening or night—or multivitamin in the morning and C and B vitamins in the evening or night—is recommended to ensure you are getting all the nutrients you need. Note that this is merely a recommendation, as you can certainly get all the vitamins and minerals you need through a healthy diet alone; however, taking vitamins is a good way to ensure this is the case.

While there a many different brands, Table 15 highlights five that were studied: Kirkland Signature Daily Multi, Rainbow Light Men's One, GNC Mega Men, Optimum Nutrition Opti-Men and One a Day Men's Health Formula. Quantitative percent rankings scores were taken from Labdoor, a company based in San Francisco that specializes in supplement ratings.

Table 15: Selected multivitamin overview.

	Kirkland	Rainbow	Men's OAD	GNC MM	Opti-Men
Price [$]	16.4	28	23	23	21
Servings [-]	500	150	300	90	50
Price/Serving [$]	0.03	0.19	0.08	0.26	0.42
Overall Rating [%[	66.1	82.2	68.4	81	80.4
Label Accuracy [%]	65*	55†	55‡	52§	33
Product Purity [%]	90	90	90	90	90
Nutritional Value [%]	61	91	65	95	90
Ingredient Safety [%]	64	84	78	76	78
Projected Efficacy [%]	61	77	56	78	82

Note that shaded values have high scores in this category. *Vitamin A is a little low, some metals a little low, everything else is higher. †Everything is over. ‡Vitamin C is only at 53% of label when tested, everything else is over. §Magnesium is 15% lower, everything else is higher.

Rainbow, GNC and Opti-Men all scored higher in overall quality, most notably in projected efficacy (i.e. how well the ingredients get absorbed by your body). However these are also more expensive than the Kirkland and Men's One a Day brands, with Kirkland's brand being the most affordable.

Table 16 breaks down the multivitamins by essential vitamins and minerals.[106] All of the brands meet most of the essential vitamin daily recommended values with varying amounts for macrominerals and trace minerals. The more expensive brands in general have a higher concentration of vitamins and minerals, although once you are saturated, having excess does not increase your body's absorption. A better methodology is the efficacy that these vitamins and minerals are absorbed, which the more expensive vitamins are rated higher in based on Labdoor's studies.

Non-essential minerals are shown in Tables 17 and 18. These categories are difficult to compare between brands, as they are non-essential, although that does not necessarily mean non-beneficial. Many of the ingredients are compared in the tables, although some specific to the brand were left off if other brands did not have similar ingredients to compare against. In general, the more expensive brands add supplemental nutrients while the less expensive ones do not. How valuable these are is somewhat unknown without quite a bit of research. Therefore I recommend focusing primarily on comparing the essential vitamins and nutrients above. As of this writing, I take GNC Mega Men once per day in the morning, supplementing with a less potent and inexpensive generic gummy multivitamin at night to replenish vitamin C and B-complex vitamins.

Note that four of the brands in the tables are specific to men; Kirkland is not, although all four brands have corresponding women's multivitamins with similar trends in terms of pricing, efficacy, safety, label accuracy and overall rating as the men's vitamins.

106 Missing from these tables are sodium and sulfur. The former is contained in most foods and individuals often are trying to reduce their sodium intake, which is likely why it is not included as a supplement. Sulfur is found in biotin and sulfate, which is why it is not included as a standalone supplement. Also note that molybdenum is included in this table since nearly all multivitamins have it.

Table 16: Selected multivitamin vitamin, micromineral and trace mineral breakdown.

Vitamin or Mineral		Kirkland [%]	Rain-bow [%]	Men's OAD [%]	GNC MM [%]	Opti-Men [%]
A	N/A	70	100	70	100	200
B1	Thiamine	100	1,667	90	3,333	5,000
B2	Riboflavin	100	1,471	100	2,951	4,412
B3	Niacin	100	125	90	250	375
B5	Pantothenic Acid	100	250	160	100	750
B6	Pyridoxine	100	1,250	150	2,500	2,500
B8	Biotin	10	50	25	100	100
B9	Folate (Folic Acid)	125	200	100	100	150
B12	Cobalamins	100	417	300	833	1,667
C	N/A	150	200	100	500	500
D	N/A	100	200	175	400	75
E	N/A	100	100	75	100	667
K	N/A	31	125	25	100	94
*	Calcium	20	5	21	20	20
†	Iron	100	0	0	0	0
*	Phosphorus	11	0	0	0	0
†	Iodine	100	0	0	100	100
*	Magnesium	25	6	35	25	25
†	Zinc	73	133	100	167	200
†	Selenium	79	286	157	286	286
†	Copper	45	100	100	100	100
†	Manganese	115	100	100	100	250
†	Chromium	29	167	100	100	100
†	Molybdenum	60	100		100	107
*	Chloride	2	0	0	0	0
*	Potassium	2	0	0	0	0

* Macromineral. †Trace mineral.

Table 17: Selected multivitamin non-essential minerals (1 of 2).

Mineral	Kirkland [mg]	Rainbow [mg]	Men's OAD [mg]	GNC MM [mg]	Opti-Men [mg]
Silicon	2	0	0	0	0
Choline	0	20	0	10	10
Inositol	0	20	0	10	10
Citrix Bioflavonoid Complex	0	25	0	0	0
Men's Strengthening Blend	0	100*	0	0	50
Vegetable Juice Complex	0	10†	0	0	0
Complete Digestive Support	0	34‡	0	0	0
Betaine HCL	0	10	0	0	0
Proprietary Amino Acid Blend	0	0	0	150	1,000
Fruit and Vegetable Blend	0	0	0	0	525
Alpha Lipoic Acid	0	0	0	25	0
Green Tea Leaf	0	0	0	40	0
Silica	0	0	0	4	5
Boron	0	0	0	2	2

*Equivalent to 150 mg of food and herbal powder: saw palmetto [berry] 2:1 extract, organic spirulina. †4:1 extract, equivalent to 40 mg of vegetable powder, kale, spinach, dandelion greens and beets. ‡Protease, amylase, lipase and cellulase.

Table 18: Selected multivitamin non-essential minerals (2 of 2).

Mineral	Kirkland [mcg]	Rainbow [mcg]	Men's OAD [mcg]	GNC MM [mcg]	Opti-Men [mcg]
Lycopene*	300	1,000‡	300	1,000	500
Lutein (Flower)†	250	0	0	0	0
Boron	150	0	0	0	0
Vanadium	10	0	0	10	100
Nickel	5	0	0	0	0
Lutemax	0	0	0	2,000	0
Zeaxanthin	0	0	0	400	28
Astaxanthin	0	0	0	50	0
Vanadium	0	0	0	10	0
Rutin	0	0	0	0	0
ProbioActive	0	25 mill CFU	0	0	0

*Carotenoid, is not converted to vitamin A and prevents prostate cancer. †Carotenoid, is not converted to vitamin A and is important for vision. ‡Natural, from tomatoes.

MOERMAN'S THERAPY

One therapy I came across during my research on nutrition during cancer treatment was the book *Dr. Moerman's Anti-Cancer Diet* by Ruth Jochems.[107] In this book, Dr. Cornelius Moerman's anti-cancer nutrition treatment is outlined. Dr. Moerman was a twentieth-century Dutch physician who perfected his therapy as a young doctor in the 1930s by treating a sick pigeon. His first experiment was to extract cancerous cells from the tumor of a sick pigeon and inject them into the breast of a healthy pigeon. As the former died and the latter did not, Moerman concluded that a healthy pigeon would be able to fend off cancer. Over the course of a 10-year period, Moerman experimented with various vitamins, minerals and acids until he concluded the following had the most anti-cancerous effects:

- Vitamins A, B, C and E.
- Minerals iodine, iron and sulfur.
- Citric acid.

Moerman's first patient was Leendert Brinkman, who had a metastatic stomach tumor that was unsuccessfully treated using the treatments available at the time. Having been told by his surgeon that there was nothing more left to do, Leendert sought out Moerman. Leendert was treated with a mixture of the eight substances identified above for roughly a year, which included high quantities of oranges and lemons. This treatment led to a full remission of his cancer and Leendert lived until the tender age of 90!

Jochems's book provides many additional case studies that came after this first patient, along with some additional data. One such interesting piece of data revolved around the cancer diagnosis rates in the Netherlands around World War II. During the Nazi occupation, Adolph Hitler instituted a strict ration of food, which replaced white bread with corn and rye bread, as well as prevented the consumption of sugar, coffee and tea. Additionally, items that were generally consumed were unavailable or scarce due to the war. This included meats, butter, margarine and alcoholic beverages; instead, citizens drank lots of fruit juice. Data shows an approximately 15% decrease in cancer diagnosis rates during the period of 1942–1946, corresponding to the Nazi occupation in 1942–1945—the extra year accounts for lingering effects of the diet.

107 Ruth Jochems, *Dr. Moerman's Anti-Cancer Diet* (New York: Avery Publishing, 1995).

During Moerman's treatment refinement, he came up with 17 symptoms that could be signs of cancer. This included dry skin, aberration of the mucous membranes, chapped lips, red spots on the skin around the nostrils, chapped hands, brown furry coating on the tongue, thinning hair, bleeding gums, black contusions from a superficial impact, slow healing wounds, formation of jelly-like regeneration tissue in a postoperative wound, fatigue, pale complexion, craving sour food, apathy, low energy and sudden weight loss. While there are a multitude of conditions that could cause one of these symptoms, these signs were red flags. Moerman used these red flags to identify patients that may have cancer or be in a weakened state that would make them more susceptible to cancer.

Throughout his career, Moerman faced pushback from the mainstay medical community. After his death, the Ministry of Health at the Hague—the Netherlands' version of the FDA—in 1987 recognized Moerman therapy as a means to treat cancer.

One point Jochems mentions in her book that aligns with other sources[108] is that it is better to get the eight Moerman therapy substances from fruits, vegetables and whole grain sources rather than vitamin supplements. While the latter are beneficial in their own right, whole grains are preferred since they contain additional nutrients such as enzymes that aid in the digestion of food. Processed foods of the modern-day diet remove many of the beneficial nutrients that are found in whole foods, which is why they should be consumed minimally as treats, or not consumed at all, rather than as a mainstay of your diet.

While I agree with the Moerman therapy diet high in fruits, vegetables and whole foods, this should be done *in addition* to standard therapies such as chemotherapy, radiation or surgery. I am a strong proponent of taking the best offerings of various therapies, both natural and medical, in order to give yourself the greatest odds of survival. Although there are negative side effects of chemotherapy, radiation and surgery, the statistics on their ability to cure cancers is undeniable. That said, a healthy diet can both aid in your treatment by strengthening your immune system and act as a lifestyle change to prevent a recurrence or getting cancer at all.

108 Columbu, *The Bodybuilder's Nutrition Book.*

Chapter 16

Exercise

Exercise is extremely important for fighting cancer, both physically and even more so mentally. Physically, radiation and chemotherapy unfortunately attack not only cancerous cells but healthy cells as well. Muscle atrophy occurs due to being sedentary and battling these unwanted toxins and high-energy photons. By following a proper exercise routine, you can reduce these undesirable side effects of cancer treatment.

Psychologically, exercising releases endorphins that can improve your mood. Having a fighting mentality carries over to your treatment and improves your chances of beating cancer. Although extremely difficult when under conditions where your body is being poisoned and dosed with radiation, you must force yourself to exercise. I maintained a steady routine of weights and cardiovascular exercises throughout my chemotherapy and radiation treatments. Part of this was ingrained in my mental makeup of being a lifelong athlete and fighter, and thus, forgoing exercise during treatment was not an option.

Doctors likely will vary in their opinions about if and how much you should exercise during treatment. Once again, this is where being your best advocate becomes imperative. The key is not to overdo it—your body is being attacked and you should not be training for a marathon; rather a steady dose of exercise is very helpful.

Before cancer I would work out roughly three times a week and play tennis three times a week. This provided a good balance of strength and conditioning. I decided to track my before, during and after pictures throughout my treatment to show the effects of the chemotherapy and radiation and track my goal of reducing muscle atrophy. All these pictures are shown in Figure 74. The first picture is a few months prior to my diagnosis while the second is right after my diagnosis and before treatment. The third figure is after my biopsy and initial chemotherapy. You can notice the patch of hair removed from my chest on my left pectoral where my biopsy was performed. This picture shows a more tired demeanor than the previous ones. The next figure was taken after weeks of chemotherapy. You see less hair on my head and body from the impact of chemotherapy as well as a decrease in weight.

Picture number five was taken after radiation. Almost all of my body hair was gone, but my facial hair and hair on my head was starting to return. While my overall muscle mass is less than at the beginning of treatment, consistent exercise kept me in decent shape considering the months of chemotherapy and radiation treatment. Finally, the last picture was taken a few months after treatment and shows my hair growing back curly and blonder, an impact of the chemotherapy that eventually would go away.

For reference of the amount of muscle lost during treatment, my bench press prior to treatment was around 300 lbm and after was around 245 lbm. Had I not exercised during treatment, I estimate it would have been well under 200 lbm by the end of my treatment (my bench was up to 350 lbm within two years after treatment).

The motivation to work out during treatment has to come from your inner strength. While you are lurching from chemo or exhausted from another round of radiation, the last thing you want to do is go to the gym. I would often sit in my car or the locker room for long durations trying to muster up the willpower to get up and work out. Having someone encourage you is great, although most of the motivation is going to come from you.

Before I was diagnosed, my workout routine consisted of rotating between chest and back one day, arms and shoulders another and legs and heavy deadlifting sprinkled in. I would do a lower amount of reps (5–8) in order to achieve maximum muscle gain. However during my treatment I decreased the weight and did higher reps (12–20) with a reduced amount of different exercises—rather than trying to isolate muscles, I attempted to preserve them. I also eliminated deadlifts or other purely power exercises and focused on maintaining (or rather reducing the regression of) my muscles.

This switch from a routine of muscle exhaustion and growth to a more conservative one of higher repetitions and sustainment is very important for two reasons. First, your body's lack of energy and overall bombardment with poisons and photons precludes it from repairing itself like it normally would when you are healthy. Therefore overstressing your muscles and powerlifting is counterproductive, as your body is not going to be able to repair proteins nearly as efficiently. Second, due to your overall lack of energy and body's plight, you are more susceptible to injury. Doing higher weights at lower reps creates more stress on your body and could lead to muscle tears, torn ligaments, sprains, etc., which is the last thing you need during your treatment.

By doing low weights at high reps, your body is able to get stimulation while not overdoing the workload and becoming counter-productive. This approach is also easier mentally as you can exert less overall energy but still let your muscles break down and repair at a more achievable pace.

Treatment Exercise Regimen

Below is my recommendation for a cancer-fighting workout, broken down into chemotherapy and radiation treatments. This is an example that you can vary depending on your pretreatment fitness level and age. The goal is to choose a weight that you struggle to complete by the last repetition (not to exhaustion like you would normally work out, but rather to a comfortable strain so as not to exhaust yourself).

Chemotherapy Exercise Plan

Note that the days of the week provided for your chemotherapy exercise plan are an example and should be adjusted for your chemotherapy day. Not all chemo treatments will be weekly; for those that are bi-weekly you can maintain the same schedule for your off-week and rest on the day you would have had chemotherapy during your on-week.

Monday (Pre-Chemo)

- 3 sets of bench press,[109] 12 to 20 reps each (note: all exercises should have a warm-up set with an easy weight that is not included in the 3 regular sets).

109 For those unfamiliar with how to do some of the exercises listed, I recommend looking for examples on YouTube where many great videos demonstrating correct techniques exist. Once you find a good channel you can bookmark it or subscribe. If you would like expert instruction, I recommend Bodybuilding.com.

Image 1: 9/12/09

Image 2: 11/24/09

Image 3: 1/4/10

Image 4: 1/30/10

Image 5: 3/5/10

Image 6: 4/12/10

Image 7: 7: 8/13/10

Figure 74: Muscle tracking throughout treatment and exercise regimen.

Image 1: 7 weeks and 3 days before 1st treatment. **Image 2:** 1 week and 2 days before 1st treatment. **Image 3:** 4 weeks and 4 days after 1st treatment. **Image 4:** 8 weeks and 4 days after 1st treatment. **Image 5:** 13 weeks and 1 day after first chemo. treatment; 1 week and 1 day after last chemo treatment; 4 days before first radiation treatment. **Image 6:** 18 weeks and 4 days after first chemo treatment; 5 weeks and 6 days after first radiation treatment; 3 days after last radiation treatment. **Image 7:** 35 weeks and 1 day after first chemo treatment; 17 weeks and 6 days after last radiation treatment.

- 3 sets of lat pulldowns, 12 to 20 reps each.
- 3 sets of triceps pulldowns, 12 to 20 reps each.
- 3 sets of biceps curls (barbell or dumbbell), 12 to 20 reps each.

TUESDAY (CHEMO)

- Rest and recover.

WEDNESDAY (POST-CHEMO DAY ONE)

- 30 minutes of cardio followed by 5 minutes of cooldown (see target heart rate (HR) listed in Table 19.

Cycling is recommended for cardio since it is the easiest on your body during treatment and you are not trying to burn excessive calories, as you are already losing weight. Before treatment, I would ramp up to level 12 resistance on a stationary bike during the first 10 minutes with a heart rate moving up toward 135 and then do 20 minutes at level 12 with a heart rate ending around 170 over the last 5 minutes, along with a 5-minute cooldown. During treatment, I ramped up to around level 6 to 10 and kept my heart rate steady at near 130 the entire time. Heart rate is age-dependent; I was 28 at the time of my diagnosis and treatment, although I believe there should not be as large of a drop-off with age as you will find listed on machines for recommended heart rates. I have seen men in their 70s able to work out with sustained heart rates in the 130s, for instance.

Table 19 provides a range of recommended heart rates provided by the American Heart Association. While these values serve as solid guidelines, I have listed my recommended values for pre and post-cancer cardio that are slightly more rigorous and do not drop off as precipitously with age. Also listed in Table 19 you will find recommended values to use during cancer treatment for both chemotherapy and radiation. Average values only are listed since during treatment the goal is a sustained heart rate at a moderate level versus pre and post-cancer workouts where you can push yourself toward the maximum range.

During treatment you will notice your body's fatigue prevents you from doing what you are used to so remember the goal is not to push yourself to exhaustion but rather to reduce muscle atrophy and exercise your heart. For me in particular it was mind-boosting to think that I was helping my heart fight the unwanted radiation being shot at it due to the tumor's close proximity.

Table 19: Target heart rate values.

Age	American Heart Association				Shepard Recommendation			
	Target HR Zone 50–85%			100%	Pre/Post Cancer		Chemo	Radia-tion
[yrs]	Min	Avg	Max	Max	Avg	Max	Avg	Avg
20	100	135	170	200	140	185	130	135
30	95	129	162	190	138	183	128	133
35	93	125	157	185	136	181	126	131
40	90	122	153	180	134	179	124	129
45	88	119	149	175	132	177	122	127
50	85	115	145	170	130	175	120	125
55	83	112	140	165	128	173	118	123
60	80	108	136	160	126	171	116	121
65	78	105	132	155	124	169	114	119
70	75	102	128	150	122	167	112	117

Thursday (Post-Chemo Day Two)

- 3 sets of back sled, 12 to 20 reps each (note: all exercises should have a warm-up set with an easy weight that is not included in the 3 regular sets).
- 3 sets of forearm curls, 12 to 20 reps each.
- 3 sets of shoulder press, 20 to 30 reps each.
- 3 sets of shrugs (barbell or dumbbell), 12 to 20 reps each.

Friday (Post-Chemo Day Three)

- Rest and recover.

Saturday (Post-Chemo Day Four)

- 3 sets of squats, 12 to 15 reps (note: all exercises should have a warm-up set with an easy weight that is not included in the 3 regular sets; if squats are too strenuous, do seated leg press).
- 3 sets of sit-ups (do as many as you comfortably can each set with 2 minutes of rest in between sets).

Sunday (Post-Chemo Day Five)

- 30 minutes of cardio followed by a 5-minute cooldown.

You will notice the regimen is biased toward more workout days at the end of the week (or days leading up to your treatment) since your body gets stronger the more days that go by after your treatment. Also, cardio is done the day after chemotherapy since in general cardio days should be slightly easier on the body than weight training days.

Radiation Exercise Plan

Exercising during radiation treatment differs from exercise during chemotherapy primarily due to a more consistent treatment schedule. For example, during my radiation treatment I came in daily for around five weeks. Therefore rather than peaks and valleys in my strength and energy, as is the case for chemotherapy where the drugs change weekly and your treatment is generally once a week or once every two weeks, my ability to exercise was more consistent.

The goals for exercising during radiation were similar to chemotherapy, although I found I was able to push myself slightly more during this phase (note my energy level was still far from what it normally was). As a result, the target cardio heart rates listed in Table 19 have five beats per minute (BPM) added to them for radiation versus chemotherapy treatment.

Below is an exercise routine to perform during your radiation treatment:

Monday (Radiation Day One)

- 3 sets of bench press, 12 to 20 reps each (note: all exercises should have a warm-up set with an easy weight that is not included in the 3 regular sets).
- 3 sets of lat pulldowns, 12 to 20 reps each.
- 3 sets of triceps pulldowns, 20 to 30 reps each.
- 3 sets of biceps curls (barbell or dumbbells), 12 to 20 reps each.

Tuesday (Radiation Day Two)

- 30 minutes of cardio followed by a 5-minute cooldown.

Wednesday (Radiation Day Three)

- 3 sets of back sled, 12 to 20 reps each (note: all exercises should have a warm-up set with an easy weight that is not included in the 3 regular sets).
- 3 sets of forearm curls, 12 to 20 reps each (if squats are too strenuous do seated leg press).
- 3 sets of shoulder press, 20 to 30 reps each.
- 3 sets of shrugs (barbell or dumbbells), 12 to 20 reps each.

Thursday (Radiation Day Four)

- 30 minutes of cardio followed by a 5-minute cooldown.

Friday (Radiation Day Five)

- 3 sets of squats, 12 to 15 reps (note: all exercises should have a warm-up set with an easy weight that is not included in the 3 regular sets).
- 3 sets of sit-ups (do as many as you comfortably can each set with 2 minutes of rest in between sets).

Saturday (Radiation Day Six)

- 30 minutes of cardio followed by a 5-minute cooldown.

Sunday (Radiation Day Seven)

- Rest and recover.

I recommend sticking to a stationary bicycle as the best compromise between getting a high heart rate and not overdoing it. Aim for a heart rate listed in Table 19 under the *radiation* column for your age. If you feel up for it you can substitute other machines like the elliptical; however be cautioned that this machine burns calories around twice as fast as a stationary bike so your time on the machine should be reduced from 30 to between 15 and 20 minutes, followed by a 5-minute cooldown.

While I did not try yoga during my treatment (I have done Bikram yoga and other forms periodically before and after my cancer diagnosis and treatment), it comes highly recommended by cancer patients who did. Yoga serves a similar role as weightlifting to preserve as best you can muscles that are not being used during treatment. Yoga also has the added benefit of practicing mental control and patience while executing various poses.

I have presented my experience and recommendation for exercise during cancer treatment and how to reduce the amount of muscle atrophy, but the key is to exercise and tailor your program toward what works best for you.

Chapter 17

Cancer Statistics

There is a plethora of statistics on cancer out there. However, finding them takes some work, so I will summarize pertinent data and provide references for further details specific to your cancer. The goal here is not to dwell on your chances of survival, but rather arm yourself and caregivers with information to let them know what you are up against. Any cancer can be beaten so you need to be optimistic and follow the nutritional, exercise and mental health approaches outlined in this book.

Per data collected by the American Cancer Society, 41% of males will develop cancer in their lifetime and 23% will die of cancer. For women, 38% will develop cancer in their lifetime and 19% will die. This shows just how significant cancer plays a role in our lives, either directly or indirectly through a loved one.

However there is an overall improvement in the treatment of cancers over time, which should further reinforce your optimistic outlook. Figures 75 and 76 show the mortality rates for cancers that cause the most deaths in men and women. This data from the American Cancer Society is dependent on both diagnosis and mortality rates, so while indicative, does not isolate the efficacy of cancer treatment with time. It does show that through a combination of preventative steps and treatment, the overall cancer mortality rates have decreased over the last few decades.

Specifically lung cancer, which has the highest mortality rate in the United States, has seen a 51% decrease in men (from 90 to 44 deaths per 100K people) and 21% decrease in women (from 42 to 36 deaths per 100K people). This is primarily due to the reduction in the number of individuals smoking cigarettes (especially in men) with a secondary cause due to improved cancer treatments.

Prostate cancer has the second highest cancer mortality among men, with a rate (up through 2017) of 19 per 100K people. Rates decreased by 53% over twenty years from a high of 40 per 100K people in 1992. This can be attributed both to increased screening and awareness, as well as an improvement in cancer treatment. Note that prostate cancer is diagnosed at a higher rate than lung cancer in men (21.4 vs. 13.0% of all male cancer diagnoses projected for 2020 per Table 20); however, the survival rate is much higher, leading to an overall mortality rate that lags lung cancer (lagging in mortality rate is positive, meaning fewer deaths).

For women, breast cancer has the second largest mortality rate and like prostate and lung cancer, has an appreciable decrease over the last few decades. With a peak value of 33 per 100K in 1987 (the most recent peak, with the absolute peak from 1930–2017 being 35 per 100K in 1948), as of 2017 this value has fallen to 20 per 100K, a decrease of 42%. Like prostate cancer, breast cancer far exceeds lung cancer in terms of diagnosis rate among women (30.3 vs. 12.3% of female cancer diagnoses projected for 2020); however due to the higher survival rate, breast cancer lags lung cancer in overall mortality rate. The decrease in mortality can be attributed both to improved screening methods and awareness as well as improvements in cancer treatment.

All other cancers in Figures 75 and 76 show decreased mortality rates over time, apart from pancreatic and liver cancer (uterus cancer also has a small upward trend over the last eight years, but less pronounced). These two cancers have gone up slightly over time, which is an indicator of a higher frequency of occurrence in both as medical treatment only improves over time. While pancreatic cancer mortality rates are only slightly increased and have been fairly steady over the past 40 years in both men and women, liver cancer mortality rates show a significant upward trend over that time period, especially in men. This may be an indicator of alcoholic consumption rates increasing over time.

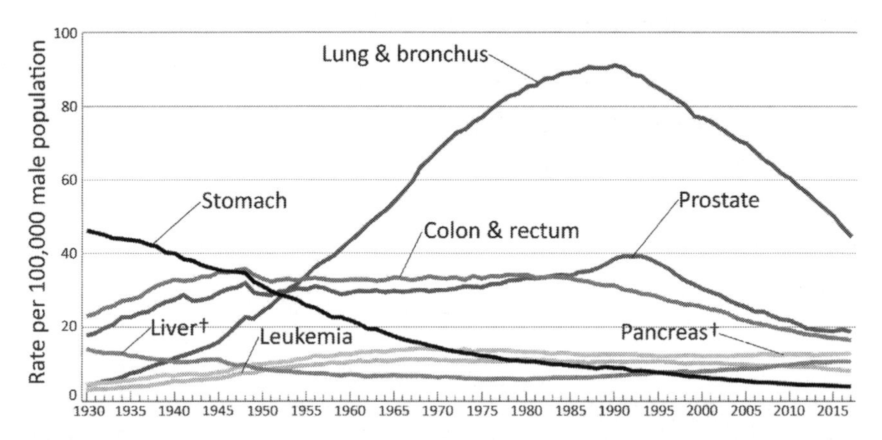

*Per 100,000, age adjusted to the 2000 US standard population. †Mortality rates for pancreatic and liver cancers are increasing. Note: Due to changes in ICD coding, numerator information has changed over time. Rates for cancers of the liver, lung and bronchus, and colon and rectum are affected by these coding changes. **Source**: US Mortality Volumes 1930 to 1959, US Mortality Data 1960 to 2017, National Center for Health Statistics, Centers for Disease Control and Prevention.

Figure 75: US male cancer mortality rates, 1930–2017.[110]

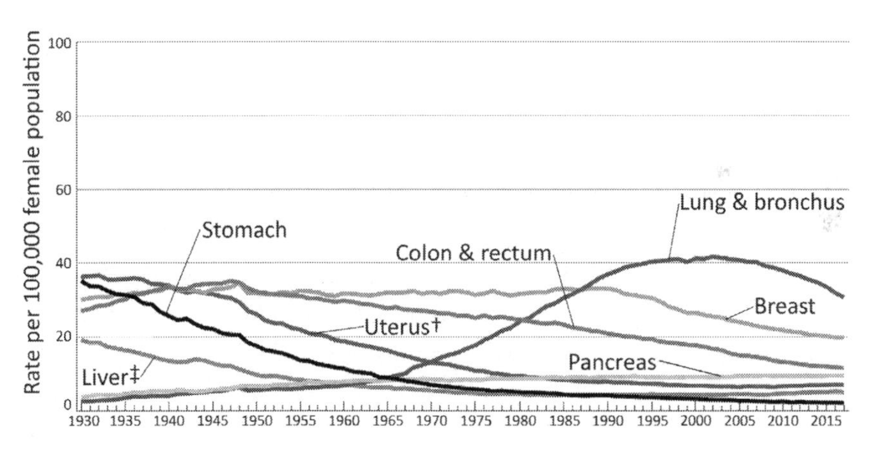

*Per 100,000, age adjusted to the 2000 US standard population. Rates exclude deaths in Puerto Rico and other US territories. †Uterus refers to uterine cervix and uterine corpus combined. ‡The mortality rate for liver cancer is increasing. Note: Due to changes in ICD coding, numerator information has changed over time. Rates for cancers of the liver, lung and bronchus, colon and rectum, and uterus are affected by these coding changes. **Source**: US Mortality Volumes 1930 to 1959, US Mortality Data 1960 to 2017, National Center for Health Statistics, Centers for Disease Control and Prevention.

Figure 76: US female cancer mortality rates, 1930–2017.[111]

110 "Cancer Facts and Figures 2020," *American Cancer Society, Inc.*

111 "Cancer Facts and Figures 2020," *American Cancer Society, Inc.*

DIAGNOSIS RATES

Table 20 shows how many new cancer cases are estimated to be diagnosed in the United States during 2020, broken down by major cancer types for both men and women. Projections are based on incidence data reported by the National Cancer Institute's (NCI's) Surveillance, Epidemiology and End Results (SEER) registry from 2000–2016. This list shows the top ten diagnosed cancers only, with the many additional forms wrapped into the remainder.

The top three diagnosed cancers for men are prostate cancer at 21.4%, lung and bronchus cancer at 13.0%, and colon and rectum cancer at 8.8%. For women, breast cancer dominates at 30.3% followed by lung and bronchus cancer at 12.3% and colon and rectum cancer at 7.6%. As of January 2020, the US population is around 329.2 million, which allows us to compute the percentage of the US population diagnosed with a specific type of cancer each year. This comes out to roughly 0.27% for men and 0.28% for women, for a combined value of 0.54%. Although this may seem low, remember this is per year and it is compared against the entire US population, which includes infants, children and adolescents.

Table 21 provides statistics on diagnosis rates by age group for both males and females. As expected, cancer diagnoses increase with age, with a 3.5% chance of men being diagnosed before 50, 6.2% chance in your 50s, 13.3% chance in your 60s and 32.7% chance if you are over 70. The overall chance of males developing cancer over their lifetime is 40.1%. Note that the percentage by age group is not additive to get the percentage over your lifetime since death rate must be considered. For women, the numbers are a 5.8% chance before 50, 6.4% chance in your 50s, 10.2% chance in your 60s and 26.7% chance over 70. At 38.7%, the overall likelihood of a woman developing cancer over her lifetime is slightly less than her male counterpart's value of 40.1%.

This data also demonstrates that women are 66% more likely to develop cancer before 50 than men, while men are 30% more likely to develop cancer in their 60s and 22% more likely when over 70. Other interesting takeaways from the data are that prostate cancer increases dramatically for men over 50, which is why screening is so important. For women, breast cancer has a steadier rise in incident rate with age, meaning screening should start rather early (most doctors recommend around age 40).

Table 20: Projected percentage of US population that will be diagnosed with cancer in 2020.[112]

Males

Cancer Type	Diagnosed	% of Cancers
Prostate	191,360	21.4
Lung & Bronchus	116,300	13.0
Colon & Rectum	78,300	8.8
Urinary Bladder	62,100	6.9
Melanoma of the Skin	60,190	6.7
Kidney & Renal Pelvis	45,520	5.1
Non-Hodgkin's Lymphoma	42,380	4.7
Oral Cavity & Pharynx	38,380	4.3
Leukemia	35,470	4.0
Pancreas	30,400	3.4
Other	193,350	21.6
Total	**893,660**	**100.0**

Females

Cancer Type	Diagnosed	% of Cancers
Breast	276,480	30.3
Lung & Bronchus	112,520	12.3
Colon & Rectum	69,650	7.6
Uterine Corpus	65,620	7.2
Thyroid	40,170	4.4
Melanoma of the Skin	40,160	4.4
Non-Hodgkin's Lymphoma	34,860	3.8
Kidney & Renal Pelvis	28,230	3.1
Pancreas	27,200	3.0
Leukemia	25,060	2.7
Other	192,980	21.1
Total	**912,930**	**100.0**

Bold cancers are in the top ten for men or women only, whereas non-bold cancers are in the top ten for both men and women. Estimates are rounded to the nearest 10 and cases exclude basal cell and squamous cell skin cancers and in-situ carcinoma except urinary bladder. Estimates do not include Puerto Rico or other US territories. Ranking is based on modeled projections and may differ from the most recent observed data.

112 "Cancer Facts and Figures 2020," *American Cancer Society, Inc.*

Table 21: Probability [%] of developing cancer based on age and sex in the US, 2014–2016.*[113]

Cancer Type	Sex	Birth to 49	50 to 59	60 to 69	70 and Older	Birth to Death
All Sites†	Male	3.5 (1/29)	6.2 (1/16)	13.3 (1/8)	32.7 (1/3)	40.1 (1/2)
	Female	5.8 (1/17)	6.4 (1/16)	10.2 (1/10)	26.7 (1/4)	38.7 (1/3)
Breast	Female	2.0 (1/49)	2.4 (1/42)	3.5 (1/28)	7.0 (1/14)	12.8 (1/8)
Colon & Rectum	Male	0.4 (1/262)	0.7 (1/143)	1.1 (1/90)	3.3 (1/30)	4.4 (1/23)
	Female	0.4 (1/274)	0.5 (1/190)	0.8 (1/ 126)	3.0 (1/33)	4.1 (1/25)
Kidney & Renal Pelvis	Male	0.2 (1/415)	0.4 (1/266)	0.7 (1/153)	1.4 (1/74)	2.2 (1/46)
	Female	0.2 (1/661)	0.2 (1/551)	0.3 (1/317)	0.7 (1/136)	1.2 (1/82)
Leukemia	Male	0.3 (1/391)	0.2 (1/550)	0.4 (1/249)	1.5 (1/69)	1.9 (1/54)
	Female	0.2 (1/499)	0.1 (1/838)	0.2 (1/433)	0.9 (1/109)	1.3 (1/77)
Lung & Bronchus	Male	0.1 (1/730)	0.6 (1/158)	1.8 (1/57)	6.0 (1/17)	6.7 (1/15)
	Female	0.2 (1/659)	0.6 (1/169)	1.4 (1/70)	4.8 (1/21)	6.0 (1/17)
Melanoma of the Skin‡	Male	0.4 (1/228)	0.5 (1/197)	0.9 (1/109)	2.6 (1/38)	3.6 (1/28)
	Female	0.6 (1/156)	0.4 (1/245)	0.5 (1/194)	1.2 (1/86)	2.5 (1/41)
Non-Hodgkin's Lymphoma	Male	0.3 (1/367)	0.3 (1/340)	0.6 (1/176)	1.9 (1/53)	2.4 (1/41)
	Female	0.2 (1/529)	0.2 (1/463)	0.4 (1/238)	1.4 (1/72)	1.9 (1/52)
Prostate	Male	0.2 (1/441)	1.8 (1/57)	4.7(1/21)	8.2 (1/12)	11.6 (1/9)
Thyroid	Male	0.2(1/449)	0.1 (1/694)	0.2 (1/558)	0.2 (1/405)	0.7 (1/144)
	Female	0.9 (1/112)	0.4 (1/252)	0.4 (1/273)	0.4 (1/251)	1.9 (1/52)
Uterine Cervix	Female	0.3 (1/367)	0.1 (1/831)	0.1 (1/921)	0.2 1/595)	0.6 (1/159)
Uterine Corpus	Female	0.3 (1/323)	0.6 (1/157)	1.0 (1/95)	1.5 (1/69)	3.1 (1/33)

*For those who are free of cancer at the beginning of each age interval. †All sites exclude basal and squamous cell skin cancers and in-situ cancers except urinary bladder. ‡Statistic is for non-Hispanic whites. Note: The probability of developing cancer for additional sites, as well as the probability of cancer death, can be found in Supplemental Data at cancer.org/research/cancerfactsstatistics/index. **Source**: DevCan: Probability of Developing or Dying of Cancer Software, Version 6.7.7. Statistical Research and Applications Branch, *National Cancer Institute,* 2019. surveillance.cancer .gov/devcan.

Table 22 breaks down cancer statistics by race between whites, blacks, Asians/Pacific Islanders, Native Americans and Hispanics/Latinos. These are the amount of diagnosed cancers between the years 2012 and 2016 per 100,000 in that ethnicity's population. Men of black ethnicity have the highest cancer rate at 0.54% over the five-year range, followed by white males at 0.50%. Native Americans are third among males at 0.40%, followed by Hispanics/Latinos at 0.37%.

113 "Cancer Facts and Figures 2020," *American Cancer Society, Inc.*

Table 22: US cancer incidence rates* based on sex and ethnicity, 2012–2016. [114]

Cancer Type	All races	Non-Hispanic White	Non-Hispanic Black	Asian/ Pacific Islander	American Indian/ Alaska Native†	Hispanic/ Latino
All Sites	448.4	464.6	460.4	288.4	380.7	346.4
Male	489.4	501.2	540	292.3	399.2	372.9
Female	421.1	440.7	407.2	289.5	370.9	333.4
Breast (Female)	125.2	130.8	126.7	93.3	94.7	93.9
Colon & Rectum	38.7	38.6	45.7	30	43.3	34.1
Male	44.4	44	53.8	35.3	48.5	40.8
Female	33.9	33.9	39.9	25.7	39.1	28.7
Kidney & Renal Pelvis	16.6	16.8	18.7	7.7	23.1	16.4
Male	22.5	22.8	25.9	11	29.7	21.6
Female	11.5	11.5	13.2	5.1	17.5	12.2
Liver & Intrahepatic Bile Duct	8.3	6.9	10.9	12.7	15.1	13.4
Male	12.7	10.5	17.9	19.4	21.6	20
Female	4.4	3.7	5.4	7.3	9.4	7.8
Lung & Bronchus	59.3	63.5	62.4	34.4	53.6	30.2
Male	69.3	72.4	82.7	43.5	60.1	37.9
Female	51.7	56.7	48.6	27.6	48.8	24.6
Prostate	104.1	97.1	173	52.9	68	86.8
Stomach	6.6	5.4	10.1	10.3	8.8	9.6
Male	9	7.6	13.9	13.3	11.6	12.1
Female	4.6	3.5	7.5	7.9	6.7	7.7
Uterine Cervix	7.6	7.1	9.1	6	8.7	9.6

Hispanic origin is not mutually exclusive from Asian/Pacific Islander or American Indian/Alaska Native. *Rates are per 100,000 population and age adjusted to the 2000 US standard population and exclude data from Puerto Rico. †Data based on Purchased/Referred Care Delivery Area (PRCDA) counties. **Source**: Incidence – *North American Association of Central Cancer Registries*, 2019. Mortality – National Center for Health Statistics, Centers for Disease Control and Prevention, 2019.

114 "Cancer Facts and Figures 2020," *American Cancer Society, Inc.*

Asian and Pacific Islanders are the least likely to develop cancer, with a 0.29% chance for males. Trends are similar for females, except for white females being more likely than black females to develop cancer at 0.44 versus 0.41%. Both remain below their male counterparts. Female Native Americans are next at 0.37%, followed by Hispanic/Latino women at 0.33 and finally Asian and Pacific Islander women at 0.29%, the least likely of either sex or ethnicity to develop cancer.

Theories as to these differences are discussed in *Cancer Facts and Figures 2020*, by The American Cancer Society, Inc. and include factors such as Asians and Hispanics smoking less, thereby decreasing their lung cancer rates. Also, Asians and Hispanics have genetic factors that likely reduce their risk of developing breast cancer. While not part of the overall cancer rate trends, differences in individual cancer rate types are discussed, such as immigrant Asians and Hispanics having higher stomach cancer rates because of poorer standards of living in their native countries.

Tables 23 and 24 provide projected cancer incident rates by state for 2020. For the most part this data falls as expected, with more populous states such as California, Florida, Texas and New York having the highest quantities. In order to examine the data closer, the number of cancers diagnosed must be compared against the population of the state, which is provided in Table 24.

All states are relatively similar with a minimum of 0.37% cancer diagnoses per state population (Utah), a maximum of 0.70% (Florida) and an average of 0.56% across all states; however when compared against one another, these differences are significant and interesting. For instance, in comparing these two extremes, you are 88% more likely to get cancer if you live in Florida versus Utah. Before everyone from Florida gets carried away and moves to Utah, it is worth additional research to understand why these trends exist.

The top five most likely states to get cancer (in order) are Florida, Delaware, West Virginia, Pennsylvania and South Carolina. Note that four out of five of these states are on the East Coast, with West Virginia being close to the East Coast. If you continue for six through ten, all five are also on the East Coast or in the Midwest (in order Michigan, Ohio, Missouri, Maine and Wisconsin). Further study would be of interest to determine what factors in these regions of the country contribute to higher cancer rates. For instance the weather is colder in many of these climates, but there are other regions of the country (such as Alaska) that are cold where cancer rates are low. A larger percentage of whites (particularly in the New England and Midwest regions) live in these regions and whites have higher incidence rates than Hispanics/Latinos, Asian/Pacific Islanders and Native Americans/Alaska Natives. This part of the country also tends to have a higher smoking rate and restaurants that allow smoking when compared to the West Coast.

When looking at the healthy end of the spectrum, the states with the lowest cancer rates are (in order) Utah, Alaska, California, Texas, New Mexico, Washington, Hawaii, Colorado, Idaho and Wyoming. No East Coast or Midwest state shows up in this list of top 10, with only Georgia (12th), the District of Columbia (13th) and Massachusetts (16th) falling in the top 20. The states with lower cancer diagnosis rates lean toward healthier lifestyles (diet, exercise and no smoking) and the states in general have greener laws when it comes to pollution. A much further in-depth study should be made to examine these hypotheses and how to reduce the chances of getting cancer.

Worldwide, there is a plethora of data summarized in *Global Cancer Facts & Figures, 4th edition*, by the American Cancer Society, Inc. that readers are encouraged to study, especially if you live outside of the United States. Figure 77 shows an excerpt from this reference summarizing the most common cancers diagnosed in each country around the globe. For men in North and South America, Australia, Western Europe and large portions of Central and South Africa, prostate cancer dominates. In Russia, China, Eastern Europe and Northern Africa, lung, bronchus and trachea cancers are the largest offenders. Other cancers such as liver, stomach, oral cavity, sarcoma, colon, esophagus and leukemia show up in pockets around the globe.

The segregation and geographic groupings make it likely there are valid explanations for the distribution. Without additional research into the topic, however, it is difficult to make conclusions. Countries such as Russia and China do have a larger percentage of the population smoking than the United States, with Russia having 40.9% of adults 15 years of age or older smoking in 2016, China at 25.2% and the United States at 21.9% (data courtesy of the World Health Organization).[115] However Chile shows up at 37.9% (tenth in the world), yet prostate cancer remains

115 "Prevalence of Tobacco Smoking," *World Health Organization*, https://apps.who.int/gho/data/node.sdg.3-a-viz?lang=en.

more prevalent than lung cancer. Therefore while smoking may partially tell the story, additional factors such as diet, cancer treatment efficacy, environmental pollution and cancer screening play a role.

For women the trend is much more straightforward. Breast cancer dominates most of the world, followed by cervical cancer and high rates of lung, thyroid and liver cancer in North Korea, South Korea and Mongolia, respectively. It is not clear why these few countries differ from the rest of the world, but the same variables as discussed above likely play a role. Some suggest Eastern Asian women have lower rates of breast cancer due to a lower fat diet; however, the majority of Eastern Asian countries such as China, Vietnam and Japan have breast cancer as their leading cancer and there are thus likely multiple factors that play a role.

Table 23: Cancer type diagnosis by state, 2020 projections.[116]

State	All Sites	Female Breast	Uterine Cervix	Colon & Rectum	Uterine Corpus	Leukemia	Lung & Bronchus	Melanoma of the Skin	Non-Hodgkin's Lymphoma	Prostate	Urinary Bladder
AL	28,570	4,120	240	2,460	780	810	4,230	1,550	1,000	3,530	1,090
AK	2,960	510	†	320	120	90	400	120	120	340	160
AZ	36,730	5,630	260	3,010	1,240	990	4,200	2,380	1,500	3,830	1,810
AR	17,200	2,430	140	1,540	500	630	2,760	800	650	1,860	760
CA	172,040	30,650	1,630	15,530	7,030	6,060	18,040	10,980	8,200	20,160	7,780
CO	27,290	4,530	190	2,040	920	910	2,550	1,920	1,150	3,140	1,250
CT	20,300	3,590	130	1,520	910	400	2,650	1,110	930	2,320	1,080
DE	6,660	960	†	470	220	230	890	420	260	770	320
D.C.	3,600	510	†	250	120	110	300	90	130	370	80
FL	150,500	19,900	1,130	11,310	4,460	3,370	18,150	8,750	7,170	13,950	6,780
GA	55,190	8,340	440	4,660	1,710	1,550	7,240	3,190	2,280	6,840	2,110
HI	6,800	1,300	60	730	330	230	870	520	290	700	300
ID	8,540	1,340	60	730	310	340	990	740	390	1,160	470
IL	71,990	11,020	540	6,240	2,850	2,400	9,210	3,700	2,920	8,000	3,310
IN	37,940	5,410	270	3,410	1,430	1,290	5,700	2,370	1,590	3,570	1,720
IA	18,460	2,710	110	1,600	700	840	2,440	1,150	800	1,920	870
KS	16,170	2,390	110	1,320	560	620	2,020	890	650	1,730	640
KY	26,500	3,800	200	2,440	870	920	4,890	1,330	1,040	2,440	1,130
LA	26,480	3,910	260	2,370	690	930	3,700	1,030	1,110	2,970	1,050
ME	8,180	1,370	50	670	390	160	1,430	520	390	800	520
MD	34,710	5,500	250	2,570	1,300	820	3,930	1,780	1,330	4,410	1,360

116 "Cancer Facts and Figures 2020," *American Cancer Society, Inc.*

State	All Sites	Female Breast	Uterine Cervix	Colon & Rectum	Uterine Corpus	Leukemia	Lung & Bronchus	Melanoma of the Skin	Non-Hodgkin's Lymphoma	Prostate	Urinary Bladder
MA	36,990	6,690	220	2,650	1,630	580	5,150	2,190	1,670	3,890	1,970
MI	61,770	8,800	360	4,620	2,380	2,060	8,140	3,290	2,450	6,820	2,890
MN	33,210	4,670	140	2,320	1,200	1,600	3,580	1,750	1,350	2,880	1,460
MS	17,190	2,390	160	1,730	450	500	2,510	620	570	2,050	630
MO	37,540	5,360	270	3,090	1,290	1,370	5,540	1,820	1,410	3,540	1,580
MT	5,850	960	†	500	220	250	770	450	250	680	330
NE	10,560	1,580	70	940	390	480	1,270	610	450	980	470
NV	16,540	2,310	130	1,480	480	520	1,850	840	650	1,780	780
NH	8,060	1,350	†	590	370	180	1,220	530	370	910	510
NJ	53,340	8,260	440	4,250	2,240	2,100	6,100	2,770	2,340	6,010	2,640
NM	9,800	1,570	80	890	370	340	1,040	610	410	920	410
NY	117,910	17,540	930	8,910	4,840	4,600	13,370	4,980	5,120	11,470	5,590
NC	59,620	9,340	430	4,540	2,030	1,640	8,470	3,680	2,480	7,200	2,510
ND	4,060	590	†	360	140	190	460	230	170	400	200
OH	71,850	10,350	440	5,910	2,790	2,280	10,110	4,100	2,820	7,030	3,190
OK	20,530	3,130	170	1,870	620	860	3,200	940	860	2,130	920
OR	23,330	3,880	160	1,740	910	740	2,930	1,730	1,000	2,470	1,150
PA	80,240	12,180	530	6,520	3,390	3,050	10,710	4,410	3,480	8,300	4,350
RI	5,930	1,020	†	430	260	100	920	340	270	650	320
SC	31,710	4,790	230	2,550	970	1,220	4,460	1,900	1,300	3,390	1,270
SD	4,960	720	†	430	170	230	590	270	200	520	240
TN	39,360	5,760	330	3,540	1,220	1,280	6,300	2,110	1,580	3,990	1,700
TX	129,770	19,590	1,410	11,430	4,120	5,260	14,830	4,530	5,650	12,110	4,590
UT	11,900	1,780	80	840	450	500	730	1,230	550	1,380	460
VT	3,740	630	†	270	170	90	570	270	170	330	210
VA	47,550	7,410	320	3,530	1,660	1,370	5,960	2,920	1,940	6,200	2,010
WA	36,290	6,690	250	2,970	1,480	1,430	4,790	2,800	1,740	4,040	1,930
WV	12,380	1,680	80	1,040	440	480	2,030	680	500	1,110	620
WI	35,280	5,120	200	2,540	1,410	1,420	4,290	2,190	1,460	3,560	1,740
WY	2,880	430	†	260	100	110	320	220	120	400	150
US Total	1,806,590	276,480	13,800	147,950	65,620	60,530	228,820	100,350	77,240	191,930	81,400

*Rounded to the nearest 10. Excludes basal and squamous cell skin cancers and in-situ carcinomas except urinary bladder. Estimates for Puerto Rico are unavailable. †Estimate is fewer than 50 cases. These estimates are offered as a rough guide and should be interpreted with caution. State estimates may not sum to US total due to rounding and exclusion of state estimates fewer than 50 cases. Note: Estimated cases for additional cancer sites by state can be found in Supplemental Data at cancer.org/statistics or via the Cancer Statistics Center (cancerstatisticscenter.cancer.org).

Table 24: Projected 2020 cancer statistics by state.[117]

Pop. Rank	State	Population	Diagno-ses	Deaths	Diagno-ses	Deaths	Survival	Diag. Rank	Death Rank	Surv. Rank
[-]	[-]	[-]	[-]	[-]	[%]	[%]	[%]	[-]	[-]	[-]
1	CA	39,512,223	172,040	60,660	0.44	0.15	63.7	3	6	21
2	TX	29,206,997	129,770	41,810	0.44	0.14	63.5	4	2	24
3	FL	21,477,737	150,500	45,300	0.70	0.21	N/A*	51	37	N/A
4	NY	19,453,561	117,910	34,710	0.61	0.18	65.8	41	16	7
5	PA	12,801,989	80,240	27,860	0.63	0.22	64.7	48	43	15
6	IL	12,671,821	71,990	24,220	0.57	0.19	65.5	26	25	8
7	OH	11,689,442	71,850	25,380	0.61	0.22	63.2	45	42	26
8	GA	10,617,423	55,190	17,990	0.52	0.17	62.1	13	9	36
9	NC	10,488,084	59,620	20,410	0.57	0.19	64.5	29	27	16
10	MI	9,986,857	61,770	21,000	0.62	0.21	63.3‡	46	36	25
11	NJ	8,882,190	53,340	15,710	0.60	0.18	65.3	40	13	12
12	VA	8,535,519	47,550	15,220	0.56	0.18	N/A*	22	15	N/A
13	WA	7,614,893	36,290	13,020	0.48	0.17	66.4	6	10	2
14	AZ	7,278,717	36,730	12,580	0.50	0.17	62.5	11	11	30
15	MA	6,902,149	36,990	12,430	0.54	0.18	N/A*	16	22	N/A
16	TN	6,770,010	39,360	14,780	0.58	0.22	63.6	33	44	23
17	IN	6,691,878	37,940	13,630	0.57	0.20	62.4	25	33	32
18	MO	6,126,452	37,540	13,010	0.61	0.21	62.8	44	39	27
19	MD	6,042,718	34,710	10,790	0.57	0.18	65.4	31	17	10
20	WI	5,813,568	35,280	11,610	0.61	0.20	65.5	42	30	9
21	CO	5,695,564	27,290	8,220	0.48	0.14	65.9	8	3	5
22	MN	5,611,179	33,210	10,040	0.59	0.18	N/A*	36	20	N/A
23	SC	5,084,127	31,710	10,780	0.62	0.21	62.4	47	38	34
24	AK	4,887,871	28,570	10,530	0.58	0.22	61.8	34	41	38

117 "Cancer Facts and Figures 2020," *American Cancer Society, Inc.* was used for diagnosis and death data; Population data is from "Annual Estimates of the Resident Population for the United States, Regions, States and Puerto Rico: April 1, 2010 to July 1, 2019 (NST-EST2019-01)," US Census Bureau, Population Division. Survival data is from "Cancer in North America: 2011-2015 Volume Four: Cancer Survival in the United States and Canada 2008-2014, pg. 49 and 50," North American Association of Central Cancer Registries (NAACCR), Inc.

Pop. Rank [-]	State [-]	Population [-]	Diagnoses [-]	Deaths [-]	Diagnoses [%]	Deaths [%]	Survival [%]	Diag. Rank [-]	Death Rank [-]	Surv. Rank [-]
25	LA	4,659,978	26,480	9,300	0.57	0.20	61.3	28	29	39
26	KY	4,468,402	26,500	10,540	0.59	0.24	62.6	37	49	29
27	OR	4,190,713	23,330	8,280	0.56	0.20	65.1	21	28	14
28	OK	3,943,079	20,530	8,430	0.52	0.21	61.9	14	40	37
29	CT	3,572,665	20,300	6,390	0.57	0.18	66.1	27	19	3
30	UT	3,193,694	11,900	3,350	0.37	0.10	N/A*	1	1	N/A
N/A	PR	3,205,958	N/A	N/A	N/A	N/A	64.3	N/A	N/A	17
31	IA	3,156,145	18,460	6,440	0.58	0.20	63.6	35	34	22
32	NV	3,034,392	16,540	5,460	0.55	0.18	N/A*	17	21	N/A
33	AR	3,013,825	17,200	6,730	0.57	0.22	62.4	30	46	31
34	MS	2,986,530	17,190	6,700	0.58	0.22	60.8	32	47	42
35	KS	2,911,505	16,170	5,520	0.56	0.19	N/A*	20	24	N/A
36	NM	2,095,428	9,800	3,730	0.47	0.18	61.3	5	14	40
37	NE	1,929,268	10,560	3,520	0.55	0.18	64.0	18	23	19
38	WV	1,805,832	12,380	4,750	0.69	0.26	62.3	49	51	35
39	ID	1,754,208	8,540	3,100	0.49	0.18	64.2	9	12	18
40	HI	1,420,491	6,800	2,540	0.48	0.18	62.6	7	18	28
41	NH	1,356,458	8,060	2,830	0.59	0.21	67.0	38	35	1
42	ME	1,338,404	8,180	3,350	0.61	0.25	65.3	43	50	11
43	MT	1,062,305	5,850	2,140	0.55	0.20	62.4	19	32	33
44	RI	1,057,315	5,930	2,120	0.56	0.20	65.9	23	31	6
45	DE	967,171	6,660	2,130	0.69	0.22	63.9	50	45	20
46	SD	882,235	4,960	1,690	0.56	0.19	N/A*	24	26	N/A
47	ND	760,077	4,060	1,260	0.53	0.17	N/A*	15	7	N/A
48	AK	737,438	2,960	1,090	0.40	0.15	60.8†	2	5	41
49	DC	702,455	3,600	1,020	0.51	0.15	N/A*	12	4	N/A
50	VT	626,299	3,740	1,450	0.60	0.23	66.0†	39	48	4
51	WY	577,737	2,880	960	0.50	0.17	65.2†	10	8	13

*State/district not included in source survey. †Age-standardized data unavailable so unstandardized data used. ‡Data is for Detroit only.

Most common cancer site

- Breast
- Cervix uteri
- Colon, rectum & anus
- Esophagus
- Lung, bronchus & trachea
- Leukemia
- Lip, oral cavity
- Liver
- Kaposi sarcoma
- Non-Hodgkin lymphoma
- Prostate
- Stomach
- Thyroid
- No data

Source: GLOBOCAN 2018.

Figure 77: World cancer prevalence by type in 2018.[118]

118 "Global Cancer Facts & Figures, 4th Edition," *American Cancer Society, Inc.*

SURVIVAL AND DEATH RATES

Cancer survival rates are shown in Tables 25 and 26 as well as in Appendix A. To reiterate, the purpose here is not to get caught up in what your chances are or bring about depression, but rather to get a feel for what you are up against and compare with what you are told by your oncologist team. Table 25 shows cancer survival rates by stage of diagnosis for the most common forms of cancer. Note that a five-year survival rate is standard to measure the effectiveness of cancer treatments. Beyond five years is also important; however, other health factors come into play. This, coupled with the fact that many cancer patients are older, make five-year survivability a convenient barometer.

Table 25: US 5-year cancer survival rates* by cancer type, 2009–2015.[119]

Cancer Type	All Stages	Local	Regional	Distant	Cancer Type	All Stages	Local	Regional	Distant
Breast (Female)	90	99	86	27	Oral Cavity & Pharynx	65	84	66	39
Colon & Rectum	64	90	71	14	Ovary	48	92	75	29
Colon	63	90	71	14	Pancreas	9	37	12	3
Rectum	67	89	71	15	Prostate	98	>99	>99	31
Esophagus	20	47	25	5	Stomach	32	69	31	5
Kidney†	75	93	70	12	Testis	95	99	96	73
Larynx	60	77	45	33	Thyroid	98	>99	98	56
Liver‡	18	33	11	2	Urinary Bladder§	77	70	36	5
Lung & Bronchus	19	57	31	5	Uterine Cervix	66	92	56	17
Melanoma of the Skin	92	99	65	25	Uterine Corpus	81	95	69	17

*Rates are adjusted for normal life expectancy and are based on cases diagnosed in the SEER 18 areas from 2009–2015, all followed through 2016. †Includes renal pelvis. ‡ Includes intrahepatic bile duct. §Rate for in-situ cases is 96%. **Local**: an invasive malignant cancer confined entirely to the organ of origin. **Regional**: a malignant cancer that 1) has extended beyond the limits of the organ of origin directly into surrounding organs or tissues; 2) involves regional lymph nodes; or 3) has both regional extension and involvement of regional lymph nodes. **Distant**: a malignant cancer that has spread to parts of the body remote from the primary tumor either by direct extension or by discontinuous metastasis to distant organs, tissues, or via the lymphatic system to distant lymph nodes. **Source**: Howlader N, Noone AM, Krapcho M, et al (eds). SEER Cancer Statistics Review, 1975–2016, National Cancer Institute, Bethesda, MD, https://seer.cancer.gov/csr/1975_2016/, based on November 2018 SEER data submission, posted to the SEER website, April 2019.

119 "Cancer Facts and Figures 2020," *American Cancer Society, Inc.*

Observe that there is a large drop-off for all cancers once they have metastasized. For example, breast cancer has a very high five-year survival rate of 99% when the cancer is local. When the cancer is regional (e.g. it has spread to another adjacent organ or tissue), the survival rate drops to 86%. Once metastasized and spread to far away organs, tissues, lymph nodes or bone marrow, the rate drops precipitously to 27%. This highlights the importance of detecting cancers as early as possible, which is why specifically for breast cancer, screening starts around the age of 40 unless you are considered high-risk, at which point screening may start earlier.

Prostate cancer is also worth examining due to it being the highest cancer for men. At 99% for both local and regional, prostate cancer has a very high treatment success rate. However once metastasized, the value falls to 31%. Unfortunately, Hodgkin's and non-Hodgkin's lymphoma are not included in the data so I am unable to state what my five-year survival rate was based on staging, although average values regardless of stage are in Appendix A.

Table 26 demonstrates five-year survival rates in three distinct periods of time for all races combined, whites and blacks. Time periods where data were taken are broken into the first range (1975–1977), second range (1987–1989) and the most recent range (2008–2015). Almost all cancers summarized in this table show improvements in survival rates, which is very positive! The exceptions are larynx cancer and cancers of the uterus, which have seen decreases of between 1 and 8% for all races (although the uterine cervix, as opposed to the uterine corpus, is unchanged for all sights and has a small increase for whites). While ovarian cancer fits the trend of increased efficacy for treatment for all races and the subset of whites, it did decrease slightly for blacks. Also of note for blacks is melanoma, which increased by 39% from the 1975–1977 data to the 1987–1989 data, only to reduce by 11% from the latter range to the 2008–2015 data.

Melanoma for blacks is particularly singled out as having between a 5 and 10% error, which may minimize this surprising decrease, but not enough to make it an increase. However, the majority of trends are relatively consistent for white and black ethnicities with similar improvements over time. Although there remains much work to do in order to continue improving cancer survival rates, these trends show the payoff of modern science and medicine with improved treatments leading to higher survival rates.

Table 26: US 5-year cancer survival rates* [%] by race, 1975–2015.[120]

Cancer Type	All Races			White			Black		
	1975-77	1987-89	2008-15	1975-77	1987-89	2008-15	1975-77	1987-89	2008-15
All Sites	49	55	69	50	57	70	39	43	64
Brain & Other Nervous System	23	29	34	22	28	33	25	32	40
Breast (female)	75	84	91	76	85	92	62	71	83
Colon & Rectum	50	60	66	50	60	67	45	52	60
Colon	51	60	65	51	61	66	45	52	56
Rectum	48	58	69	48	59	69	44	52	68
Esophagus	5	9	21	6	11	23	4	7	12
Hodgkin's Lymphoma	72	79	89	72	80	89	70	72	85
Kidney & Renal Pelvis	50	57	76	50	57	75	49	55	77
Larynx	66	66	62	67	67	64	58	56	50
Leukemia	34	43	66	35	44	67	33	35	60
Liver Intrahepatic Bile Duct	3	5	20	3	6	19	2	3	16
Lung & Bronchus	12	13	21	12	13	21	11	11	19
Melanoma	82	88	94	82	88	94	57†	79†	70†
Myeloma	25	27	54	24	27	53	29	30	55
Non-Hodgkin's Lymphoma	47	51	75	47	51	76	49	46	70
Oral Cavity & Pharynx	53	54	68	54	56	70	36	34	51
Ovary	36	38	48	35	38	48	42	34	40
Pancreas	3	4	10	3	3	10	2	6	9
Prostate	68	83	99	69	84	99	61	71	97
Stomach	15	20	32	14	18	31	16	19	33
Testis	83	95	97	83	96	97	73†‡	88†	93
Thyroid	92	94	99	92	94	99	90	92	97
Urinary Bladder	72	79	78	73	80	79	50	63	65
Uterine Cervix	69	70	69	70	73	71	65	57	58
Uterine Corpus	87	82	83	88	84	86	60	57	64

*Rates are adjusted for normal life expectancy and are based on cases diagnosed in the SEER 9 areas from 1975 to 1977, 1987 to 1989 and 2009 to 2015, all followed through 2016. †The standard error is between 5 and 10 percentage points. ‡Survival rate is for cases diagnosed from 1978 to 1980. **Source**: Howlader N, Noone AM, Krapcho M, et al. (eds). SEER Cancer Statistics Review, 1975–2016, National Cancer Institute, Bethesda, MD, https://seer.cancer.gov/csr/1975_2016/, based on November 2018 SEER data submission, posted to the SEER website, April 2019.

120 "Cancer Facts and Figures 2020," *American Cancer Society, Inc.*

Table 27 shows death rates for the same five ethnic groups discussed in the *Diagnosis Rates* section broken down by gender and various common forms of cancer.

Table 27: US cancer death rates* based on sex and ethnicity, 2013–2017.[121]

Cancer Type	All races	Non-Hispanic White	Non-Hispanic Black	Asian/ Pacific Islander	American Indian/ Alaska Native†	Hispanic/ Latino
All Sites	158.2	162.9	186.4	98.1	144	111.8
Male	189.3	193.8	233.2	116.4	172.6	135.6
Female	135.5	139.9	157.5	85	122.9	95.1
Breast (Female)	20.3	20.3	28.4	11.4	14.6	14
Colon & Rectum	13.9	13.8	19	9.5	15.8	11.1
Male	16.6	16.3	23.8	11.4	19.4	14.1
Female	11.7	11.7	15.6	8.1	13	8.7
Kidney & Renal Pelvis	3.7	3.8	3.7	1.7	5.5	3.4
Male	5.4	5.6	5.6	2.6	8.2	5
Female	2.3	2.4	2.3	1.1	3.4	2.2
Liver & Intrahepatic Bile Duct	6.6	5.8	8.6	9	10.6	9.3
Male	9.6	8.4	13.5	13.4	14.8	13.2
Female	4	3.5	4.9	5.6	7.1	6
Lung & Bronchus	40.2	43.4	43.5	22	33.3	17.5
Male	49.3	51.8	60.4	29	40	24.1
Female	33.2	36.8	31.9	16.8	28.3	12.6
Prostate	19.1	18	38.7	8.6	18.7	15.7
Stomach	3.1	2.3	5.4	5.1	4.8	5
Male	4.1	3.2	8	6.6	6.3	6.4
Female	2.2	1.6	3.7	4	3.6	4
Uterine Cervix	2.3	2.1	3.6	1.7	2.5	2.6

Hispanic origin is not mutually exclusive from Asian/Pacific Islander or American Indian/Alaska Native. *Rates are per 100,000 population and age adjusted to the 2000 US standard population and exclude data from Puerto Rico. Data based on Purchased/Referred Care Delivery Area (PRCDA) counties. **Source**: Incidence – North American Association of Central Cancer Registries, 2019. Mortality – National Center for Health Statistics, Centers for Disease Control and Prevention, 2019.

121 "Cancer Facts and Figures 2020," *American Cancer Society, Inc.*

Looking at the summary for all sites, black men have the highest number of deaths per 100K sample followed by whites, Native Americans, Hispanics and Latinos and finally, Asians. These same trends hold for women as well. Note that these rates are over a five-year period from 2013–2017, meaning how many individuals died of a particular cancer over that timeframe.

The number of cancer deaths projected for 2020 is presented in Table 28. While prostate cancer was shown to have the highest projected 2020 cancer diagnosis rate for men (Table 20), it falls to second in terms of projected deaths while lung and bronchus cancer moves from second in diagnosis to first in cancer deaths. Comparing these two tables provides readers further data on the efficacy of cancer treatments, where clearly in this example lung cancer is more difficult to treat than prostate cancer. For women, although breast cancer is 2.5 times more likely to occur in women than any other cancer based on the data, it projects to have only the second highest number of fatalities in the United States for 2020. Like prostate cancer for men, breast cancer for women has high success rates (when caught early).

With lung cancer being number one in fatalities for women as well as men, this needs to be a focus area for cancer researchers and lawmakers, where aggressive measures on non-smoking policies for restaurants should continue. Marijuana laws have recently been passed in many states, which may lead to a spike in lung cancer diagnosis rates in the future. Containing many of the same carcinogens and tar as cigarettes, marijuana smoke is unhealthy for the respiratory system.[122]

Table 29 summarizes diagnosis, death and survival rates for whites, blacks, Asians, Hispanics/Latinos and Native Americans in an attempt to see trends. The data demonstrates that while Asians are the least likely to be diagnosed with and die of cancer, they only have the highest survival rates for women at 71.0%. For men, the highest survival rate is for whites, with a value of 67.2%. Survival rates for Hispanics/Latinos come in second for both men and women at values of 65.5% and 70.5%, respectively. This corresponds to Hispanic/Latino men and women likewise being second in diagnosis and death rates. Black men's survival rate comes in at third with a value of 65.2% despite being last in both diagnosis and death rates.

122 "Marijuana and Lung Health," *American Lung Association*, https://www.lung.org /stop-smoking/smoking-facts/marijuana-and-lung-health.html.

Table 28: Projected percentage of US males that will die of cancer in 2020.[123]

Males

Cancer Type	Deaths	% of Cancers
Lung & Bronchus	72,500	22.6
Prostate	33,330	10.4
Colon & Rectum	28,630	8.9
Pancreas	24,640	7.7
Liver & Intrahepatic Bile Duct	20,020	6.2
Leukemia	13,420	4.2
Esophagus	13,100	4.1
Urinary Bladder	13,050	4.1
Non-Hodgkin's Lymphoma	11,460	3.6
Brain & Other Nervous System	10,190	3.2
Other	80,820	25.2
Total	**321,670**	**100.0**

Females

Cancer Type	Deaths	% of Cancers
Lung & Bronchus	63,220	22.2
Breast	42,170	14.8
Colon & Rectum	24,570	8.6
Pancreas	22,410	7.9
Ovary	13,940	4.9
Uterine Corpus	12,590	4.4
Liver & Intrahepatic Bile Duct	10,140	3.6
Leukemia	9,680	3.4
Non-Hodgkin's Lymphoma	8,480	3.0
Brain & Other Nervous System	7,830	2.7
Other	70,330	24.6
Total	**285,210**	**100.0**

Bold cancers are in the top ten for men or women only, whereas non-bold cancers are in the top ten for both men and women. Estimates are rounded to the nearest 10 and cases exclude basal cell and squamous cell skin cancers and in-situ carcinoma, except urinary bladder. Estimates do not include Puerto Rico or other US territories. Ranking is based on modeled projections and may differ from the most recent observed data.

123 "Cancer Facts and Figures 2020," *American Cancer Society, Inc.*

Table 29: Summary of cancer statistics by race and sex.[124]

	White	Black	Asian	Native American	Hispanic/ Latino
Diagnosis Rates (Table 22: US cancer incidence rates based on sex and ethnicity, 2012–2016)					
Per/100K	[-]	[-]	[-]	[-]	[-]
Male	501.2	540	292.3	399.2	372.9
Female	440.7	407.2	289.5	370.9	333.4
Percent	[%]	[%]	[%]	[%]	[%]
Male	0.50	0.54	0.29	0.40	0.37
Female	0.44	0.41	0.29	0.37	0.33
Rank	[-]	[-]	[-]	[-]	[-]
Male	4	5	1	3	2
Female	5	4	1	3	2
Death Rates (Table 27: US cancer death rates based on sex and ethnicity, 2013–2017)					
Per/100K	[-]	[-]	[-]	[-]	[-]
Male	193.8	233.2	116.4	172.6	135.6
Female	139.9	157.5	85	122.9	95.1
Percent	[%]	[%]	[%]	[%]	[%]
Male	0.19	0.23	0.12	0.17	0.14
Female	0.14	0.16	0.09	0.12	0.10
Rank	[-]	[-]	[-]	[-]	[-]
Male	4	5	1	3	2
Female	4	5	1	3	2
Survival Rates, 2009-2015[125]					
Percent	[%]	[%]	[%]	[%]	[%]
Male	67.2	65.2	61.3	56.1	65.5
Female	69.3	62.1	71.0	66.4	70.5
Rank	[-]	[-]	[-]	[-]	[-]
Male	1	3	4	5	2
Female	3	5	1	4	2

Diagnosis and mortality rates are per 100,000 individuals.

124 Diagnosis and death rates are from "Cancer Facts and Figures 2020," *American Cancer Society, Inc.*

125 N. Howlader, A.M. Noone, M. Krapcho, D. Miller, A. Brest, M. Yu, J. Ruhl, Z. Tatalovich, A. Mariotto, D.R. Lewis, H.S. Chen, E.J. Feuer, K.A. Cronin (eds), "SEER Cancer Statistics Review 1975-2016: Tables 33.2 and 33.3," National Cancer Institute. Bethesda, MD, https://seer.cancer.gov/csr/1975_2016, based on November 2018 SEER data submission, posted to the SEER website, April 2019.

White women are third in survival rates with a value of 69.3%, compared with being last and fourth in diagnosis and death rates. Asian men are fourth in survival rates with a value of 61.3%, even though they are first in both diagnosis and death rates. Native American women are fourth in survival rates as well with a value of 66.4% compared with their diagnosis and death rates ranking of third. Finally, Native American men come in fifth for survivability at 56.1%, versus third in diagnosis and death rates. Black women are fifth for survival rates at 62.1%, versus fourth and fifth for their diagnosis and death rates.

One conclusion from the data is that women consistently have higher survival rates compared to their male counterparts, except for black women. To compare further across races, one can divide the delta[126] between diagnosis and death rates by diagnosis rates over similar time periods (the diagnosis data is from a year earlier, 2011–2015 versus 2012–2016, which satisfies the majority of cases with survivals greater than a year). While survival rate by definition is survival over a five-year window and would include diagnoses before 2011 and deaths after 2016, comparing average diagnosis and death rates over the given period of 2011–2016 is indicative of survival rates for a given ethnicity and therefore yields approximate survival rates.

The resulting approximate survival rate rankings for males (from highest to lowest) are Hispanics/Latinos, whites, Asians and a tie between blacks and Native Americans. For women, the rankings are Hispanics/Latinos, Asians, whites, Native Americans and blacks. Comparing the approximate survival data to the survival data in Table 30, which is from 2009–2015 and a different source (NCI's 2000–2016 SEER registry), shows the rankings to be similar but not exact. All are within +/-1, demonstrating that in general, survival rates are linked to a comparison of death rates and diagnosis rates, as one might expect.

Table 30 provides death rates by state and cancer type in the United States. However, as was done when looking at statistical rates by ethnicity, it is more useful to compare death rates to diagnosis rates and back out survival rates by state. Although this was done, the data was compared

126 Delta used in this context means difference, deriving from the Greek symbol Δ used in mathematics for the difference between to values.

against state survival rates from "Cancer in North America: 2011–2015 Volume Four: Cancer Survival in the United States and Canada 2008–2014" NAACCR. The comparison had quite a few dissimilarities, likely due to the number of individuals moving between states and the data being taken at slightly different times. Therefore the NAACCR data is used for our survival rate comparison and shown in Table 24, although eight states (and a district) were not included in the survey (Florida, Virginia, Massachusetts, Minnesota, Nevada, Kansas, South Dakota, North Dakota and the District of Colombia).

The states with the top ten highest survival rates are New Hampshire, Washington, Connecticut, Vermont, Colorado, Rhode Island, New York, Illinois, Wisconsin and Maryland. Conversely the lowest ten states are Mississippi, Alaska, New Mexico, Louisiana, Alabama, Oklahoma, Georgia, West Virginia, South Carolina and Montana. Of note is that while the East Coast and Midwest have in general higher diagnosis rates than the rest of the country (discussed in the previous section), these regions fare better with survival rates. Furthermore, the states with the lowest rates are in traditionally rural areas. Generally, top cancer treatment facilities are in urban centers such as New York, Boston, Washington D.C. and Chicago, versus rural areas like Mississippi or Oklahoma. Those living farther away from top cancer treatment centers will have a more difficult time traveling or perhaps receive treatment that is not as cutting-edge as they would receive at centers with more funding.

Now this does not mean that there are not excellent treatment facilities in all of the states, but patients may need to travel farther to receive this care. This hypothesis based on the data reinforces that researching and finding the best oncology team is an important part of your treatment therapy.

While the summaries in Table 24 are for an average of all cancers, readers can study the specific cancer death rate data in Table 30 and look for trends in their cancer of interest. Specifically, lung cancer can be found geographically in *Cancer Facts and Figures 2017*, American Cancer Society, where a prevalence in the Midwest and South is shown (Figure 78). States around the Rocky Mountains have the lowest death rates, most likely due to fewer instances of tobacco use and cleaner air.

Appendix A has additional data for specific cancers, showing survivability rates over a 40-year time period going back to 1975. This shows the general upward trend in cancer treatment and provides a more detailed survival summary for your specific type of cancer. For instance Hodgkin's lymphoma, which I had, shows a one-year survival rate of 86.9% in the time period of 1975-1979. This has increased to 93.0% in the year 2015, the last available time period for the data.

Looking at five-year values, the change over the same duration was 71.4 to 89.5%. Going back to 2005 you can see the 10-year survival rate to 2015 was 85.0% and the 20-year survival rate from 1995 to 2015 was 73.4%. This data is for all races, male and female. Additional information is provided in *SEER Cancer Statistics Review 1975–2016*, National Cancer Institute. Note that this data is also for all levels of staging—so that must also be considered, as well as factors such as physical health and age at the time of diagnosis. Although I was young and in good shape, my advanced staging would result in lower percentages than those presented above.

Table 30: US cancer deaths by state, 2020 projections.[127]

State	All Sites	Nervous System	Female Breast	Colon & Rectum	Leukemia	Liver‡	Lung & Bronchus	Non-Hodgkin's Lymphoma	Ovary	Pancreas	Prostate
AL	10,530	340	690	960	370	520	2,790	290	230	790	520
AK	1,090	†	70	120	†	50	190	†	†	90	60
AR	12,580	400	900	1,120	520	680	2,590	410	310	1,070	760
AK	6,730	190	410	610	240	290	1,890	190	140	450	280
CA	60,660	1,980	4,620	5,480	2,400	3,880	10,210	2,140	1,590	4,840	3,890
CO	8,220	290	640	700	330	410	1,450	260	210	620	590
CT	6,390	210	430	460	260	310	1,370	230	160	520	480
DW	2,130	60	150	160	90	120	510	80	50	190	90
D.C.	1,020	†	100	100	†	80	180	†	†	90	70
FL	45,300	1,290	3,040	3,930	1,800	2,200	10,580	1,500	1,000	3,570	2,800
GA	17,990	540	1380	1,730	600	760	4,210	530	400	1,300	990
HA	2,540	60	160	240	90	180	520	90	†	240	130
ID	3,100	100	230	260	110	160	590	120	90	260	210
IL	24,220	670	1720	2,160	900	1,080	5,710	750	560	1,780	1,560
IN	13,630	370	880	1,170	510	550	3,570	450	290	990	640
IA	6,440	190	380	560	250	260	1,530	240	150	500	340

127 "Cancer Facts and Figures 2020," *American Cancer Society, Inc.*

State	All Sites	Nervous System	Female Breast	Colon & Rectum	Leuke-mia	Liver‡	Lung & Bronchus	Non-Hodgkin's Lymph_oma	Ovary	Pancreas	Prostate
KA	5,520	170	350	500	240	250	1,300	180	120	410	290
KY	10,540	290	630	870	370	440	2,910	330	180	670	430
LA	9,300	240	640	880	320	580	2,330	280	160	750	450
ME	3,350	100	180	240	120	120	870	110	70	240	180
MD	10,790	300	850	920	410	580	2,310	340	260	870	580
MA	12,430	410	780	910	480	640	2,810	390	310	1,020	660
MI	21,000	600	1380	1,700	770	890	5,220	720	480	1,720	1,030
MN	10,040	330	630	790	430	420	2,210	390	210	820	590
MS	6,700	180	460	670	220	320	1,740	160	120	520	360
MO	13,010	340	850	1,090	480	570	3,250	390	250	940	570
MT	2,140	70	140	190	70	100	460	70	50	160	150
NE	3,520	120	240	320	150	120	800	120	80	280	190
NV	5,460	210	400	590	200	240	1,230	170	150	400	310
NH	2,830	90	170	290	110	120	700	90	70	200	150
NJ	15,710	480	1230	1,440	620	700	3,230	560	390	1,340	810
NM	3,730	110	280	360	120	250	670	120	110	280	230
NY	34,710	960	2430	2,950	1,370	1,610	6,510	1,230	870	2,890	1,850
NC	20,410	570	1440	1,640	710	850	5,020	610	430	1,500	1,010
ND	1,260	†	80	110	60	†	280	50	†	100	70
OH	25,380	700	1710	2,170	930	1,090	6,460	850	550	1,930	1,200
OK	8,430	230	560	800	330	410	2,180	270	190	570	430
OR	8,280	260	550	660	310	480	1,750	270	240	680	500
PA	27,860	780	1910	2,440	1,070	1,270	6,460	950	640	2,270	1,390
RI	2,120	60	120	160	80	110	540	70	†	170	110
SC	10,780	310	750	910	390	520	2,610	320	210	830	590
SD	1,690	60	110	170	70	70	400	60	†	130	90
TN	14,780	380	950	1,260	530	730	3,990	460	310	1,010	660
TX	41,810	1,260	3,060	4,070	1,620	2,740	8,420	1,350	930	3,130	2,310
UT	3,350	140	290	300	170	160	430	130	110	280	240
VT	1,450	60	70	130	50	50	350	50	†	110	70
VA	15,220	450	1140	1,400	540	730	3,450	490	370	1,180	800
WA	13,020	440	900	1,050	490	720	2,740	450	330	1,000	750

*Rounded to the nearest 10. †Estimate is fewer than 50 deaths. ‡Liver includes intrahepatic bile duct. These estimates are offered as a rough guide and should be interpreted with caution. State estimates may not sum to US total due to rounding and exclusion of state estimates fewer than 50 deaths. Estimates are not available for Puerto Rico.

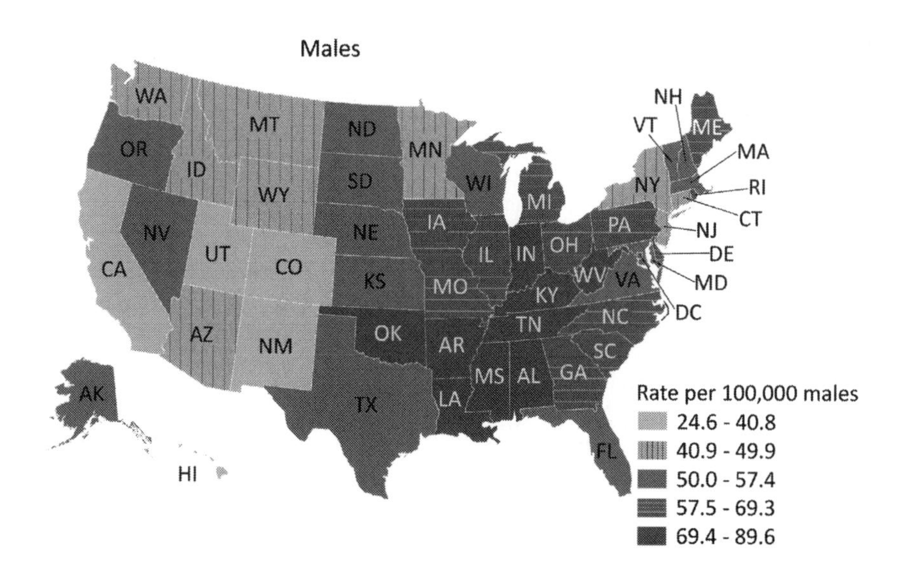

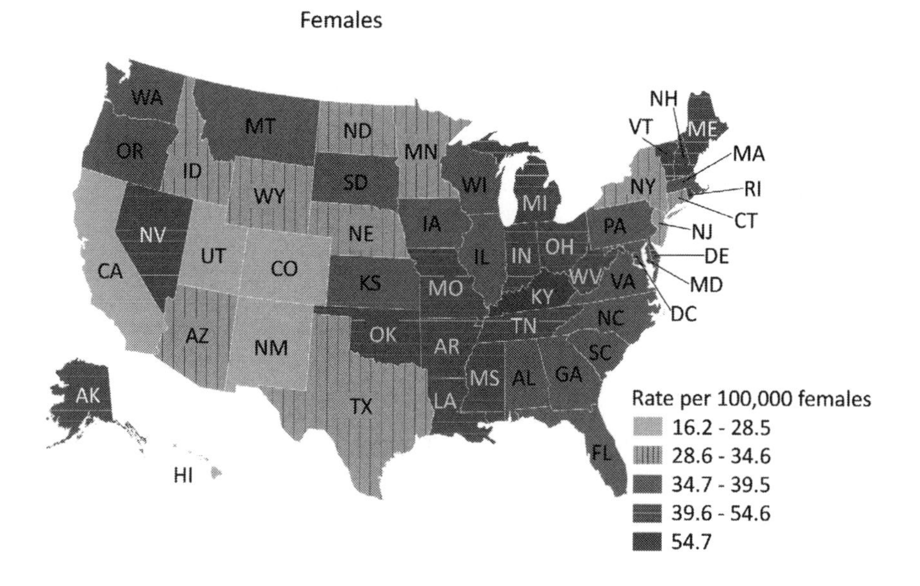

Source: Incidence – North American Association of Central Cancer Registries, 2019. Mortality – National Center for Health Statistics, Centers for Disease Control and Prevention, 2019.

Figure 78: US lung cancer death rates by state, 2010–2014.[128]

128 "Cancer Facts and Figures 2017," *American Cancer Society, Inc.*

Chapter 18

Causes and Prevention

There is an abundance of research regarding what causes cancer and how you can prevent it. While there are behaviors that can be directly linked to cancer (such as tobacco use, alcohol and obesity), there remains a lot to learn. In my case specifically, oncologists do not know what caused my lymphoma. Certain types of cancers have stronger links to causes than others. We will discuss what is known and how you can reduce your risk of getting cancer or having a recurrence. Additionally, we will look at some theories on why cancer may form without some of the direct known links.

Mutation Overview

How does cancer begin? Inside the nucleus of most human cells, chromosomes consist of DNA that makes an identical copy as part of the cell division process. Recall RBCs do not have a nucleus and DNA, which is why WBCs proliferate in leukemia since they do have a nucleus with DNA. During cell division, mutations may occur. A healthy cell will attempt to detect mutations and order a controlled death if one is found via a process called apoptosis. However a cell may miss the mutations and divide anyways, creating two new cells with one having more and the other having less DNA than is needed, both of which cause the cell to behave abnormally. Furthermore, mutated cells that become cancer cells will ignore their instructions to die, causing a buildup of unwanted cancer cells that form tumors. These tumors will grow and impinge on your organs and bodily systems, eventually leading to death if not taken care of.

Note that each cell has thousands of genes and it takes more than one mutation to create a cancerous cell. These mutations often occur by chance and your body's WBCs will kill off mutated cells before they multiply and turn cancerous. But when your immune system fails to remove a mutated cell that begins to multiply, these cells become cancerous. Why your immune system fails after removing mutated cells (daily) is a focus area for research.

MUTATION CAUSES

Cell mutations can be divided into two subtypes:

- Gene mutations that you are born with. This does not mean you will get cancer, but that you are more likely to contract cancer as additional gene mutations combine with the one(s) you are born with.
- Gene mutations after you are born (which are more likely to cause cancer than hereditary mutations). Causes of these gene mutations include.
    - Carcinogens.
    - Obesity (due to diet or lack of exercise).
    - Insomnia.

As multiple mutations are generally required for cancer cells to develop, having mutated genes that you are born with will make you more likely to get cancer, but they still need to combine with gene mutations later in life to cause cancer. Breast cancer is a good example. While women with a history of breast cancer in their family may have a gene mutation that puts them at higher risk for breast cancer, not all female offspring will develop breast cancer.

CARCINOGENS

A carcinogen is any substance that can cause gene mutations and abnormal cancerous cells to form from healthy cells (termed carcinogenesis). There are ongoing studies to determine carcinogenic substances, along with a substantial list at present. The two most reputable organizations for defining carcinogenic substances (either from laboratory testing or studying human populations) are the International Agency for Research on Cancer (IARC) and the National Toxicology Program (NTP). The former is part of the World Health Organization (WHO) while the latter has contributions from several United States agencies, i.e. the National

Institute of Health (NIH), the Centers for Disease Control and Prevention (CDC) and the FDA.

The IARC has tested over 900 substances and classifies them into one of the following groups:

- ***Group 1:*** Carcinogenic to humans.
- ***Group 2A:*** Probably carcinogenic to humans.
- ***Group 2B:*** Possibly carcinogenic to humans.
- ***Group 3:*** Unclassifiable as to carcinogenicity in humans.
- ***Group 4:*** Probably not carcinogenic to humans.

The NTP lists only substances that are known or are likely carcinogenic and does not include substances that may have been tested and deemed non-carcinogenic. Their list contains over 250 substances categorized into two groups:

- ***Group 1:*** Known to be human carcinogens.
- ***Group 2:*** Reasonably anticipated to be human carcinogens.

Lists for Group 1 and 2A for the IARC and Group 1 and 2 for the NTP are listed in Appendix C/Carcinogens at the back of this book. Some common carcinogens[129] to be aware of from these lists include:

- Tobacco.
- Alcoholic beverages.
- Processed meats.
- Radiation (UV, X-ray, gamma-ray, ionizing and neutron).
- Soot.
- Coal processing.
- Rubber processing.
- Salted fish (Chinese-style).[130]
- Plutonium and radium.
- Exposure as a painter.

129 "Known and Probable Human Carcinogens," *American Cancer Society,* https://www.cancer.org/cancer/cancer-causes/general-info/known-and-probable-human-carcinogens.html.

130 A traditional Chinese dish where fish is preserved or cured with a high quantity of salt.

- Air pollution.
- Isopropyl alcohol manufacturing.
- Iron and steel founding (workplace exposure).
- HIV, HPV, Hepatitis B/C, HTLV-1, KSHV and MCV.[131]
- Asbestos.
- Vinyl chloride (used in the production of PVC).
- Nickel compounds.

OBESITY

There have been numerous studies regarding obesity and its link to cancer, beginning relatively recently at the onset of the 1990s. While this is another field of ongoing research (like the carcinogenic substance studies), there are useful conclusions that can be made.

There is a plethora of literature on this topic and specifically for which type of cancers are caused by obesity. One study by the GLOBOCAN project[132] estimated that in 2012, 3.5% of male cancer diagnoses and 9.5% of female cancer diagnoses were due to obesity. The World Cancer Research Fund estimates that 20% of all cancers diagnosed in the United States are due to obesity, lack of exercise, poor nutrition or alcohol consumption (stating that obesity is the largest component). A 2014 Cancer Progress Report provided by the American Association for Cancer Research (AACR) states that 25% of cancers are related to obesity, behind only tobacco use as the leading cause of cancer. While the statistics vary depending on the source, there is clearly a link between obesity and cancer.

Table 31 summarizes seven sources that were found to have reputable data regarding the link between cancer and obesity. These sources are listed in the table along with a link to their websites. Since there is not a consensus, data is organized and the number of obesity citations for that particular cancer is summarized. If six to seven citations were received, green text is used. For four to five, orange; two to three, blue, and for one, black. Cancers not listed means they did not show up in any of the seven sources as being linked to obesity.

131 HPV: Human papillomavirus; HTLV-1: human T-cell leukemia virus; KSHV: Kaposi's sarcoma herpesvirus; MCV: molluscum contagiosum virus.

132 The GLOBOCAN project is led by the IARC and summarizes cancer statistics using data available from multiple countries, segregating by cancer site and sex.

Looking particularly at the green and orange-colored cancers, there are twelve that are cited by more than half of the sources as being linked to cancer (with nine of these being cited by six to seven sources). The nine with 85–100% citations are breast (post-menopausal), colorectal, esophageal, gallbladder, kidney, liver, ovarian, pancreatic and uterine cancer. Those with citations of 57–71% include multiple myeloma, stomach and thyroid cancer.

Fat cells influence cancer in the following ways:

- They produce hormones such as insulin (a peptide hormone) and estrogen (a steroid hormone) that interact with nearby cells and can cause cell mutations. The latter can inhibit or stimulate cell growth. Specifically, estrogen increase can lead to breast (post-menopausal), ovarian and endometrial (lining of the uterus) cancers. High levels of insulin can result in colon, kidney, prostate and endometrial cancers.
- They produce adipokines, which are cytokines (cell-signaling proteins) that can stimulate or inhibit cell growth and lead to tumorigenesis (the creation of a tumor).
- They create body inflammation. Recall that inflammation is the body's natural response to injured tissue, whereby WBCs are recruited to the area to cause local swelling. Chronic inflammation resulting from obesity causes inflammation to occur without an injury and to be prolonged when there is an injury. Local cell DNA can be damaged with excess inflammation, leading to cancer.
- They impact the mammalian target of rapamycin (mTOR) and AMP-activated protein kinase (AMPK), which are protein enzyme cell growth regulators that can cause DNA mutations and rapidly dividing cancer cells.

Although studies are ongoing, the links already found between obesity and cancer demonstrate that individuals will reduce their chances of contracting cancer with decreased obesity. The *Nutrition* and *Exercise* chapters presented in this book are meant not only for patients with cancer, but also for people of all ages and fitness levels to promote a healthy lifestyle and decrease your risk of cancer.

Table 31: Obesity link to various types of cancers.

Cancer	Source							Prevalence	
	A	B	C	D	E	F	G	Citations	Weight
Bowel					x			1	4
Breast*	x	x	x	x	x	x	e	7	**1**
Cervix‡			e				e	2	3
Colorectal*	x	x	x	x		x	e	6	**1**
Esophageal*	x	x	x	x	x	x		6	**1**
Head and Neck (Includes Mouth, Pharynx and Larynx)‡	e					x		2	3
Gallbladder*	e	x	e	x	x	x	x	7	**1**
Kidney*	x	x	x	x	x	x	x	7	**1**
Leukaemia							e	1	4
Liver*	x	x	e	x	x		x	6	**1**
Meningioma (Form of Brain Cancer)†		x	e	x	x			4	2
Multiple Myeloma (Form of Blood Cancer)			e					1	4
Non-Hodgkin's Lymphoma*	e	x	e	x	x		e	6	**1**
Ovarian*	x	x	x	x	x	x		6	**1**
Pancreatic‡	e		e			x		3	3
Prostate†	e	x		x	x			4	2
Stomach (Includes Gastric Cardia)†		x		x	x	x	e	5	2
Thyroid*	x	x	x	x	x	x	x	7	**1**
Uterine (Includes Endometrium, a Lining of the Uterus)					x			1	4
Total	**12**	**13**	**13**	**13**	**13**	**10**	**10**	**NA**	**NA**

Key:

A: "Diet, Nutrition, Physical Activity and Cancer: a Global Perspective, Continuous Update Project Expert Report 2018" (London/Washington D.C.: *World Cancer Research Fund* and *American Institute for Cancer Research*, 2018), https://www.wcrf.org/dietandcancer/resources-and-toolkit#accordion-1.

B: Beatrice Lauby-Secretan, Chiara Scoccianti, Danna Loomis et al., "Body Fatness and Cancer–Viewpoint of the IARC Working Group," *New England Journal of Medicine* 375, no. 8 (Aug. 2016): 794-798, https://www.nejm.org/doi/full/10.1056/NEJMsr1606602.

C: "Does Body Weight Affect Cancer Risk?," *American Cancer Society*, https://www.cancer.org/cancer/cancer-causes/diet-physical-activity/body-weight-and-cancer-risk/effects.html.

D: "Obesity and Cancer," *National Cancer Institute*, https://www.cancer.gov/about-cancer/causes-prevention/risk/obesity/obesity-fact-sheet.

E: "Does Obesity Cause Cancer?," *Cancer Research UK*, https://www.cancerresearchuk.org/about-cancer/causes-of-cancer/bodyweight-and-cancer/how-being-overweight-causes-cancer.

F: "Obesity, Weight and Cancer Risk," *American Society of Clinical Oncology*, https://www.cancer.net/navigating-cancer-care/prevention-and-healthy-living/obesity-and-cancer/obesity-weight-and-cancer-risk.

G: Krishnan Bhaskaran, Ian Douglas, Harriet Forbes, et al., "Body-Mass Index and Risk of 22 Specific Cancers: A Population-Based Cohort Study of 5.24 Million UK Adults," *The Lancet* 384, no. 9945 (Aug. 30 2014): 755-765, https://www.ncbi.nlm.nih.gov/pmc/articles/PMC4151483.

x: Causes cancer.
e: Evidence supporting cause of cancer.
*** Shows up in 6–7 of 7 sources.**
† Shows up in 4–5 of 7 sources.
‡ Shows up in 2–3 of 7 sources.
Shows up in 1 of 7 sources (cancers not shown do not show up at all).

Sleep

Getting a good night's sleep is often taken for granted, yet is one of the most important lifestyle changes you can make to improve your chances of not contracting cancer. As described earlier, your body is constantly creating mutated cells that must be removed before they can multiply and become cancerous. When your sleep is disrupted, studies have shown that the number of WBCs produced by your body decreases. With this decrease comes a statistically higher chance that a mutated cell is not ingested and destroyed. Additionally, insomnia can cause inflammation in your body, which can also lead to an increase in cancer risk as discussed in the *Inflammation* section of this book.

I believe this may have been the cause or certainly contributed to my cancer. Not having ever used tobacco or drugs, exercising regularly and being in great shape, having a decent diet and generally adhering to a healthy lifestyle, I was an atypical cancer patient on many levels. Still, I went through bouts of insomnia, which likely reduced the efficacy of my immune system. As there are no direct links to lymphoma (as opposed to some cancers such as lung cancer that is linked to tobacco usage), insomnia certainly may have contributed.

I adopted many of the strategies and habits discussed in the *Insomnia* section throughout the course of my adult life, which has led to improved sleep. There are additional books you can read on fighting insomnia, as well as courses you can take; I took a class from my health provider.[133]

Cancer Prevention

While we have discussed many causes of cancer, there remain ongoing studies to understand the various precursors of cancer and therefore how one can prevent being diagnosed with cancer. However based on the data and research already done to date, there are clear lifestyle changes that can be made to help prevent cancer.

- Get adequate sleep. Adults should shoot for 7 to 9 hours every night.
- Do not smoke or use other forms of tobacco products.
- Monitor your alcohol intake or do not drink.
- Exercise regularly and maintain your fat percentage below the normal value for your age and sex.
- Eat healthy, focusing on fruits and vegetables with lean forms of protein, along with controlled portions of carbohydrates.
- Take a multi-vitamin.
- Adopt anti-anxiety exercises, both physically and mentally, such as yoga, meditation, etc.
- Maintain a strong support network of family and friends to improve your mental health.
- Get annual health checkups and screenings for cancer when appropriate based on your age and sex.[134]

---

133 Michael J. Breus, "What is the Connection Between Sleep and Cancer," *Psychology Today*, https://www.psychologytoday.com/us/blog/sleep-newzzz/201911/what-is-the-connection-between-sleep-and-cancer;
Carol A. Everson, Christopher J. Henchen, Aniko Szabo and Neil Hogg, "Cell Injury and Repair Resulting from Sleep Loss and Sleep Recovery in Laboratory Rats," *Sleep* 37, no. 12 (Dec. 2004): 1929-1940, https://doi.org/10.5665/sleep.4244;
"Does Insomnia Increase Risk of Cancer?," *YogaU,* https://www.yogauonline.com/yogau-wellness-blog/does-insomnia-increase-risk-cancer;
"Long-Term Study Links Chronic Insomnia to Increased Risk of Death," *Mercola.* https://articles.mercola.com/sites/articles/archive/2010/06/24/longterm-study-links-chronic-insomnia-to-increased-risk-of-death.aspx;
S. Sephton and D. Spiegel, "Circadian Disruption in Cancer: A Neuroendocrine-Immune Pathway From Stress to Disease," *Brain, Behavior, and Immunity* 17, no. 5 (Oct. 2003): 321–8, https://www.ncbi.nlm.nih.gov/pubmed/12946654?dopt=Abstract.

134 Elaine R. Mardis et al., AACR Cancer Progress Report 2019 (Philadelphia: *American Association for Cancer Research*, 2019), http://www.cancerprogressreport.org.

Specific cancer screening guidelines for men and women based on age are provided in Table 32.

Table 32: Cancer screening recommendations.[135]

Cancer Site	Population	Test or Procedure	Recommendation*
Breast	Women, ages 40–54	Mammography	Women should have the opportunity to begin annual screening between the ages of 40 and 44. Women should undergo regular screening mammography starting at age 45. Women ages 45 to 54 should be screened annually.
	Women, ages 55+	N/A	Transition to biennial screening or have the opportunity to continue annual screening. Continue screening as long as overall health is good and life expectancy is 10+ years.
Cervix	Women, ages 21–29	Pap test	Screening should be done every 3 years with conventional or liquid-based Pap tests.
	Women, ages 30–65	Pap test & HPV DNA test	Screening should be done every 5 years with both the HPV test and the Pap test (preferred), or every 3 years with the Pap test alone (acceptable).
	Women, ages 66+	Pap test & HPV DNA test	Women ages 66+ who have had ≥ 3 consecutive negative Pap tests or ≥ 2 consecutive negative HPV and Pap tests within the past 10 years, with the most recent test occurring in the past 5 years should stop cervical cancer screening.
	Women who have had a total hysterectomy	N/A	Stop cervial cancer screening.
Colorectal†	Men and women, ages 45+	Guaiac-based fecal occult blood test (gFOBT) with at least 50% sensitivity or fecal immunochemical test (FIT) with at least 50% sensitivity, **OR**	Annual testing of spontaneously passed stool specimens. Single stool testing during a clinician office visit is not recommended, nor are "throw in the toilet bowl" tests. In comparison with guaiac-based tests for the detection of occult blood, immunochemical tests are more patient-friendly and are likely to be equal or better in sensitivity and specificity. There is no justification for repeating FOBT in response to an initial positive finding.
		Multi-target stool DNA test, **OR**	Every 3 years.
		Flexible sigmoidoscopy (FSIG), **OR**	Every 5 years alone, or consideration can be given to combining FSIG performed every 5 years with a highly sensitive gFOBT or FIT performed annually.
		Colonoscopy, **OR**	Every 10 years.
		CT Colonography	Every 5 years.

135 "Cancer Facts and Figures 2020," *American Cancer Society, Inc.*

Cancer Site	Population	Test or Procedure	Recommendation
Endometrial	Women at menopause	N/A	Women should be informed about risks and symptoms of endometrial cancer and encouraged to report unexpected bleeding to a physician.
Lung	Current or former smokers ages 55-74 in good health with 30+ pack-year history	Low-dose helical CT (LDCT)	Clinicians with access to high-volume, high-quality lung cancer screening and treatment centers should initiate a discussion about annual lung cancer screening with apparently healthy patients ages 55-74 who have at least a 30 pack-year smoking history, and who currently smoke or have quit within the past 15 years. A process of informed and shared decision making with a clinician related to the potential benefits, limitations, and harms associated with screening for lung cancer with LDCT should occur before any decision is made to initiate lung cancer screening. Smoking cessation counseling remains a high priority for clinical attention in discussions with current smokers, who should be informed of their continuing risk of lung cancer. Screening should not be viewed as an alternative to smoking cessation.
Prostate	Men, ages 50+	Prostate-specific antigen test with or without digital rectal examination	Men who have at least a 10-year life expectancy should have an opportunity to make an informed decision with their health care provider about whether to be screened for prostate cancer, after receiving information about the potential benefits, risks, and uncertainties associated with prostate cancer screening. Prostate cancer screening should not occur without an informed decision-making process. African-American men should have this conversation with their provider beginning at age 45.

CT stands for computed tomography. *All individuals should become familiar with the potential benefits, limitations and harms associated with cancer screening. †All positive tests (other than colonoscopy) should be followed up with a colonoscopy.

Men should be checked regularly starting at age 50 for prostate cancer and colorectal cancer, while women should begin screening as early as 21 years of age for cervical cancer. Once age 40 is reached, women are recommended to regularly be checked for signs of breast cancer followed by endometrial cancer screening post-menopause. After 50 years of age, women are also recommended for colorectal cancer screening. Finally, former smokers should be screened for lung cancer (both men and women) starting at age 55.

It is imperative that individuals follow cancer screening guidelines, as discovering cancer early can save your life, reduce the amount of time required for treatment and limit the potential side effects associated with your cancer treatment. Unfortunately, my cancer was inside my chest and I did not exhibit symptoms until my tumor had grown very large and I had reached a Stage IV diagnosis. Being 28, there are no cancer screening guidelines that would have caught this.

So while cancer screening recommendations exist for the most common types of cancers, they will not cover all possibilities. Here it is best to follow the cancer prevention guidelines and be attuned to your body so that you see a physician if any abnormal behavior occurs. Had I not visited the doctor after my symptoms worsened, my tumor would have killed me.

Chapter 19

Research

There is a large amount of cancer research ongoing across the United States and the world. With such a broad an important topic requiring volumes of books to cover adequately, I will present a brief summary of some promising ongoing research.[136] Readers are encouraged to delve into this constantly changing topic.

Precision Medicine via Targeted Therapy

Precision medicine is based on developing cancer treatment plans that are unique to an individual, based on that individual's genetic factors, environment and lifestyle. Researchers have found that patient tumors have unique genetics that cause cancers to spread. Furthermore, two patients with the same cancer and stage may have genetic differences that cause their cancers to behave differently.

The goal of precision medicine is to better understand how genetics factor into a patient's cancer and find ways to tailor treatments, rather than have blanket treatment plans prescribed to all patients with a given cancer and stage. Targeted therapy is presently used for this tailoring, often in conjunction with standard chemotherapy and radiation treatments. These targeted therapies can aid the immune system in killing your cancer by better identifying cancerous cells for destruction or boosting your immune system to better attack the cancer. They can also prevent cancer cells from growing, block signals that create the blood vessels required for tumors to survive, cause cancer death, starve a cancer

136 E. Kiesler and M. Begley, "The Future of Cancer Research: Five Reasons for Optimism," *Memorial Sloan Kettering Cancer Center,* https://www.mskcc.org/blog/future-five-reasons-optimism.

from vital hormones required to grow and directly deliver cancer-killing substances to the cancerous cells.

While targeted therapy is presently being used, it is still in early stages. I did not have any targeted therapy for my cancer treatment (in 2009 and 2010), nor was I given the option; however, there is information regarding this treatment online. As precision medicine continues to grow and researchers are better able to decipher how these drugs can be catered to an individual, targeted therapy will likely become more prevalent.[137]

Presently the NIH is undergoing the All of Us Research Program, in which they are asking for a minimum of one million volunteers to study how differences in individuals such as their age, gender and ethnicity impact their genetic predispositions for cancers and diseases, as well as determine how better to create user-specific treatments based on this information.

Additionally, Memorial Sloan Kettering (MSK) researchers (including molecular pathologists, genome scientists and bioinformaticians) are performing the integrated mutation profiling of actionable cancer targets test (MSK-IMPACT), which is a targeted tumor-sequencing test that detects aberrations and gene mutations in both rare and common cancerous tumors. With this information, oncologists can use precision medicine to tailor patient treatment plans, improving their efficacy while reducing side effects.[138]

President Barack Obama announced the Precision Medicine Initiative in his 2015 State of the Union with the goal of galvanizing federal research to make advances in precision medicine. In addition to the NIH initiative, other agencies under the Department of Health and Human Services (HHS) actively participating are the FDA, Office of the National Coordinator (ONC) for Health Information Technology and the Office for Civil Rights (OCR). The Department of Veterans Affairs (VA) and Department of Defense (DOD) are also expanding the Million Veteran Program to study how genes impact health.

137 "Understanding Targeted Therapy," *Cancer.net,* http://www.cancer.net/navigating-cancer-care/how-cancer-treated/personalized-and-targeted-therapies/understanding-targeted-therapy.

138 "MSK-IMPACT: A Targeted Test for Mutations in Both Rare and Common Cancers," *Memorial Sloan Kettering Cancer Center,* https://www.mskcc.org/msk-impact.

CHECKPOINT INHIBITORS

WBC T-cells are charged with killing your cancer cells. In order to do so, they utilize *checkpoint proteins* or *immune checkpoints*, which are protein receptors on the T-cell that identify proteins on healthy or cancerous cells to identify if they are a threat. Unfortunately cancerous cells often have proteins on their surface that replicate healthy cell proteins, causing T-cells to erroneously believe these are healthy cells and thus ignore them. Checkpoint inhibitors are a class of immunotherapy drugs that essentially go after and block checkpoint proteins (either on the T-cell or cancer cell side) to prevent them from recognizing one another, thereby enabling T-cells to decipher cancerous cells from healthy cells.

As an example, PD-1 is a checkpoint protein on T-cells that looks for PD-L1, a checkpoint protein found on many healthy cells. PD-L1 is also found on cancerous cells and checkpoint inhibitors will therefore target the production of PD-L1. Drugs that target PD-1 include pembrolizumab (Keytruda) and nivolumab (Opdivo). Cancers including bladder, head and neck, Hodgkin's lymphoma, kidney, melanoma and non-small cell lung cancer have utilized these drugs for successful treatment. Drugs that inhibit PD-L1 include atezolizumab (Tecentriq), avelumab (Bavencio) and durvalumab (Imfinzi). These drugs have shown promise in treating bladder, non-small lung and Merkel cell skin cancer.

An additional protein that acts as a switch to prevent T-cells from recognizing cancerous cells is CTLA-4. Ipilimumab (Yervoy) is a checkpoint inhibitor that blocks this protein, which is often used to treat melanoma and is being studied for use with other cancers.[139]

CELL-BASED THERAPY

Cell-based therapy (CBT) is another type of immunotherapy that utilizes the body's own immune system to fight cancer. Specifically, CBT enhances the body's immune system by removing T-cells from a patient's body, genetically altering them and re-inserting them back into the body.

139 "What is a Checkpoint Inhibitor? Immune Checkpoint Inhibitor Definition," *Dana-Farber Cancer Institute,* http://blog.dana-farber.org/insight/2015/09/what-is-a-checkpoint-inhibitor;
"Checkpoint Inhibitors," *Cancer Treatment Centers of America,* http://www.cancercenter.com/treatments/checkpoint-inhibitors.

Scientists at research institutes such as MSK are using Chimeric Antigen Receptor (CAR) T-cell therapy, a form of CBT that injects a virus inside of removed T-cells that carries a gene inside of the cells. This gene programs the cell to detect a specific kind of protein present in a cancer cell. As the type of proteins present depend on the type of cancer, CAR T-cell therapy can be tailored toward a specific cancer.

Presently, CAR T-cell therapy has been used for types of leukemias and lymphomas, with promising results. For these cancers the T-cells are genetically engineered to recognize the protein CD19, which is found on the surface of B-cells (relevant to blood cancers). In one study performed by researchers at MSK, this type of therapy, when utilized with patients having ALL, resulted in 88% of patients responding positively to the therapy. This therapy was given a "breakthrough therapy" designation by the FDA in 2014.

The goal is increasing the potency of this treatment and to expand it to other types of cancers. Active research is presently ongoing for prostate cancer and sarcomas. Additional trials are underway for breast, lung, mesothelioma and ovarian cancers.[140]

EPIGENETIC THERAPY

Recent research has found that in addition to genetic mutations (which have primarily been the theory behind how cancer cells form), an additional modification to cells deemed *epigenetic* can create cancer cells. While the former is a process in which a biochemical modification attaches directly to DNA, the latter has to do with turning a gene on or off. Specifically DNA is wrapped around protein histones, which when wound tightly, can block a DNA gene sequence from view of the cell. This effectively turns off the gene sequence. Normally this happens normally throughout the body; for instance, skin cells and eye cells have genetically the same instructions in their DNA, yet their differences result from certain genes being hidden and effectively turned off. Researchers are finding that these epigenetic changes are also partially to blame for

140 Julie Grisham, "CAR T Cell Therapies Are a Growing Area of Research," *Memorial Sloan Kettering Cancer Center,* https://www.mskcc.org/blog/car-t-cell-therapy-growing-area-research;
Eric Bender. "Cell-Based Therapy: Cells on Trial," *Nature* 540 (2016), http://www.nature.com/nature/journal/v540/n7634_supp/full/540S106a.html?foxtrotcallback=true.

cancerous mutations. Cancers are believed to be a combination of genetic mutations and epigenetic modifications, with certain cancers having a higher percentage of one type of modification than the other.

Based on this research, the idea is that rather than destroying cancerous cells using traditional means (including chemotherapy, radiation and surgery), epigenetic therapy can reverse the abnormal epigenetic modifications causing the cells to be cancerous. By doing so, genes that should have been turned on or off can be reversed and the cancerous cells can once again become healthy.

Various research institutes have been undergoing these forms of study and using trials to judge its efficacy. MSK is performing trials on patients with AML and MDS. A 38% positive response rate was achieved as of September 2015 based on a 159-patient sampling. Trials are also underway for additional blood cancers. Work at the Anderson Cancer Center is focused on MDS, with a trial of over 100 individuals. Results have shown complete remissions in over half of patients, with an additional 25% showing some positive response. [141]

METASTASIS RESEARCH

Approximately nine out of ten cancer deaths are caused by metastasis, where cancerous cells travel from the primary tumor into other parts of the body through either the lymph system or bloodstream. As such, many institutes are researching this phenomenon to better understand how these cells are distributed and survive transportation throughout the body (and subsequently re-attachment to vital organs). This research also includes how the cells are formed, how they lay dormant and how they end up in specific distant organs (are there any trends into where they land when spreading, etc.).

MSK has a Metastasis and Tumor Ecosystems Center (MTEC), which brings together scientists to study metastatic traits, brain metastasis, tumor ecosystems and the metastatic environment. This center also focuses on latent metastasis, which is when a cancer is cured, and latent cancer cells show up years later; a recent study showed latent metastasis in a patient 29 years after initial cancer cells were destroyed. Unfortunately many

141 "Epigenetic Therapy," *PBS*, http://www.pbs.org/wgbh/nova/body/epigenetic-therapy.html.

patients see this recurrence, although 29 years is extreme. For example, roughly 25% of women with human epidermal growth factor receptor 2 (HER2)-positive breast cancer, which promotes the growth of cancer cells, and 50% of patients with lung cancer, experience latent metastasis.

Research to date has found that metastatic cancer cells are adept at clinging to blood vessels, a trait that likely aids their transport and survival. Additionally, researchers have found that some of these cells produce a protein inhibitor (WNT inhibitor) that blocks cell division. When natural killer cells are searching for cancerous cells, this trait can prevent the metastatic cells from being detected, as they are less conspicuous when compared with traditional non-metastatic cancer cells. Furthermore, chemotherapy agents generally attack dividing cells, making metastatic cancer cells less likely to die. Researchers at MSK and other cancer institutes (such as the National Cancer Institute (NCI)) lead ongoing research initiatives to target metastatic cells and prevent their proliferation.

CHAPTER 20

FUNDRAISING

THERE ARE MANY cancer organizations out there with goals such as cancer research and aiding cancer patients in need. I am most familiar with the Leukemia and Lymphoma Society, (LLS) due to their notoriety and because I had lymphoma. Doing some research into the charities listed below is recommended so you can ensure your money or raised money goes toward a cause of your choice.

Many cancers funds have races (running, biking, triathlons) to raise money, which can be a liberating way to help a loved one and people you do not even know battle cancer, all while getting in shape! Not only are you helping others, but you also can feel good about yourself knowing that you are making a positive difference in the lives of others. Research takes money so the more raised, the better our chances are of improving cancer treatment and one day eradicating cancer once and for all!

Below is a list of the largest cancer-related charities in the United States for 2019,[142] ranked by annual donations, with total annual revenue also listed. Note that the rank is compared against all types of charities in the United States, not just cancer-related. This is not to say that smaller charities are not worthwhile. They are. But it makes you aware of some of the larger charities that are easy to get involved with, having local chapters throughout the United States. If you would prefer to make monetary donations rather than raise money, it is as simple as donating online.

Largest (by donations) cancer charities in the United States:

- American Cancer Society (ACS) (#17, $728M, $724M): https://www.cancer.org.

142 "The 100 Largest U.S. Charities, 2019 rankings," *Forbes,* https://www.forbes.com/top-charities/list.

- Leukemia and Lymphoma Society (LLS) (#36, $421M, $298M): http://www.lls.org.
- Dana-Farber Cancer Institute (#37, $418M, $1.8B): http://www.dana-farber.org.
- Memorial Sloan Kettering Cancer Center (MSKCC) (#41, $392M, $5.1B): https://www.mskcc.org.
- Make-A-Wish Foundation of America (#42, $367M, $388M): https://worldwish.org.
- Susan G. Komen (#87, $176M, $191M): http://www.komen.org.

Here are some additional cancer charities that, while not as high in yearly donations and revenue as those listed above, have strong ratings as charitable organizations:

- Breast Cancer Research Foundation (BCRF): https://www.bcrf.org.
- Cancer Research Institute (CRI): https://www.cancerresearch.org.
- Livestrong Foundation: http://www.livestrong.com.
- Multiple Myeloma Research Foundation (MMRF): https://www.themmrf.org.
- National Breast Cancer Coalition (NBCC): http://www.breastcancerdeadline2020.org/homepage.html.
- Ovarian Cancer Research Fund Alliance (OCRFA): https://ocrfa.org.
- Pancreatic Cancer Action Network (PanCAN): https://www.pancan.org.
- Prostate Cancer Foundation (PCF): https://www.pcf.org.

There are various charity rating services available to cross-reference the above charities or compare others against:

- Charity Navigator: https://www.charitynavigator.org.
- Charity Watch: https://www.charitywatch.org/home.
- Consumer Reports: https://www.consumerreports.org/charities/best-charities-for-your-donations.

Charity Navigator is the most comprehensive charity rating site I found, as it provides quantitative data for financial statements broken down by categories of interest to show how it determines its ratings.

Appendix A

Cancer Survivability Rates

This appendix is a good resource for understanding cancer survivability rates and how they have improved over time. Data comes from the following source, where additional cancer statistics can be found:

N. Howlader, A.M. Noone, M. Krapcho, D. Miller, A. Brest, M. Yu, J. Ruhl, Z. Tatalovich, A. Mariotto, D.R. Lewis, H.S. Chen, E.J. Feuer, K.A. Cronin (eds), "SEER Cancer Statistics Review 1975-2016," *National Cancer Institute*. Bethesda, MD, https://seer.cancer.gov/csr/1975_2016/, based on November 2018 SEER data submission, posted to the SEER website, April 2019.

All data in this appendix is based on the Surveillance, Epidemiology and End Results (SEER) 9 areas (San Francisco, Connecticut, Detroit, Hawaii, Iowa, New Mexico, Seattle, Utah and Atlanta). The exceptions are the first figure's mortality rates, which come from a different source, as well as the second figure's data, which is from SEER 18. SEER 18 includes the SEER 9 locations but also these additional regions: San Jose-Monterey, Los Angeles, Alaska Native Registry, Rural Georgia, California (excluding San Francisco, San Jose-Monterey and Los Angeles), Kentucky, Louisiana, New Jersey and Georgia (excluding Atlanta and rural Georgia).

Table A1: Summary of changes in cancer mortality (1950–2017) and five-year relative survival rate (1950–2016); males and females, by primary cancer site.[143]

	Whites			
	U.S. Mortality [%] Change (1950–2016)†	5-Year Relative Survival (Percent)‡		
Primary Site	Total	APC	1950–1954	2009–2016
Oral Cavity And Pharynx	-49.0	-1.2*	46	67.7
Esophagus	20.6	0.7*	4	21.1
Stomach	-89.3	-3.3*	12	29.0
Colon And Rectum	-59.3	-1.4*	37	67.5
Colon	-54.7	-1.2*	41	67.2
Rectum	-69.7	-2.1*	40	68.3
Liver And Intrahepatic Bile Duct	66.6	0.9*	1	16.8
Pancreas	29.8	0.1*	1	7.8
Larynx	-47.5	-0.9*	52	65.1
Lung And Bronchus	147.4	0.9*	6	18.8
Males	79.4	0.2*	5	16.2
Females	439.1	2.1*	9	21.6
Melanoma Of The Skin	115.3	1.0*	49	93.4
Breast (Females)	-40.6	-0.8*	60	92.2
Cervix Uteri	-82.1	-3.0*	59	71.4
Corpus And Uterus, NOS	-62.3	-1.3*	72	85.7
Ovary	-21.0	-0.4*	30	45.5
Prostate	-39.0	-0.7*	43	99.8
Testis	-70.1	-2.6*	57	97.4
Urinary Bladder	-30.4	-0.6*	53	79.9
Kidney And Renal Pelvis	22.0	0.3*	34	74.8
Brain And Nervous System	55.8	0.4*	21	34.2
Thyroid	-42.6	-0.8*	80	98.4
Hodgkin's Lymphoma	-85.6	-3.3*	30	88.7
Non-Hodgkin's Lymphoma	58.7	0.5*	33	73.3
Myeloma	194.7	0.9*	6	48.3
Leukemia	-10.7	-0.4*	10	63.9
Childhood (Ages 0-14)	-75.4	-2.6*	20	84.9
All Sites	-21.7	-0.3*	35	70.0

†The APC is the Annual Percent Change over the time interval. Rates used in the calculation of the APC are age-adjusted to the 2000 U.S. standard population (18 age groups - Census P25-1130). U.S. Mortality Files, National Center for Health Statistics, Centers for Disease Control and Prevention. Due to coding changes throughout the years: Colon excludes other digestive tract; Rectum includes anal canal; Liver & intrahepatic bile duct includes gallbladder & biliary tract, NOS; Lung & bronchus includes trachea & pleura; Ovary includes fallopian tube; Urinary bladder includes other urinary organs; Kidney & Renal pelvis includes ureter; NHL and myeloma each include a small number of leukemias; NHL includes a small number of ill-defined sites. ‡Survival estimates for 1950-54 are from NCI Survival Report 5 with the exception of All Sites, Oral cavity & pharynx, Colon & rectum, Non-Hodgkin lymphoma and Childhood cancers which come from historical Connecticut data. Survival estimates for 2009-2015 are from the SEER 9 areas (San Francisco, Connecticut, Detroit, Hawaii, Iowa, New Mexico, Seattle, Utah, and Atlanta). Rates are based on follow-up of patients into 2016. * The APC is significantly different from zero ($p<0.05$).

143 "SEER Cancer Statistics Review, 1975-2017," *National Cancer Institute.*

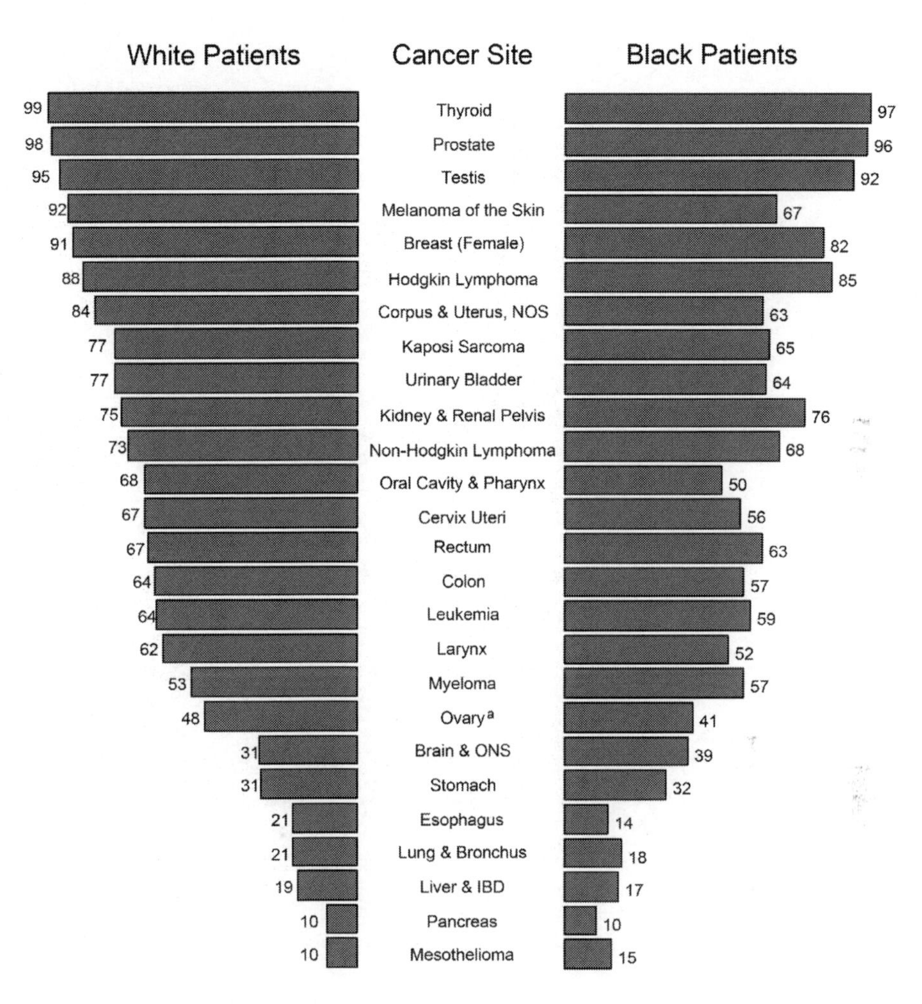

Source: SEER 18 areas (San Francisco, Connecticut, Detroit, Hawaii, Iowa, New Mexico, Seattle, Utah, Atlanta, San Jose-Monterey, Los Angeles, Alaska Native Registry, Rural Georgia, California excluding SF/SJM/LA. Kentucky, Louisiana, New Jersey and Georgia excluding ATL/RG. [a]Ovary excludes borderline cases or histologies 8442, 8451, 8462, 8472, and 8473.

Figure A1: Five-year relative survival (%) SEER 18 program (2009–2016). Both sexes, by race and cancer site.[144]

144 "SEER Cancer Statistics Review, 1975-2017," *National Cancer Institute.*

Year of Diagnosis

Survival Time	1975-1979	1980-1984	1985-1989	1990-1994	1995	1996	1997	1998	1999	2000	2001	2002	2003	2004	2005	2006	2007	2008	2009	2010	2011	2012	2013	2014	2015	2016
1-year	80.6	84.5	86.6	89.5	90.4	92.6	90.7	90.7	91.1	92.5	92.0	91.3	93.4	92.6	92.4	93.9	92.3	92.6	92.2	93.3	95.2	93.8	93.7	92.3	93.3	95.1
2-year	70.9	75.6	78.4	82.9	84.6	86.4	84.7	84.4	85.4	86.1	86.9	87.5	89.3	88.3	87.8	89.3	87.8	88.1	86.6	89.9	91.1	89.3	89.3	88.5	87.9	
3-year	65.4	70.7	74.4	79.7	81.0	83.6	82.3	81.6	83.1	83.3	84.1	84.2	86.3	85.5	84.9	87.9	86.2	86.1	84.0	87.4	89.7	87.7	88.0	86.3		
4-year	62.1	67.6	72.0	77.6	78.9	80.8	80.5	78.9	81.5	81.8	82.6	82.4	85.3	84.0	83.4	86.5	84.9	84.3	83.2	86.4	88.3	86.8	86.7			
5-year	59.9	65.9	70.7	76.4	77.4	80.4	79.3	78.1	79.7	80.8	81.9	81.4	84.3	83.1	82.1	85.8	83.8	83.5	82.7	85.7	87.7	86.1				
6-year	58.3	64.8	69.7	75.8	76.4	79.7	78.4	76.7	79.2	79.7	81.2	80.6	83.9	82.8	81.1	85.2	82.4	83.2	82.0	85.5	87.1					
7-year	57.2	63.9	68.8	75.2	75.9	79.1	78.0	76.3	78.5	79.1	80.3	80.2	83.3	82.2	80.8	84.8	81.9	83.1	81.6	85.1						
8-year	56.6	63.2	68.3	74.8	75.0	78.7	76.9	75.8	78.2	78.4	80.2	79.6	82.7	81.5	80.1	83.9	81.1	82.6	80.9							
9-year	56.2	62.7	67.9	74.3	74.6	78.3	76.7	75.1	77.7	78.4	79.8	79.4	82.2	81.1	79.8	83.4	80.8	82.3								
10-year	55.9	62.1	67.5	74.2	74.6	78.0	76.2	74.8	77.3	78.1	79.6	79.0	82.0	80.9	79.6	83.3	80.3									
11-year	55.6	61.9	67.2	73.8	74.3	77.9	76.1	74.4	77.0	77.9	79.4	78.3	81.5	80.8	79.6	83.2										
12-year	55.3	61.4	66.9	73.6	74.1	77.8	75.7	74.1	76.9	77.8	79.3	77.7	81.1	80.6	79.2											
13-year	55.0	61.0	66.7	73.4	74.0	77.7	75.5	73.7	76.5	77.4	79.2	77.4	81.0	80.5												
14-year	54.7	60.7	66.3	73.1	74.0	77.5	75.4	73.6	76.5	77.3	79.1	77.3	80.8													
15-year	54.4	60.5	66.2	72.9	73.9	77.3	75.4	73.5	76.5	77.0	79.1	76.8														
16-year	54.1	60.2	65.9	72.6	73.2	77.3	75.0	73.2	76.4	76.8	78.9															
17-year	53.7	60.0	65.7	72.4	72.9	77.1	74.8	73.2	76.1	76.5																
18-year	53.5	59.7	65.5	72.1	72.8	76.8	74.7	73.2	76.0																	
19-year	53.2	59.5	65.4	72.0	72.6	76.6	74.4	73.2																		
20-year	53.1	59.3	65.1	71.9	72.5	76.6	74.2																			

Figure A2: All cancer sites (invasive). Relative survival (%) by year of diagnosis. All races, males and females. Ages 0-14.

Year of Diagnosis

Survival Time	1975-1979	1980-1984	1985-1989	1990-1994	1995	1996	1997	1998	1999	2000	2001	2002	2003	2004	2005	2006	2007	2008	2009	2010	2011	2012	2013	2014	2015	2016
1-year	82.4	85.7	88.2	90.1	90.1	93.0	91.2	91.7	91.0	91.3	92.3	92.1	93.5	92.2	92.9	94.0	93.2	93.9	93.6	93.2	95.4	94.8	94.2	93.5	93.8	95.0
2-year	73.0	77.4	80.4	83.3	84.6	87.5	85.0	85.4	85.9	84.4	87.3	87.4	89.2	88.0	88.6	89.1	88.4	89.3	88.5	89.7	91.5	90.1	90.4	90.0	88.8	
3-year	68.0	72.5	76.5	80.0	81.3	85.0	82.0	82.5	83.3	82.0	84.4	84.3	86.3	84.8	86.1	87.4	86.6	87.2	86.1	87.2	89.3	87.6	88.9	87.8		
4-year	64.9	69.6	74.0	78.1	79.8	82.3	80.4	80.0	81.5	80.5	82.6	82.3	85.3	83.6	84.6	85.9	85.0	85.2	85.1	86.3	88.1	86.4	87.4			
5-year	62.9	67.9	72.9	76.8	78.4	81.8	78.8	78.9	80.0	79.3	81.9	81.2	84.3	82.4	83.4	85.1	83.7	84.0	84.5	85.5	87.2	85.7				
6-year	61.3	66.5	72.0	76.1	77.4	81.2	77.7	77.8	79.6	78.5	81.3	80.3	83.9	82.1	82.2	84.1	82.5	83.4	83.6	85.3	86.6					
7-year	60.3	65.8	71.2	75.4	76.9	80.7	77.5	77.4	79.1	77.8	80.4	79.9	83.2	81.5	81.8	83.6	82.0	82.9	83.3	84.8						
8-year	59.6	65.0	70.5	75.0	76.0	80.2	76.7	77.1	78.9	77.3	80.2	79.1	82.6	81.0	81.1	82.8	81.2	82.5	82.8							
9-year	59.1	64.4	70.1	74.5	75.7	79.8	76.5	76.5	78.5	77.1	79.8	78.8	82.3	80.5	80.7	82.5	80.8	82.2								
10-year	58.7	63.9	69.7	74.3	75.5	79.5	76.0	76.1	78.2	76.8	79.7	78.5	82.1	80.2	80.5	82.3	80.3									
11-year	58.4	63.6	69.4	74.0	75.4	79.3	75.8	75.8	77.9	76.7	79.5	77.9	81.8	80.1	80.5	82.2										
12-year	58.0	63.2	69.2	73.8	75.2	79.2	75.6	75.3	77.6	76.4	79.3	77.4	81.4	79.9	80.1											
13-year	57.8	62.8	68.9	73.5	75.1	79.1	75.3	74.9	77.4	76.1	79.1	77.2	81.2	79.8												
14-year	57.6	62.5	68.4	73.2	75.0	79.0	75.3	74.9	77.4	75.8	79.0	77.1	81.0													
15-year	57.3	62.2	68.2	73.1	74.9	78.8	75.1	74.8	77.3	75.6	79.0	76.9														
16-year	57.0	61.8	67.9	72.8	74.3	78.7	74.7	74.5	77.2	75.4	78.9															
17-year	56.7	61.6	67.7	72.5	74.1	78.4	74.6	74.5	77.0	75.1																
18-year	56.4	61.4	67.4	72.3	74.0	78.0	74.5	74.4	76.8																	
19-year	56.2	61.1	67.2	72.1	74.0	77.9	74.2	74.4																		
20-year	55.9	60.8	66.9	72.0	73.7	77.7	73.9																			

Figure A3: All cancer sites (invasive). Relative survival (%) by year of diagnosis. All races, males and females. Ages 0-19.

Year of Diagnosis

Survival Time	1975-1979	1980-1984	1985-1989	1990-1994	1995	1996	1997	1998	1999	2000	2001	2002	2003	2004	2005	2006	2007	2008	2009	2010	2011	2012	2013	2014	2015	2016
1-year	69.1	70.6	73.2	76.3	76.6	77.0	77.6	77.9	78.8	79.0	79.5	79.7	79.6	80.0	80.3	80.9	81.5	81.6	81.9	81.8	82.2	81.8	82.1	81.9	82.7	82.8
2-year	59.8	61.3	64.4	68.8	69.4	70.1	70.9	71.5	72.3	72.7	73.4	74.0	73.5	74.0	74.3	75.4	75.8	75.9	76.1	76.3	76.5	75.9	76.3	76.3	77.0	
3-year	54.8	56.2	59.8	64.9	65.8	66.7	67.5	68.1	69.0	69.5	70.3	70.9	70.4	71.0	71.3	72.3	72.7	72.7	73.1	73.2	73.4	72.6	73.0	73.1		
4-year	51.4	52.8	56.8	62.4	63.4	64.3	65.1	65.8	66.9	67.5	68.3	68.8	68.3	68.9	69.3	70.2	70.8	70.7	71.1	71.0	71.3	70.6	70.9			
5-year	48.9	50.2	54.4	60.4	61.6	62.5	63.3	64.0	65.2	66.0	66.8	67.3	66.8	67.3	67.7	68.7	69.4	69.1	69.6	69.5	69.7	69.0				
6-year	47.0	48.2	52.6	58.9	60.2	61.2	61.9	62.7	63.9	64.8	65.7	66.2	65.6	66.1	66.5	67.7	68.2	68.0	68.4	68.5	68.5					
7-year	45.4	46.5	51.1	57.6	59.1	60.1	61.0	61.7	62.9	63.7	64.7	65.2	64.5	65.1	65.6	66.7	67.2	66.9	67.5	67.5						
8-year	44.1	45.1	49.8	56.5	58.1	59.2	60.1	60.9	62.0	62.7	63.8	64.4	63.7	64.3	64.8	65.9	66.4	66.1	66.7							
9-year	42.9	43.9	48.6	55.5	57.2	58.2	59.2	59.9	61.2	62.0	63.1	63.6	62.9	63.5	64.0	65.2	65.8	65.3								
10-year	41.8	42.9	47.7	54.6	56.5	57.4	58.5	59.1	60.5	61.3	62.3	63.1	62.0	62.8	63.3	64.5	65.1									
11-year	40.9	41.9	46.8	53.7	55.6	56.8	57.8	58.4	59.8	60.5	61.7	62.4	61.4	62.1	62.7	63.9										
12-year	40.1	41.1	45.9	53.0	54.9	56.2	57.2	57.8	59.2	59.9	61.1	61.7	60.8	61.6	62.1											
13-year	39.3	40.3	45.1	52.3	54.0	55.5	56.5	57.1	58.6	59.2	60.4	61.2	60.3	61.0												
14-year	38.7	39.6	44.5	51.7	53.3	54.8	55.9	56.6	57.9	58.6	59.9	60.7	59.6													
15-year	38.0	39.0	43.8	51.1	52.7	54.3	55.4	56.1	57.4	58.0	59.4	60.3														
16-year	37.5	38.3	43.2	50.5	52.2	53.8	54.7	55.5	56.8	57.6	58.9															
17-year	36.9	37.7	42.7	49.8	51.6	53.2	54.2	55.0	56.3	57.1																
18-year	36.4	37.2	42.2	49.3	51.0	52.7	53.6	54.8	55.7																	
19-year	36.0	36.6	41.8	48.7	50.5	52.0	53.2	54.3																		
20-year	35.5	36.1	41.3	48.2	49.8	51.4	52.8																			
21-year	35.0	35.6	40.9	47.7	49.2	50.9																				
22-year	34.6	35.2	40.5	47.1	48.7																					
23-year	34.2	34.7	40.0	46.7																						
24-year	33.7	34.3	39.5																							
25-year	33.2	33.9	39.1																							
26-year	32.8	33.5	38.6																							
27-year	32.5	33.1	38.2																							
28-year	32.2	32.8	37.7																							
29-year	31.8	32.5																								
30-year	31.5	32.1																								

Figure A4: All cancer sites (invasive). Relative survival (%) by year of diagnosis. All races, males and females.

Year of Diagnosis

Survival Time	1975-1979	1980-1984	1985-1989	1990-1994	1995	1996	1997	1998	1999	2000	2001	2002	2003	2004	2005	2006	2007	2008	2009	2010	2011	2012	2013	2014	2015	2016
1-year	63.7	66.6	69.8	75.4	75.4	76.2	76.8	76.7	78.1	78.6	79.1	79.1	79.1	79.7	79.9	80.8	81.5	81.3	81.6	81.3	81.5	80.5	80.7	80.5	81.3	81.5
2-year	53.4	56.4	60.4	67.7	68.1	69.4	70.2	70.4	71.6	72.5	73.0	73.5	73.1	73.9	74.1	75.5	76.0	75.6	75.6	75.9	75.8	74.4	74.6	74.5	75.2	
3-year	48.2	51.2	55.5	64.1	64.7	66.1	67.1	67.2	68.5	69.5	70.4	70.8	70.2	71.0	71.3	72.6	73.3	72.7	73.0	73.0	72.7	71.2	71.2	71.3		
4-year	44.8	47.7	52.3	61.7	62.5	63.9	64.9	65.0	66.6	67.7	68.5	69.1	68.4	69.3	69.6	70.8	71.5	70.9	71.2	71.1	70.8	69.3	69.2			
5-year	42.2	45.0	49.9	59.8	60.9	62.3	63.2	63.5	65.2	66.3	67.4	67.9	67.2	68.0	68.2	69.5	70.3	69.6	69.8	69.7	69.5	67.8				
6-year	40.2	43.0	48.1	58.4	59.6	61.1	61.9	62.3	64.1	65.2	66.4	67.0	66.1	66.9	67.0	68.6	69.2	68.5	68.9	68.8	68.5					
7-year	38.5	41.3	46.4	57.1	58.6	60.1	61.1	61.3	63.2	64.3	65.6	66.3	65.2	66.0	66.1	67.8	68.3	67.7	68.2	68.0						
8-year	37.1	39.9	45.1	55.9	57.6	59.3	60.3	60.5	62.4	63.3	64.9	65.5	64.5	65.3	65.3	67.0	67.7	67.0	67.6							
9-year	35.8	38.6	43.8	54.9	56.9	58.4	59.4	59.6	61.6	62.7	64.4	64.9	63.6	64.5	64.7	66.3	67.1	66.4								
10-year	34.6	37.6	42.8	54.0	56.0	57.6	58.8	59.0	61.0	62.0	63.6	64.4	62.8	63.8	64.1	65.7	66.5									
11-year	33.5	36.6	41.9	53.1	55.3	56.8	58.2	58.3	60.5	61.1	63.0	63.6	62.2	63.2	63.6	65.2										
12-year	32.6	35.7	40.9	52.4	54.4	56.2	57.6	57.5	59.8	60.6	62.6	63.2	61.6	62.7	63.1											
13-year	31.8	34.9	40.1	51.7	53.6	55.7	56.8	56.9	59.3	59.8	61.9	62.6	61.2	62.2												
14-year	31.1	34.1	39.3	51.0	52.7	55.1	56.2	56.5	58.6	59.2	61.5	62.2	60.6													
15-year	30.4	33.4	38.6	50.3	52.0	54.5	55.7	56.1	58.0	58.6	61.0	61.8														
16-year	29.7	32.7	37.9	49.7	51.4	54.0	55.0	55.6	57.6	58.3	60.5															
17-year	29.0	32.0	37.4	49.0	50.8	53.4	54.3	55.1	57.2	57.9																
18-year	28.5	31.5	36.8	48.5	50.4	53.1	53.6	54.9	56.6																	
19-year	28.0	30.9	36.3	47.9	49.8	52.5	53.3	54.6																		
20-year	27.6	30.4	35.9	47.5	48.8	51.9	53.0																			
21-year	27.1	29.8	35.5	46.8	48.2	51.6																				
22-year	26.6	29.3	35.1	46.2	47.6																					
23-year	26.1	28.9	34.5	45.7																						
24-year	25.6	28.5	34.1																							
25-year	25.1	28.2	33.6																							
26-year	24.6	27.7	33.2																							
27-year	24.3	27.4	32.9																							
28-year	24.0	27.1	32.4																							
29-year	23.6	26.8																								
30-year	23.2	26.5																								

Figure A5: All cancer sites (invasive). Relative survival (%) by year of diagnosis. All races, males.

Year of Diagnosis

Survival Time	1975-1979	1980-1984	1985-1989	1990-1994	1995	1996	1997	1998	1999	2000	2001	2002	2003	2004	2005	2006	2007	2008	2009	2010	2011	2012	2013	2014	2015	2016
1-year	74.6	74.8	76.6	77.4	78.0	77.8	78.5	79.2	79.5	79.4	79.9	80.3	80.1	80.4	80.7	80.9	81.4	81.9	82.2	82.3	82.9	83.2	83.5	83.4	84.1	84.1
2-year	66.1	66.2	68.7	70.1	70.8	71.0	71.6	72.6	73.1	73.0	73.8	74.4	74.0	74.2	74.6	75.2	75.7	76.1	76.5	76.7	77.3	77.3	78.0	78.1	78.8	
3-year	61.3	61.2	64.2	65.9	67.0	67.3	67.9	69.1	69.5	69.6	70.2	70.9	70.5	70.9	71.3	72.0	72.1	72.7	73.2	73.3	74.0	74.0	74.8	74.8		
4-year	58.0	57.9	61.4	63.2	64.3	64.8	65.4	66.6	67.1	67.3	68.0	68.5	68.2	68.4	69.1	69.7	70.0	70.6	71.0	71.0	71.8	71.9	72.6			
5-year	55.6	55.3	59.0	61.2	62.4	62.8	63.5	64.6	65.3	65.7	66.2	66.6	66.3	66.6	67.3	67.9	68.3	68.6	69.3	69.3	70.0	70.1				
6-year	53.6	53.4	57.3	59.6	61.0	61.3	62.0	63.2	63.8	64.4	64.9	65.3	65.0	65.3	66.0	66.7	67.0	67.3	68.0	68.1	68.6					
7-year	52.1	51.7	55.8	58.2	59.6	60.2	60.9	62.1	62.6	63.2	63.7	64.1	63.8	64.2	65.0	65.6	66.0	66.1	66.8	67.0						
8-year	50.8	50.4	54.6	57.1	58.6	59.1	60.0	61.2	61.6	62.2	62.7	63.2	62.9	63.4	64.2	64.7	65.1	65.1	65.8							
9-year	49.7	49.2	53.5	56.1	57.7	58.1	58.9	60.2	60.8	61.3	61.7	62.3	62.1	62.5	63.3	63.9	64.2	64.2								
10-year	48.7	48.1	52.5	55.3	57.0	57.3	58.1	59.3	60.1	60.6	61.1	61.7	61.2	61.7	62.5	63.2	63.4									
11-year	47.8	47.1	51.7	54.4	56.1	56.7	57.4	58.6	59.2	59.8	60.4	61.1	60.6	61.0	61.8	62.6										
12-year	47.1	46.4	50.9	53.7	55.4	56.1	56.9	58.1	58.5	59.2	59.5	60.3	60.0	60.5	61.1											
13-year	46.4	45.6	50.1	53.0	54.5	55.4	56.3	57.3	57.9	58.5	58.8	59.7	59.4	59.7												
14-year	45.8	44.9	49.5	52.4	54.0	54.7	55.7	56.8	57.2	58.0	58.2	59.2	58.7													
15-year	45.1	44.3	48.9	51.9	53.5	54.1	55.1	56.1	56.7	57.4	57.8	58.7														
16-year	44.6	43.6	48.3	51.3	52.9	53.6	54.4	55.6	56.0	56.9	57.3															
17-year	44.1	43.0	47.8	50.7	52.4	52.9	54.1	55.1	55.5	56.4																
18-year	43.6	42.5	47.3	50.2	51.6	52.4	53.6	54.8	54.9																	
19-year	43.2	42.0	46.9	49.6	51.2	51.7	53.1	54.2																		
20-year	42.7	41.6	46.5	49.1	50.7	51.1	52.5																			
21-year	42.2	41.0	46.0	48.6	50.2	50.3																				
22-year	41.8	40.6	45.6	48.0	49.6																					
23-year	41.4	40.2	45.1	47.6																						
24-year	40.9	39.7	44.7																							
25-year	40.4	39.2	44.3																							
26-year	40.1	38.9	43.7																							
27-year	39.7	38.4	43.2																							
28-year	39.3	38.0	42.7																							
29-year	39.0	37.7																								
30-year	38.6	37.2																								

Figure A6: All cancer sites (invasive). Relative survival (%) by year of diagnosis. All races, females

Survival Time	Year of Diagnosis																									
	1975-1979	1980-1984	1985-1989	1990-1994	1995	1996	1997	1998	1999	2000	2001	2002	2003	2004	2005	2006	2007	2008	2009	2010	2011	2012	2013	2014	2015	2016
1-year	46.3	47.4	50.7	52.6	52.6	53.2	51.2	53.8	55.3	55.6	57.1	59.2	57.2	56.9	58.5	61.5	62.1	62.4	62.4	62.0	61.5	63.1	64.5	63.5	62.6	59.0
2-year	32.3	33.9	36.9	38.9	39.2	40.7	38.0	40.8	40.4	41.8	42.4	46.6	43.6	43.0	45.8	47.3	47.4	46.7	47.0	46.8	45.4	46.4	48.5	45.8	45.5	
3-year	27.9	29.4	32.7	34.9	36.0	36.6	34.1	35.7	36.1	38.4	39.0	40.6	38.1	38.2	40.0	40.6	41.0	41.0	41.4	39.9	39.8	38.7	40.2	38.2		
4-year	25.2	26.6	30.0	32.5	34.3	34.6	31.9	33.0	33.8	36.1	36.7	38.2	36.0	36.0	37.2	38.2	37.5	37.6	38.2	37.0	36.2	35.5	37.5			
5-year	23.1	24.6	28.0	30.8	33.3	32.5	30.3	31.1	32.4	34.9	34.8	36.1	34.2	35.0	34.9	35.9	35.5	36.1	36.3	34.3	33.9	33.4				
6-year	21.5	23.1	26.6	29.6	32.3	31.2	28.2	29.8	30.8	33.4	33.7	34.5	32.9	33.4	32.9	34.2	33.5	34.5	34.7	32.5	32.7					
7-year	20.5	21.9	25.5	28.3	31.3	30.3	27.5	28.8	30.1	31.7	32.6	33.6	31.4	32.0	31.9	33.4	32.5	33.1	33.6	31.3						
8-year	19.4	20.9	24.2	27.4	30.4	28.9	26.8	27.9	29.4	31.0	31.7	32.7	30.4	31.0	30.6	31.8	31.7	32.1	33.0							
9-year	18.4	19.9	23.4	26.6	29.9	27.7	25.6	27.2	28.3	30.1	31.1	31.7	29.6	30.2	29.8	30.7	30.7	31.0								
10-year	17.8	18.9	22.6	25.9	28.9	27.1	24.6	26.4	27.8	29.4	30.6	30.9	29.2	29.3	29.1	30.1	29.9									
11-year	17.2	18.2	22.0	25.2	28.1	26.1	23.9	25.9	26.9	28.3	29.7	30.0	28.9	28.4	28.1	29.4										
12-year	16.6	17.6	21.5	24.6	27.8	25.2	23.1	25.2	26.1	27.5	29.0	29.5	28.3	27.9	27.7											
13-year	16.2	17.0	20.9	24.1	26.6	24.8	22.8	24.4	25.6	26.8	28.4	29.0	27.6	27.3												
14-year	15.9	16.6	20.3	23.5	26.0	24.6	22.5	23.9	25.1	26.4	27.8	28.4	26.9													
15-year	15.4	16.1	19.6	22.9	25.3	24.0	22.4	22.6	24.9	26.1	27.4	28.0														
16-year	15.1	15.6	19.3	22.5	24.2	23.3	21.6	22.3	24.7	25.2	26.5															
17-year	14.7	15.3	18.8	22.0	24.0	23.2	21.2	22.1	24.4	25.0																
18-year	14.4	14.9	18.3	21.5	23.8	22.8	21.0	21.8	23.9																	
19-year	14.1	14.4	17.9	21.3	23.1	22.5	20.5	21.2																		
20-year	13.8	14.0	17.6	20.7	23.0	22.5	20.1																			

Figure A7: Cancer of the brain and nervous system (invasive). Relative survival (%) by year of diagnosis. All races, males and females.

Year of Diagnosis

Survival Time	1975-1979	1980-1984	1985-1989	1990-1994	1995	1996	1997	1998	1999	2000	2001	2002	2003	2004	2005	2006	2007	2008	2009	2010	2011	2012	2013	2014	2015	2016
1-year	94.4	95.1	96.3	97.1	97.1	96.9	97.3	97.5	97.5	97.5	97.3	97.5	97.8	97.8	97.7	97.7	97.7	97.7	97.8	97.6	97.8	98.1	98.0	97.6	98.1	97.7
2-year	88.7	89.9	92.3	93.8	94.0	93.6	94.3	95.0	95.2	95.3	95.0	95.5	95.5	95.6	95.7	95.7	96.1	95.7	96.0	95.8	96.1	96.0	96.2	96.0	96.4	
3-year	83.1	84.6	88.3	90.5	91.4	90.9	92.0	92.9	93.1	93.4	92.8	93.6	93.3	93.3	93.8	93.8	94.0	93.8	94.0	94.0	94.5	94.1	94.5	94.3		
4-year	78.5	80.2	85.1	87.8	88.8	88.9	89.9	91.1	91.3	91.6	91.0	91.8	91.6	91.4	91.9	92.3	92.5	92.4	92.6	92.3	92.8	92.6	93.1			
5-year	74.6	76.3	82.2	85.6	86.8	86.6	88.4	89.5	89.6	90.1	89.4	90.2	89.8	90.0	90.5	90.8	91.2	90.7	91.4	91.0	91.5	91.2				
6-year	71.4	73.4	79.8	83.8	85.4	85.0	86.9	88.3	88.1	89.0	88.2	88.8	88.6	88.7	89.2	89.7	89.9	89.5	90.5	90.1	90.3					
7-year	68.7	70.7	77.9	82.1	83.8	83.8	85.9	87.1	87.0	87.7	86.9	87.7	87.5	87.5	88.1	88.8	89.0	88.4	89.4	89.3						
8-year	66.5	68.4	76.3	81.0	82.6	82.4	85.0	85.8	85.9	86.4	86.0	86.5	86.4	86.6	87.4	87.8	88.1	87.4	88.6							
9-year	64.3	66.4	74.6	79.8	81.4	81.3	83.7	84.6	85.0	85.3	85.2	85.6	85.9	85.7	86.6	86.9	87.1	86.5								
10-year	62.5	64.5	73.3	78.7	80.6	80.4	82.6	83.5	84.2	84.5	84.4	85.1	84.8	84.9	85.8	86.0	86.2									
11-year	60.8	62.7	72.1	77.7	79.5	79.5	81.8	82.7	83.0	83.6	83.7	84.5	84.0	84.4	85.1	85.3										
12-year	59.3	61.6	70.9	76.7	78.5	78.6	81.1	82.1	82.5	82.6	82.5	83.5	83.4	83.9	84.2											
13-year	58.1	60.4	69.8	75.8	77.5	77.6	80.1	81.1	81.8	81.8	81.9	82.7	82.9	83.0												
14-year	57.1	59.2	69.0	75.0	76.9	76.9	79.4	80.5	80.8	81.2	81.1	82.2	82.2													
15-year	56.1	58.3	68.1	74.2	76.0	76.2	78.5	79.5	80.3	80.5	80.5	81.7														
16-year	55.2	57.3	67.3	73.3	75.0	75.6	77.5	78.9	79.5	80.0	79.9															
17-year	54.2	56.3	66.5	72.4	74.2	74.9	77.1	78.2	79.0	79.4																
18-year	53.4	55.6	65.7	71.7	73.0	74.2	76.4	77.7	78.1																	
19-year	52.6	55.0	65.1	71.1	72.5	73.2	75.5	76.7																		
20-year	51.9	54.2	64.5	70.3	71.9	72.4	74.6																			
21-year	51.1	53.5	63.8	69.4	71.2	71.2																				
22-year	50.6	52.8	63.2	68.7	70.5																					
23-year	50.1	52.2	62.6	68.2																						
24-year	49.4	51.6	62.0																							
25-year	48.7	51.1	61.4																							
26-year	48.4	50.7	60.6																							
27-year	48.0	50.1	59.7																							
28-year	47.6	49.6	59.1																							
29-year	47.1	48.8																								
30-year	46.7	48.2																								

Figure A8: Cancer of the female breast (invasive). Relative survival (%) by year of diagnosis. All races, females.

	Year of Diagnosis																									
Survival Time	1975-1979	1980-1984	1985-1989	1990-1994	1995	1996	1997	1998	1999	2000	2001	2002	2003	2004	2005	2006	2007	2008	2009	2010	2011	2012	2013	2014	2015	2016
1-year	86.5	87.2	87.0	87.7	89.3	88.6	89.1	88.0	89.7	88.3	85.8	87.8	86.7	86.2	87.1	86.3	88.3	87.1	88.8	87.8	85.1	87.1	86.5	86.9	88.3	87.2
2-year	77.9	77.3	77.9	79.5	82.2	81.8	80.0	81.4	82.8	81.0	79.0	79.7	79.1	77.3	77.2	78.6	78.7	79.5	79.9	80.0	76.3	78.0	79.8	80.5	81.3	
3-year	73.2	72.7	73.1	75.2	78.8	77.8	75.9	76.8	78.4	76.5	75.5	74.9	75.8	73.1	72.3	75.8	74.8	75.7	74.5	73.9	72.4	73.9	74.8	76.3		
4-year	70.3	69.7	70.5	72.2	76.0	74.7	73.3	74.1	76.1	74.3	72.4	72.5	72.5	70.2	69.9	73.7	73.0	72.8	71.4	71.4	69.6	70.9	71.3			
5-year	68.3	67.6	68.5	70.4	74.9	73.6	71.8	72.6	74.4	72.2	70.4	69.9	70.4	67.5	68.0	72.0	71.2	70.1	68.8	69.5	67.9	69.3				
6-year	67.1	65.7	67.2	69.0	73.6	72.9	70.2	71.0	73.4	71.1	69.1	68.9	68.8	66.0	67.1	70.1	70.5	67.7	67.8	68.2	65.5					
7-year	66.0	64.1	66.1	68.1	72.5	71.7	69.8	69.9	73.0	70.3	67.8	68.2	67.5	65.3	66.8	69.1	69.4	66.6	66.3	67.3						
8-year	65.1	62.8	65.0	67.2	71.8	70.8	69.3	69.5	72.6	70.0	67.4	67.6	66.2	63.6	65.8	68.6	68.2	65.8	65.9							
9-year	64.3	62.2	64.4	66.2	70.8	69.3	67.7	68.8	72.2	68.9	66.9	67.3	65.8	62.7	65.1	67.7	67.5	65.0								
10-year	63.1	61.4	63.7	65.6	70.0	68.8	67.1	67.8	70.9	68.3	66.7	66.6	65.5	61.8	64.5	67.5	67.0									
11-year	62.5	60.8	63.1	65.0	69.6	68.2	66.2	67.1	69.5	67.9	65.9	66.2	64.6	61.3	63.8	66.8										
12-year	61.7	60.0	62.3	64.4	68.6	67.1	66.0	66.1	68.8	66.9	64.9	65.5	64.3	60.7	63.6											
13-year	61.1	59.5	61.6	63.8	67.1	66.5	65.7	64.8	67.8	66.0	64.0	65.2	64.1	59.9												
14-year	60.5	58.7	61.0	63.2	66.6	65.8	65.4	64.6	67.2	65.6	63.6	64.7	63.5													
15-year	59.8	57.9	60.4	62.7	66.3	65.1	65.0	64.3	67.0	64.9	62.7	63.8														
16-year	59.3	57.2	59.7	62.0	65.4	64.7	63.4	64.1	66.7	63.9	62.2															
17-year	58.6	56.3	58.9	61.6	64.6	63.8	62.6	63.8	66.5	63.5																
18-year	58.2	55.5	58.7	60.8	64.1	62.9	62.6	63.2	65.9																	
19-year	57.4	54.9	58.3	60.3	63.7	62.2	62.0	63.1																		
20-year	56.6	54.2	57.3	59.9	63.3	61.0	61.4																			

Figure A9: Cancer of the cervix uteri (invasive). Relative survival (%) by year of diagnosis. All races, females.

Year of Diagnosis

Survival Time	1975-1979	1980-1984	1985-1989	1990-1994	1995	1996	1997	1998	1999	2000	2001	2002	2003	2004	2005	2006	2007	2008	2009	2010	2011	2012	2013	2014	2015	2016
1-year	73.9	77.0	79.9	80.8	79.5	81.1	80.2	81.9	82.4	81.9	83.0	83.2	82.5	83.4	83.9	83.5	84.1	84.6	84.3	84.2	83.8	84.0	84.1	84.5	85.0	84.5
2-year	63.0	66.8	71.1	72.5	71.4	73.3	72.3	74.6	74.7	74.4	75.5	76.9	75.8	76.6	76.9	77.2	77.3	78.3	77.3	77.4	76.9	76.5	77.5	78.1	78.0	
3-year	56.9	60.5	65.7	67.2	65.9	68.5	67.3	69.5	70.1	70.2	71.1	72.2	70.8	71.5	72.6	72.2	72.4	73.4	72.6	72.3	72.0	71.6	72.3	72.9		
4-year	53.0	56.5	61.9	63.3	62.0	65.0	64.2	66.3	66.9	67.0	68.2	69.0	67.7	68.4	68.9	68.5	69.2	70.1	69.1	68.9	68.5	68.3	68.6			
5-year	50.4	53.7	59.3	60.7	59.8	62.5	61.4	63.2	64.6	64.8	66.0	66.2	65.4	65.8	66.2	66.2	66.8	67.5	66.6	66.4	65.9	65.5				
6-year	48.6	51.8	57.5	59.0	57.3	61.2	59.7	61.4	63.1	63.1	64.1	64.4	63.6	63.9	64.6	64.7	64.8	65.6	64.8	64.2	64.2					
7-year	47.4	50.3	56.0	57.5	55.6	60.2	58.8	60.1	62.0	61.7	62.3	63.1	62.1	62.7	63.0	63.0	63.4	63.9	63.2	62.8						
8-year	46.4	49.3	55.0	56.3	54.9	59.5	58.0	59.3	60.7	60.3	61.3	62.1	61.4	61.8	62.0	61.8	62.2	62.8	61.9							
9-year	45.8	48.4	54.1	55.4	54.6	58.7	57.1	58.0	59.8	59.5	60.4	60.8	60.2	61.0	61.3	60.8	61.6	61.7								
10-year	45.1	47.8	53.3	54.6	53.9	58.1	56.4	57.3	59.2	59.0	59.6	60.5	59.1	60.1	60.4	60.4	60.4									
11-year	44.5	47.2	52.9	54.1	53.1	57.5	55.6	56.5	58.6	58.1	59.2	59.8	58.3	58.9	59.9	60.2										
12-year	44.1	46.8	52.5	53.6	52.4	57.4	55.3	56.2	57.9	57.7	58.3	58.8	58.0	58.9	59.1											
13-year	43.7	46.4	51.7	53.0	51.7	56.8	55.0	55.7	57.7	57.2	57.7	58.4	57.0	57.9												
14-year	43.2	45.9	51.4	52.6	51.2	55.9	54.1	55.6	57.0	56.8	57.1	57.5	56.2													
15-year	42.8	45.6	50.9	52.2	50.6	55.3	53.6	55.5	56.6	56.6	56.6	56.6														
16-year	42.4	45.2	50.7	52.2	50.1	55.0	52.8	55.2	55.7	56.0	55.9															
17-year	42.0	44.7	50.0	51.6	49.6	53.9	52.3	54.8	54.7	55.6																
18-year	41.5	44.3	49.7	51.1	49.0	53.8	51.7	54.5	54.2																	
19-year	41.5	43.9	49.5	50.5	48.2	53.1	51.5	54.4																		
20-year	41.4	43.7	49.0	49.9	47.3	52.3	51.1																			
21-year	40.9	42.9	49.0	49.4	46.5	51.9																				
22-year	40.6	42.4	48.9	48.7	45.9																					
23-year	40.1	42.3	48.4	48.3																						
24-year	39.8	41.9	47.8																							
25-year	39.5	41.4	47.5																							
26-year	39.0	41.0	46.9																							
27-year	38.5	40.6	46.4																							
28-year	38.4	40.2	45.5																							
29-year	38.0	40.0																								
30-year	37.5	39.7																								

Figure A10: Cancer of the colon and rectum (invasive). Relative survival (%) by year of diagnosis. All races, males and females.

Year of Diagnosis

Survival Time	1975-1979	1980-1984	1985-1989	1990-1994	1995	1996	1997	1998	1999	2000	2001	2002	2003	2004	2005	2006	2007	2008	2009	2010	2011	2012	2013	2014	2015	2016
1-year	93.7	91.3	92.2	92.1	92.9	91.5	93.1	93.3	93.9	93.6	92.1	92.0	92.8	92.8	92.8	92.8	93.1	93.3	92.8	93.3	92.2	93.1	92.7	93.2	93.0	92.1
2-year	90.1	86.3	87.5	88.1	88.4	88.0	88.9	89.7	89.5	89.5	88.1	88.2	88.6	88.3	89.0	89.5	89.5	88.8	88.3	89.0	88.0	88.8	87.9	88.4	89.4	
3-year	87.8	83.4	84.9	85.6	86.0	85.7	86.2	86.6	87.2	87.4	85.7	85.6	86.1	85.4	85.8	87.2	87.3	86.1	85.8	87.0	85.5	85.7	85.5	85.5		
4-year	86.3	82.0	83.4	84.1	84.7	84.6	85.2	84.5	85.5	86.0	84.2	83.7	84.6	83.8	84.2	85.6	85.7	84.5	84.5	84.9	83.8	84.2	84.0			
5-year	85.8	80.9	82.3	83.3	84.1	83.2	84.1	83.2	84.3	85.2	83.1	82.8	83.4	82.6	82.9	84.9	83.9	82.9	83.2	83.5	82.6	83.0				
6-year	85.1	80.3	81.7	82.2	83.4	82.1	83.3	82.5	83.7	84.4	82.2	82.4	82.5	81.7	81.9	84.4	83.3	82.0	81.9	82.9	81.4					
7-year	84.7	79.8	81.3	81.9	82.9	81.3	82.6	82.2	82.6	83.9	81.5	81.3	82.0	80.7	81.3	83.8	82.4	81.3	81.3	82.3						
8-year	84.5	79.1	80.8	81.2	82.5	80.4	81.6	81.9	82.0	83.3	80.8	81.0	81.9	80.3	80.9	83.5	81.9	80.8	80.8							
9-year	84.5	78.7	80.5	81.0	81.5	79.7	81.0	81.2	82.0	83.2	80.1	80.3	81.4	79.9	80.4	83.2	81.5	80.4								
10-year	84.3	78.4	80.3	80.5	81.2	79.5	81.0	81.0	81.3	82.9	80.0	80.1	80.3	79.6	79.8	82.5	81.3									
11-year	84.1	78.1	80.0	80.4	81.1	79.5	81.0	81.0	81.2	81.9	79.2	80.1	80.1	79.1	78.9	82.3										
12-year	83.9	77.8	79.6	79.9	80.9	79.5	81.0	80.8	80.9	81.8	78.6	80.0	79.4	78.9	78.8											
13-year	83.9	77.5	79.2	79.6	80.7	79.2	81.0	80.6	80.4	81.5	78.1	80.0	78.8	78.2												
14-year	83.7	77.1	78.9	79.2	80.2	78.9	81.0	79.7	80.0	81.4	78.1	79.6	78.1													
15-year	83.7	76.8	78.6	79.1	79.7	78.4	81.0	79.7	80.0	80.2	78.1	79.4														
16-year	83.4	76.2	78.1	78.8	79.4	77.5	80.8	78.8	79.2	79.8	78.1															
17-year	83.4	75.9	78.0	78.3	79.0	76.5	80.2	78.0	78.9	79.3																
18-year	83.4	75.5	77.8	77.8	78.1	75.7	79.4	78.0	78.4																	
19-year	83.4	74.8	77.7	77.2	77.5	75.4	78.9	77.7																		
20-year	83.4	74.7	77.2	76.5	76.6	74.5	78.9																			

Figure A11: Cancer of the corpus and uterus (invasive). Relative survival (%) by year of diagnosis. All races, females.

	Year of Diagnosis																									
Survival Time	1975-1979	1980-1984	1985-1989	1990-1994	1995	1996	1997	1998	1999	2000	2001	2002	2003	2004	2005	2006	2007	2008	2009	2010	2011	2012	2013	2014	2015	2016
1-year	25.8	30.0	34.9	38.2	40.6	42.1	40.0	40.8	44.3	45.9	46.2	41.8	47.0	47.9	45.3	47.4	50.2	47.7	48.5	49.6	49.4	50.8	51.9	50.9	50.7	54.8
2-year	11.5	14.3	19.0	21.9	22.6	24.4	23.9	23.9	26.7	28.3	29.3	26.0	30.4	30.2	31.0	30.8	33.1	32.3	30.7	33.9	32.4	34.3	36.2	35.2	33.6	
3-year	7.7	10.0	13.5	16.3	16.8	18.7	17.5	18.4	20.7	23.7	23.0	20.9	24.0	25.8	24.3	25.1	27.1	24.5	24.6	27.8	24.7	26.9	29.0	27.8		
4-year	6.0	8.4	10.8	13.7	13.2	16.0	14.3	14.2	18.3	20.6	19.2	17.9	20.1	22.3	21.4	21.8	23.6	21.9	21.7	24.9	22.1	23.3	25.1			
5-year	4.9	7.0	9.5	12.2	11.6	15.1	12.4	12.6	17.4	18.9	17.8	16.8	17.2	20.2	18.9	20.1	21.8	18.8	20.7	22.6	18.9	20.6				
6-year	4.3	6.2	8.5	11.1	10.8	14.0	11.3	11.6	15.9	17.9	15.9	15.3	14.6	19.6	17.3	18.7	19.6	18.0	19.6	20.3	17.8					
7-year	3.8	5.6	7.7	10.2	10.2	13.0	10.6	11.3	14.7	16.6	14.8	14.9	13.3	18.3	15.7	17.9	18.4	16.8	18.7	19.3						
8-year	3.4	4.9	7.0	9.5	9.6	12.2	9.8	9.8	14.2	15.8	13.8	13.7	13.0	17.9	14.7	16.7	18.4	16.1	17.5							
9-year	3.2	4.5	6.4	9.0	8.9	11.5	9.3	9.6	13.5	15.6	13.6	13.2	12.9	16.5	13.7	15.9	17.6	15.7								
10-year	2.9	4.3	6.0	8.3	8.6	10.7	9.2	9.4	12.9	15.1	13.3	13.0	11.9	15.5	13.3	15.2	16.8									
11-year	2.7	4.0	5.4	7.8	7.6	9.7	9.0	9.1	12.7	14.3	12.3	13.0	11.9	14.9	13.2	14.0										
12-year	2.6	3.8	4.9	7.1	7.2	9.4	7.6	8.5	11.9	14.0	12.0	12.4	11.5	14.6	12.8											
13-year	2.4	3.5	4.6	7.0	6.8	9.4	7.0	8.5	11.0	12.2	11.8	11.4	11.3	13.8												
14-year	2.4	3.4	4.3	6.6	5.7	9.0	6.4	8.4	10.4	12.0	11.2	10.9	10.6													
15-year	2.4	3.4	4.2	6.2	5.5	8.8	6.0	7.6	10.2	11.6	11.2	10.9														
16-year	2.3	3.3	3.6	5.8	4.9	8.0	5.9	7.5	9.9	10.5	11.2															
17-year	2.3	3.0	3.2	5.5	4.9	7.1	5.9	7.5	9.7	10.1																
18-year	2.2	2.7	3.1	5.3	4.9	7.1	5.6	7.5	9.2																	
19-year	2.2	2.5	2.9	4.9	3.8	6.3	5.0	7.5																		
20-year	2.1	2.5	2.6	4.8	3.0	5.3	5.0																			

Figure A12: Cancer of the esophagus (invasive). Relative survival (%) by year of diagnosis. All races, males and females.

Year of Diagnosis

Survival Time	1975-1979	1980-1984	1985-1989	1990-1994	1995	1996	1997	1998	1999	2000	2001	2002	2003	2004	2005	2006	2007	2008	2009	2010	2011	2012	2013	2014	2015	2016
1-year	86.9	89.4	90.7	92.1	92.8	91.6	93.5	94.6	91.0	92.4	94.0	92.9	93.2	91.3	93.5	92.4	93.5	94.7	92.3	91.9	93.7	94.5	93.8	93.9	93.0	94.3
2-year	80.8	83.6	86.3	87.7	88.6	88.2	90.6	90.5	87.8	90.6	90.8	90.0	92.2	89.4	91.1	90.1	92.5	92.8	89.6	90.8	92.5	92.9	92.4	93.3	92.4	
3-year	76.5	80.1	83.0	85.1	86.7	86.8	89.5	88.3	85.4	88.7	88.5	88.4	90.1	87.3	90.5	89.0	90.8	91.7	88.4	89.8	91.4	91.4	91.3	92.5		
4-year	73.7	77.4	80.6	83.2	83.5	84.6	88.6	86.6	84.2	87.9	86.6	86.2	88.1	86.1	89.6	88.5	89.2	90.8	87.3	88.7	90.1	90.7	91.0			
5-year	71.4	74.8	79.0	81.6	82.2	83.3	87.3	84.9	83.4	86.4	85.2	85.2	87.5	85.9	88.3	88.1	88.6	90.5	85.7	86.8	89.8	90.3				
6-year	69.0	72.3	77.5	80.0	80.2	81.8	86.1	83.8	82.2	85.6	84.3	84.7	87.0	84.9	87.9	86.9	87.1	89.7	84.6	85.6	88.6					
7-year	67.1	70.7	75.2	78.6	79.0	80.1	85.4	83.0	80.9	84.9	84.3	83.4	86.4	84.3	87.4	86.3	85.4	88.0	83.8	85.5						
8-year	65.0	69.3	74.1	77.8	78.5	79.4	84.9	82.4	80.0	83.9	83.8	82.3	85.7	83.4	85.8	86.2	84.2	87.4	82.9							
9-year	62.9	67.9	73.3	76.9	77.3	78.3	84.8	81.5	79.0	83.0	82.8	81.3	84.8	83.1	85.5	35.8	84.2	86.6								
10-year	61.7	66.8	72.6	76.1	76.8	76.8	84.1	80.2	78.3	83.0	81.9	80.5	84.5	82.6	85.4	85.3	84.2									
11-year	60.4	65.8	71.7	75.5	75.7	75.8	83.6	79.6	78.0	82.9	81.3	79.0	84.1	82.4	84.9	84.9										
12-year	59.5	64.7	70.9	74.8	75.6	74.9	83.2	79.2	77.6	82.4	81.3	78.3	83.5	81.9	84.6											
13-year	58.7	63.8	69.7	74.1	75.5	74.4	82.0	79.0	76.4	82.0	79.6	77.5	82.3	81.2												
14-year	57.5	62.7	68.7	73.6	75.5	73.6	80.5	78.4	76.3	81.2	79.5	77.5	81.9													
15-year	56.5	61.9	67.5	73.1	75.4	73.3	79.7	77.2	75.7	78.9	78.0	76.9														
16-year	55.7	60.9	66.9	72.6	75.2	73.3	79.5	77.1	74.1	77.9	77.2															
17-year	54.7	59.5	66.3	72.0	74.9	72.6	78.4	76.7	73.4	77.3																
18-year	53.9	58.6	65.6	71.3	73.9	72.1	77.4	76.1	73.2																	
19-year	53.2	58.1	64.9	70.4	73.9	71.8	76.6	75.6																		
20-year	52.1	57.0	64.2	69.9	73.5	70.9	76.3																			

Figure A13: Hodgkin's lymphoma (invasive). Relative survival (%) by year of diagnosis. All races, males and females.

Year of Diagnosis

Survival Time	1975-1979	1980-1984	1985-1989	1990-1994	1995	1996	1997	1998	1999	2000	2001	2002	2003	2004	2005	2006	2007	2008	2009	2010	2011	2012	2013	2014	2015	2016
1-year	94.4	65.4	59.0	59.0	61.8	69.0	72.8	73.8	76.8	74.1	80.5	78.1	75.3	82.6	82.1	81.7	82.8	80.0	82.3	83.5	85.2	80.6	86.6	87.3	83.7	82.2
2-year	89.6	44.5	32.8	29.4	46.8	60.4	65.1	65.5	65.1	66.3	73.4	71.1	72.5	73.5	78.8	79.0	80.4	76.1	79.5	77.7	81.5	80.5	83.9	80.1	79.2	
3-year	88.7	36.1	19.8	18.3	42.9	54.1	62.1	61.1	61.5	60.1	67.6	67.6	66.9	70.9	75.0	76.8	78.8	74.9	76.4	75.0	80.1	79.0	82.5	79.8		
4-year	83.8	32.0	13.8	14.0	38.8	51.6	58.9	59.0	56.7	57.0	65.6	66.4	65.0	69.3	74.2	76.1	76.8	74.9	74.7	73.4	79.2	79.0	78.1			
5-year	83.8	28.0	10.6	11.8	35.7	48.4	56.7	58.5	55.8	54.2	64.1	60.8	64.7	68.2	72.3	74.0	74.9	74.5	74.7	73.3	77.7	77.6				
6-year	76.9	24.5	8.9	10.9	35.1	45.6	54.6	55.7	53.4	52.8	63.8	59.1	64.4	67.0	70.7	73.1	73.5	73.9	74.1	71.7	76.2					
7-year	76.8	23.5	7.7	10.3	33.4	44.9	49.9	53.1	52.8	52.8	63.0	59.1	60.7	66.8	68.7	72.0	72.5	73.9	72.9	71.7						
8-year	76.8	20.8	6.8	9.9	32.2	44.5	47.2	50.4	52.4	52.4	61.6	57.3	58.5	66.7	67.5	70.2	72.5	71.8	69.6							
9-year	76.8	19.2	6.1	9.5	31.0	42.7	44.8	49.9	52.4	50.8	59.6	57.2	56.8	65.5	65.7	68.3	71.8	71.3								
10-year	76.8	17.2	5.9	9.4	29.9	42.7	43.7	49.9	51.5	48.9	56.9	57.2	56.3	63.0	65.1	66.3	68.4									
11-year	74.7	15.9	5.7	9.2	28.8	42.3	42.4	49.2	50.4	46.7	56.9	57.2	55.6	63.0	65.1	63.0										
12-year	74.7	15.6	5.6	9.2	28.0	41.9	41.2	49.2	50.4	46.1	56.1	57.2	54.7	62.5	64.1											
13-year	74.7	14.7	5.2	9.2	27.5	41.9	40.2	47.9	50.4	45.6	53.7	57.2	54.7	60.4												
14-year	74.7	14.7	5.0	8.9	26.9	40.6	40.2	45.6	49.2	45.6	53.7	57.2	54.1													
15-year	74.7	14.7	4.9	8.7	26.3	40.4	40.2	44.5	48.5	45.6	53.7	57.2														
16-year	74.7	14.7	4.8	8.5	25.8	40.1	40.2	42.9	47.9	45.6	53.2															
17-year	74.7	14.4	4.8	8.4	25.5	38.9	38.8	42.2	45.2	43.3																
18-year	74.7	13.2	4.8	8.2	25.4	38.2	37.7	42.1	45.2																	
19-year	74.7	13.2	4.6	8.0	25.4	38.2	37.7	40.5																		
20-year	74.7	13.2	4.5	7.8	24.9	37.9	37.7																			

Figure A14: Kaposi sarcoma (invasive). Relative survival (%) by year of diagnosis. All races, males and females.

	Year of Diagnosis																									
Survival Time	1975-1979	1980-1984	1985-1989	1990-1994	1995	1996	1997	1998	1999	2000	2001	2002	2003	2004	2005	2006	2007	2008	2009	2010	2011	2012	2013	2014	2015	2016
1-year	68.8	70.4	72.9	76.4	77.0	77.0	76.3	77.7	78.0	79.7	81.2	79.6	81.8	85.0	83.9	85.0	86.4	85.2	86.5	86.4	85.9	87.3	87.8	86.4	87.1	88.2
2-year	60.3	61.7	65.0	69.2	69.7	70.5	70.0	71.2	71.8	73.8	73.9	74.7	76.2	79.6	80.1	81.8	82.5	81.2	82.2	81.6	81.0	83.4	83.6	82.1	83.6	
3-year	56.3	57.0	60.8	65.8	65.5	67.3	66.1	67.5	67.6	69.3	70.8	71.3	73.4	76.4	77.7	78.5	80.3	78.9	79.4	79.3	77.9	80.7	81.3	79.6		
4-year	53.0	53.9	57.8	63.2	63.3	64.3	63.7	64.5	64.5	67.5	68.1	69.2	71.1	73.5	76.0	75.9	77.7	76.4	77.6	77.0	75.1	78.9	78.9			
5-year	50.9	52.0	55.8	60.7	61.4	62.5	61.7	63.3	62.5	65.4	66.4	67.5	69.0	71.1	74.4	73.8	75.1	74.4	76.2	74.8	73.3	77.2				
6-year	49.3	49.7	53.8	59.1	59.4	60.5	59.7	61.2	60.5	63.8	64.9	65.5	67.9	69.1	72.8	72.5	73.7	73.2	74.4	73.9	72.1					
7-year	47.5	48.6	52.3	57.5	58.0	59.0	58.1	59.5	59.1	62.2	63.1	63.6	66.2	67.8	71.9	70.9	72.6	71.7	73.2	72.0						
8-year	46.2	47.4	51.1	56.1	56.8	58.6	56.0	58.0	57.3	60.2	61.9	61.8	64.8	65.8	70.9	69.5	71.0	70.6	72.0							
9-year	45.0	46.0	49.6	55.0	55.3	57.8	54.1	56.6	55.8	58.9	60.8	60.4	63.4	64.1	69.8	68.3	70.0	69.1								
10-year	44.0	44.9	48.0	53.3	54.3	55.9	53.0	55.5	53.9	56.8	59.5	59.4	62.3	62.5	69.0	67.5	68.5									
11-year	43.1	43.6	46.5	51.8	52.3	55.0	52.1	55.0	52.8	55.8	57.6	58.5	60.9	61.4	66.9	66.3										
12-year	42.2	42.8	45.3	50.7	50.7	53.5	51.4	54.3	52.0	55.1	56.2	57.7	59.5	60.3	66.1											
13-year	41.5	42.0	44.3	49.5	49.2	51.8	50.4	53.4	51.4	54.1	54.4	56.2	58.0	59.1												
14-year	40.5	41.4	43.0	48.2	48.2	50.3	49.8	52.2	50.1	52.7	53.6	55.9	56.8													
15-year	39.5	41.0	42.1	47.5	47.2	49.0	48.9	50.7	48.8	52.2	53.0	54.9														
16-year	39.2	40.4	41.1	46.8	46.9	48.1	47.6	49.6	47.8	51.2	52.0															
17-year	38.7	39.8	40.2	46.0	45.7	47.9	46.3	48.3	46.8	49.6																
18-year	37.6	39.3	39.3	45.0	44.6	47.3	45.2	47.5	46.4																	
19-year	36.7	38.3	38.5	44.3	44.0	46.3	44.7	46.5																		
20-year	36.1	37.3	37.9	43.4	43.1	45.0	43.6																			

Figure A15: Cancer of the kidney and renal pelvis (invasive). Relative survival (%) by year of diagnosis. All races, males and females.(%) by year of diagnosis. All races, males and females.

Year of Diagnosis

Survival Time	1975-1979	1980-1984	1985-1989	1990-1994	1995	1996	1997	1998	1999	2000	2001	2002	2003	2004	2005	2006	2007	2008	2009	2010	2011	2012	2013	2014	2015	2016
1-year	88.3	88.4	88.2	87.7	89.2	88.1	85.8	86.1	85.5	85.3	85.3	84.0	84.0	85.4	86.6	86.4	88.2	84.2	86.9	85.7	86.9	87.6	84.2	85.6	85.5	84.8
2-year	79.1	79.5	79.0	78.0	80.0	78.9	77.3	76.1	74.8	76.8	75.1	74.8	74.6	75.2	78.1	74.7	75.6	76.2	76.7	77.2	73.0	78.6	76.1	77.7	77.5	
3-year	73.3	74.1	73.6	71.8	74.9	74.1	72.3	70.8	70.7	71.0	70.0	70.0	66.7	68.6	72.6	68.4	70.6	71.2	71.7	70.5	67.4	72.9	69.3	71.0		
4-year	68.9	70.0	68.8	67.4	69.3	69.7	68.8	67.0	68.6	66.7	66.6	67.0	62.6	64.4	70.0	62.7	65.9	67.0	68.1	65.5	64.0	66.6	64.7			
5-year	65.5	66.7	65.8	64.5	66.1	65.8	65.2	63.4	63.8	63.3	62.9	62.7	60.3	62.4	65.7	59.4	62.1	64.5	64.1	60.7	61.6	63.2				
6-year	63.1	63.9	62.5	61.0	63.3	63.4	62.3	59.1	60.5	60.4	60.5	60.1	56.0	59.8	61.0	56.8	59.5	61.1	61.4	58.6	57.8					
7-year	60.8	61.3	60.1	59.2	60.0	61.2	59.7	55.5	57.2	56.5	57.0	56.4	54.2	57.0	58.4	53.7	55.8	58.1	58.0	56.1						
8-year	58.8	59.7	57.2	56.5	57.4	57.9	57.0	52.7	54.1	53.1	54.6	53.1	52.0	53.9	54.8	51.2	53.4	54.2	55.1							
9-year	56.4	57.3	54.6	53.9	54.3	54.0	55.5	50.3	52.1	50.8	52.0	52.2	49.8	52.0	51.8	48.3	51.5	50.9								
10-year	54.2	55.4	52.3	51.4	51.3	51.8	53.5	48.3	50.5	48.9	50.4	48.9	47.6	51.0	48.8	45.6	50.1									
11-year	52.7	53.0	50.2	49.4	47.8	49.1	50.9	47.0	47.7	47.6	47.0	45.9	44.8	48.6	46.3	42.4										
12-year	51.0	50.3	48.2	47.4	46.1	46.5	46.3	45.0	45.0	44.4	45.2	42.4	43.1	46.0	43.3											
13-year	49.7	48.0	47.0	45.0	44.3	44.5	45.1	43.2	42.1	43.0	43.3	39.4	41.9	42.9												
14-year	48.0	46.5	44.3	43.1	41.5	41.5	43.2	41.6	41.1	40.4	41.2	37.9	39.7													
15-year	45.8	44.8	41.9	41.4	39.0	38.8	42.3	39.6	39.2	38.6	39.3	36.2														
16-year	43.7	42.1	40.3	39.5	38.0	37.6	41.1	39.0	37.1	36.2	37.3															
17-year	41.4	40.5	38.9	37.6	37.5	35.5	38.4	36.0	34.0	35.8																
18-year	40.2	38.8	37.4	35.5	36.4	34.0	37.3	33.2	33.1																	
19-year	38.4	37.3	34.8	33.8	35.8	31.2	36.9	32.8																		
20-year	36.5	35.6	33.6	31.3	35.4	28.8	36.3																			

Figure A16: Cancer of the larynx (invasive). Relative survival (%) by year of diagnosis. All races, males and females.

Year of Diagnosis

Survival Time	1975-1979	1980-1984	1985-1989	1990-1994	1995	1996	1997	1998	1999	2000	2001	2002	2003	2004	2005	2006	2007	2008	2009	2010	2011	2012	2013	2014	2015	2016
1-year	60.7	62.6	64.8	66.8	68.2	68.2	67.8	66.7	66.0	66.9	70.6	72.2	73.4	75.0	74.1	76.5	77.0	76.7	77.4	78.3	79.6	79.2	79.0	79.2	81.0	79.1
2-year	50.1	52.5	55.2	57.5	58.9	59.9	59.7	58.1	58.5	59.0	63.1	65.0	66.3	67.7	68.4	70.7	70.7	70.4	71.0	72.7	73.5	72.8	73.4	73.8	75.5	
3-year	43.8	46.5	49.6	52.9	54.3	55.4	55.0	54.1	54.8	55.1	59.1	62.0	62.4	64.2	64.9	67.5	67.6	66.6	68.6	69.6	70.6	69.5	70.3	70.5		
4-year	38.9	41.7	45.5	49.1	50.7	51.9	51.0	50.4	51.7	52.0	56.8	59.7	59.6	61.9	62.8	64.8	65.4	64.1	66.1	67.9	68.7	67.8	67.6			
5-year	34.6	37.6	42.3	46.1	48.1	48.7	48.3	48.2	48.5	49.9	54.4	57.7	57.5	59.6	61.0	62.7	63.8	62.1	63.7	66.3	66.9	66.1				
6-year	31.4	35.0	39.8	43.6	45.7	46.5	45.8	46.2	46.3	48.6	52.4	56.1	55.7	57.2	59.4	60.8	62.3	60.6	61.5	65.2	66.0					
7-year	29.1	32.5	37.2	41.0	44.3	45.1	43.9	44.7	44.5	47.3	51.1	54.5	53.8	56.1	58.2	59.3	61.6	58.8	60.4	64.1						
8-year	27.0	30.4	35.0	38.8	41.7	43.9	43.3	43.2	43.0	45.8	49.7	53.6	52.0	54.7	56.3	57.4	60.5	57.8	58.6							
9-year	25.0	28.5	33.0	37.3	40.6	42.1	41.9	42.2	42.2	44.9	48.3	52.0	50.9	53.1	54.7	56.2	59.5	56.8								
10-year	23.7	26.9	31.6	36.1	39.2	40.7	40.7	41.1	40.7	43.9	47.1	50.6	49.1	51.5	53.4	54.6	58.5									
11-year	22.4	25.7	29.9	34.7	37.8	39.8	38.9	39.8	39.3	42.7	46.0	49.3	47.5	49.9	52.9	53.5										
12-year	21.2	24.6	28.8	33.6	37.1	38.7	37.8	39.1	38.4	42.1	45.0	48.5	46.3	49.3	52.3											
13-year	20.5	23.6	28.0	32.7	36.7	38.3	37.1	38.1	37.3	40.5	44.2	46.8	45.7	48.0												
14-year	19.8	22.9	27.3	31.9	35.6	37.0	36.9	37.1	36.8	39.3	43.4	45.9	44.9													
15-year	19.2	22.3	26.7	31.4	35.3	36.2	36.4	36.0	36.1	38.5	42.4	44.9														
16-year	18.5	21.4	26.1	30.7	35.0	35.9	36.3	35.1	35.5	37.7	41.9															
17-year	18.1	20.7	25.6	30.0	34.0	35.5	35.3	34.7	34.8	37.4																
18-year	17.7	20.1	25.3	29.5	33.0	35.1	34.7	34.5	34.6																	
19-year	17.2	19.5	24.8	29.2	32.6	34.3	34.5	34.5																		
20-year	16.7	19.1	24.5	28.7	32.0	33.8	33.6																			

Figure A17: Leukemia (invasive). Relative survival (%) by year of diagnosis. All races, males and females.

	Year of Diagnosis																									
Survival Time	1975-1979	1980-1984	1985-1989	1990-1994	1995	1996	1997	1998	1999	2000	2001	2002	2003	2004	2005	2006	2007	2008	2009	2010	2011	2012	2013	2014	2015	2016
1-year	13.7	14.9	17.8	21.4	23.4	24.7	27.7	26.8	27.7	29.0	31.2	32.9	36.3	38.3	39.1	39.2	42.1	44.8	44.6	48.4	47.0	46.5	49.0	49.1	52.5	52.5
2-year	6.6	8.3	9.8	12.4	13.7	15.0	17.4	17.0	17.8	19.9	21.7	22.3	24.8	26.3	27.0	28.8	30.3	32.5	31.2	36.1	33.5	32.4	35.3	36.8	39.3	
3-year	4.8	6.1	7.3	8.7	9.5	10.9	12.5	13.1	14.3	15.4	16.0	17.6	20.5	20.7	21.2	23.4	24.0	25.8	24.3	27.9	26.3	24.8	27.3	29.5		
4-year	4.1	4.6	6.3	6.9	6.9	9.2	10.1	10.7	12.0	12.4	13.8	15.5	17.1	17.6	18.4	19.9	21.0	21.6	19.8	22.6	22.7	20.5	23.2			
5-year	3.4	3.8	5.3	5.6	5.7	8.2	8.2	9.4	10.1	11.7	12.0	14.1	15.6	14.7	16.8	17.1	18.5	18.7	17.0	19.4	20.3	18.0				
6-year	3.0	3.6	5.0	5.0	4.9	7.1	7.4	8.4	9.1	10.8	11.0	12.0	15.0	13.3	15.4	15.6	16.5	16.9	15.3	17.3	18.6					
7-year	2.8	3.1	4.8	4.6	4.4	6.7	6.7	7.0	8.3	10.0	9.9	11.4	13.3	12.7	14.1	14.6	14.6	15.6	14.5	15.8						
8-year	2.7	2.9	4.4	4.4	4.2	6.5	5.9	6.6	7.8	9.6	8.9	10.4	12.3	12.1	13.5	13.7	13.7	14.3	13.1							
9-year	2.5	2.9	4.2	4.3	4.1	6.0	5.7	5.6	7.5	9.4	8.5	10.1	11.7	11.6	12.4	13.0	13.0	13.6								
10-year	2.5	2.9	4.0	4.2	3.6	5.7	5.5	5.5	7.4	9.0	7.9	9.3	11.2	10.8	11.7	12.6	12.0									
11-year	2.5	2.7	4.0	4.0	3.2	5.6	5.5	4.8	6.9	8.7	7.8	8.7	10.9	10.3	11.6	12.1										
12-year	2.5	2.7	3.8	3.9	3.1	5.3	5.5	4.6	6.8	8.6	7.6	8.1	10.6	9.9	11.6											
13-year	2.3	2.5	3.8	3.8	3.1	5.1	5.5	4.5	6.6	8.4	7.5	7.6	10.2	9.7												
14-year	2.3	2.5	3.7	3.7	3.1	4.9	5.5	4.5	6.1	8.4	7.2	7.4	10.0													
15-year	2.2	2.5	3.7	3.7	3.0	4.1	5.1	4.5	5.9	8.3	6.8	7.2														
16-year	2.2	2.4	3.7	3.6	3.0	4.0	4.8	4.5	5.4	8.2	6.6															
17-year	2.1	2.4	3.7	3.5	2.9	4.0	4.6	4.4	5.3	8.1																
18-year	2.0	2.3	3.7	3.4	2.8	4.0	4.4	4.1	5.3																	
19-year	2.0	2.2	3.7	3.3	2.8	4.0	4.4	4.1																		
20-year	2.0	2.2	3.7	3.3	2.8	3.9	4.4																			

Figure A18: Center of the liver and intrahepatic bile duct (invasive). Relative survival (%) by year of diagnosis. All races, males and females.

Year of Diagnosis

Survival Time	1975-1979	1980-1984	1985-1989	1990-1994	1995	1996	1997	1998	1999	2000	2001	2002	2003	2004	2005	2006	2007	2008	2009	2010	2011	2012	2013	2014	2015	2016
1-year	35.3	37.1	38.4	39.3	40.5	39.8	40.1	39.7	41.6	40.7	40.0	40.5	42.0	42.0	43.3	43.8	44.5	45.0	45.6	44.7	46.9	45.8	47.9	48.0	50.2	51.8
2-year	21.1	21.9	22.6	23.7	24.9	24.6	24.6	25.2	25.9	25.8	25.5	26.5	26.6	27.7	28.4	29.5	30.1	30.9	31.4	31.2	32.9	31.8	33.6	35.0	37.2	
3-year	16.3	16.9	17.1	18.3	19.1	19.3	19.3	19.9	19.8	20.2	19.8	20.7	20.8	22.3	22.6	23.5	24.0	24.8	25.3	25.0	26.5	25.6	27.4	28.9		
4-year	14.0	14.5	14.6	15.5	16.5	16.3	16.6	17.0	17.0	17.5	16.8	17.5	17.9	18.9	19.5	20.0	20.6	21.4	21.7	21.6	23.0	22.2	24.2			
5-year	12.5	12.9	13.0	13.8	14.5	14.6	14.7	14.6	15.2	15.7	15.0	15.6	16.0	16.8	17.4	17.6	18.4	18.9	19.6	19.4	20.5	19.9				
6-year	11.4	11.7	11.9	12.7	13.2	13.2	13.3	13.3	13.7	14.2	13.5	14.1	14.5	15.4	15.7	16.2	16.8	17.3	17.9	17.8	18.7					
7-year	10.7	10.8	10.9	11.6	12.3	12.1	12.2	12.2	12.7	13.0	12.4	12.9	13.3	14.1	14.6	15.1	15.4	16.0	16.6	16.2						
8-year	10.0	10.0	10.1	10.7	11.5	11.2	11.3	11.2	11.8	12.0	11.5	11.9	12.1	13.2	13.5	14.1	14.2	14.6	15.3							
9-year	9.4	9.2	9.4	9.9	10.7	10.3	10.4	10.5	11.1	11.2	10.6	11.1	11.1	12.5	12.5	12.7	13.3	13.6								
10-year	8.8	8.6	8.8	9.2	10.0	9.5	9.8	10.0	10.5	10.4	10.1	10.4	10.3	11.6	11.6	11.7	12.4									
11-year	8.3	8.2	8.2	8.6	9.4	8.8	9.2	9.4	9.7	9.8	9.4	9.5	9.8	10.8	10.9	11.2										
12-year	7.8	7.7	7.7	8.0	8.9	8.3	8.5	8.9	9.1	9.1	8.8	8.9	9.3	10.1	10.3											
13-year	7.4	7.2	7.2	7.5	8.2	7.9	8.0	8.1	8.6	8.6	8.2	8.3	8.7	9.4												
14-year	7.1	6.8	6.8	7.1	7.7	7.6	7.5	7.6	8.0	8.1	7.7	7.9	7.9													
15-year	6.6	6.4	6.4	6.7	7.3	7.0	7.1	7.1	7.6	7.7	7.2	7.6														
16-year	6.3	6.0	6.1	6.3	6.9	6.7	6.6	6.8	7.2	7.5	6.9															
17-year	6.0	5.6	5.7	5.9	6.4	6.3	6.2	6.3	6.8	7.1																
18-year	5.7	5.3	5.3	5.5	5.9	6.0	5.7	6.1	6.4																	
19-year	5.4	5.0	5.0	5.2	5.5	5.6	5.4	5.8																		
20-year	5.2	4.6	4.7	4.9	5.2	5.3	5.0																			
21-year	4.9	4.3	4.5	4.6	5.0	5.0																				
22-year	4.7	4.1	4.2	4.4	4.7																					
23-year	4.4	3.8	4.0	4.1																						
24-year	4.2	3.6	3.8																							
25-year	4.0	3.4	3.6																							
26-year	3.8	3.2	3.5																							
27-year	3.6	3.1	3.3																							
28-year	3.5	2.9	3.1																							
29-year	3.3	2.7																								
30-year	3.2	2.6																								

Figure A19: Cancer of the lung and bronchus (invasive). Relative survival (%) by year of diagnosis. All races, males and females.

Year of Diagnosis

Survival Time	1975-1979	1980-1984	1985-1989	1990-1994	1995	1996	1997	1998	1999	2000	2001	2002	2003	2004	2005	2006	2007	2008	2009	2010	2011	2012	2013	2014	2015	2016
1-year	94.2	95.6	96.3	96.7	97.3	97.6	96.6	96.6	97.7	97.8	97.4	97.9	97.7	97.7	97.8	97.6	97.8	97.9	98.1	97.8	97.8	98.1	98.6	97.9	98.3	98.6
2-year	89.8	91.7	93.2	93.9	94.4	95.7	94.2	94.5	95.6	96.2	95.5	95.5	96.1	96.1	96.3	96.1	96.3	95.9	96.1	96.7	96.4	96.6	97.9	96.7	97.3	
3-year	86.1	88.0	90.8	92.0	92.7	93.8	92.3	92.7	94.0	94.5	94.1	94.3	94.9	94.6	95.0	94.7	94.8	94.8	94.8	95.3	95.0	96.0	96.7	96.0		
4-year	83.4	85.5	89.2	90.3	91.3	93.2	91.3	91.5	93.2	93.2	92.7	93.2	93.7	93.6	94.3	93.7	94.1	94.2	93.9	94.6	94.3	95.5	96.2			
5-year	82.0	83.5	87.7	89.4	90.2	92.2	89.8	90.9	92.4	92.1	92.1	92.9	93.3	93.1	93.3	93.0	93.5	93.6	93.4	93.9	93.8	95.1				
6-year	80.2	82.3	86.6	88.5	90.0	91.5	89.3	90.4	91.7	91.8	91.3	92.5	92.5	93.1	93.0	92.3	92.9	92.8	93.0	93.7	93.5					
7-year	79.0	81.7	85.9	87.9	90.0	91.2	88.6	90.3	91.6	91.5	91.1	92.0	91.9	92.5	92.9	92.1	92.9	92.3	92.7	93.6						
8-year	77.9	80.9	85.3	87.3	89.9	90.9	87.8	89.9	91.2	91.4	90.9	91.2	91.8	92.4	92.8	92.1	92.5	92.2	92.7							
9-year	77.1	80.3	85.1	87.1	89.5	90.4	87.6	89.7	91.0	91.4	90.8	90.7	91.8	92.2	92.8	92.1	92.3	92.0								
10-year	76.6	79.8	84.9	86.9	89.4	89.9	87.6	89.3	91.0	91.3	90.2	90.6	91.5	92.2	92.8	92.1	92.1									
11-year	76.0	79.4	84.4	86.6	89.2	89.7	87.6	89.3	91.0	90.9	90.0	90.5	91.5	92.2	92.8	92.1										
12-year	75.6	79.2	84.2	86.5	89.2	89.7	87.1	88.8	91.0	90.9	90.0	90.5	91.5	92.2	92.8											
13-year	75.3	78.8	84.0	86.5	88.7	89.6	86.9	88.8	91.0	90.9	89.9	90.5	91.4	92.2												
14-year	75.1	78.6	83.9	86.4	88.7	89.5	86.7	88.7	91.0	90.9	89.9	90.5	91.3													
15-year	74.7	78.4	83.9	86.4	88.7	89.5	86.6	88.6	91.0	90.9	89.9	90.5														
16-year	74.5	78.4	83.8	86.4	88.7	89.5	86.6	88.0	91.0	90.9	89.9															
17-year	74.5	78.1	83.8	86.4	88.7	89.2	86.6	88.0	91.0	90.9																
18-year	74.5	78.1	83.8	86.4	88.7	89.2	86.6	88.0	91.0																	
19-year	74.5	78.1	83.8	86.4	88.7	89.2	86.6	88.0																		
20-year	74.5	78.1	83.8	86.4	88.7	89.2	86.6																			

Figure A20: Melanoma of the skin (invasive). Relative survival (%) by year of diagnosis. All races, males and females.

Year of Diagnosis

Survival Time	1975-1979	1980-1984	1985-1989	1990-1994	1995	1996	1997	1998	1999	2000	2001	2002	2003	2004	2005	2006	2007	2008	2009	2010	2011	2012	2013	2014	2015	2016
1-year	34.9	37.6	35.6	35.3	36.3	41.5	35.3	37.4	31.7	40.0	40.4	32.1	47.1	39.5	40.4	44.5	43.9	45.3	45.4	33.5	47.8	40.3	41.5	40.0	43.8	44.8
2-year	16.1	18.7	16.5	15.7	13.8	25.1	16.9	21.2	18.6	14.6	24.0	16.1	22.3	16.6	19.6	28.4	18.7	21.1	26.1	16.5	26.8	24.7	25.8	21.5	22.1	
3-year	10.9	11.7	10.8	11.0	11.8	17.5	10.3	14.7	12.8	10.5	13.8	7.4	16.8	10.4	12.7	19.6	11.5	14.9	16.3	8.5	17.9	14.9	19.0	14.0		
4-year	9.2	8.4	8.0	8.2	8.9	11.4	7.8	12.6	10.5	7.0	11.4	5.2	13.5	8.5	10.2	14.8	7.1	12.5	14.3	6.0	15.8	10.5	17.3			
5-year	7.9	7.0	6.6	7.1	7.1	10.2	5.7	12.6	8.2	3.1	10.9	4.8	9.0	6.0	9.2	12.6	4.8	10.4	12.9	4.9	13.6	9.5				
6-year	6.7	5.7	5.2	6.0	6.1	10.0	4.6	11.3	7.8	1.3	9.9	4.3	8.0	6.0	9.2	11.3	4.3	8.5	11.2	3.9	10.2					
7-year	6.2	5.1	4.5	5.6	5.1	8.6	3.0	10.6	7.4	1.3	7.7	3.8	7.0	4.6	9.1	10.3	3.7	7.3	10.8	3.9						
8-year	5.6	4.8	4.0	5.0	5.1	6.5	3.0	8.5	7.4	0.7	7.7	3.8	5.7	4.1	8.2	9.2	3.7	7.3	10.8							
9-year	5.3	4.6	3.8	4.7	3.0	4.9	2.5	8.5	6.0	0.7	7.7	3.8	5.7	3.5	8.2	6.4	3.7	7.0								
10-year	5.2	4.3	3.4	4.3	3.0	4.4	2.5	7.7	6.0	0.7	6.8	3.7	4.0	3.5	8.2	6.4	3.7									
11-year	5.0	4.3	3.4	4.1	3.0	4.0	1.9	7.4	5.6	0.7	5.0	3.7	4.0	3.5	8.2	6.4										
12-year	4.9	4.1	3.4	3.8	2.5	3.5	1.9	5.6	5.6	0.7	5.0	3.3	4.0	3.1	7.2											
13-year	4.7	4.0	3.3	3.6	2.5	3.0	1.9	5.6	5.6	0.7	5.0	2.8	3.5	2.5												
14-year	4.7	4.0	3.3	3.6	2.5	3.0	1.9	5.6	5.2	0.7	4.0	2.8	3.5													
15-year	4.3	3.5	3.2	3.6	2.5	3.0	1.9	5.6	5.2	0.0	3.2	1.9														
16-year	4.1	3.5	3.2	3.6	2.5	3.0	1.9	5.6	4.7	0.0	3.2															
17-year	4.0	3.5	3.2	3.6	2.5	2.9	1.9	5.4	4.7	0.0																
18-year	4.0	3.5	3.2	3.2	2.5	2.0	1.9	4.8	4.7																	
19-year	4.0	3.5	3.2	2.5	2.5	2.0	1.9	4.1																		
20-year	4.0	3.5	3.1	2.5	2.5	2.0	1.9																			

Figure A21: Mesothelioma (invasive). Relative survival (%) by year of diagnosis. All races, males and females

	Year of Diagnosis																									
Survival Time	1975-1979	1980-1984	1985-1989	1990-1994	1995	1996	1997	1998	1999	2000	2001	2002	2003	2004	2005	2006	2007	2008	2009	2010	2011	2012	2013	2014	2015	2016
1-year	67.9	69.4	72.2	71.9	74.1	70.6	74.3	74.9	73.7	72.5	72.7	73.5	77.4	74.6	76.1	73.9	77.3	77.9	81.9	81.4	80.1	81.7	81.4	83.2	82.3	83.7
2-year	52.1	54.4	57.6	57.5	58.7	57.4	60.3	62.4	60.1	60.0	59.6	63.0	65.1	66.5	64.6	66.5	67.1	69.1	72.8	72.2	71.9	74.0	72.6	75.4	74.1	
3-year	39.9	43.0	45.4	46.3	47.8	44.8	48.6	51.2	48.0	49.8	49.7	53.6	56.0	58.7	57.6	58.3	58.5	61.9	65.6	65.7	65.2	67.1	66.7	69.1		
4-year	31.5	34.3	35.8	37.0	40.2	38.2	38.7	40.2	39.1	41.3	44.4	47.2	47.9	50.9	51.0	51.4	50.8	56.2	60.0	59.1	58.2	60.8	61.3			
5-year	25.1	27.0	27.4	29.9	33.3	31.7	31.3	34.3	31.8	34.3	37.5	40.2	42.3	45.4	46.7	46.3	46.6	50.6	53.2	53.3	51.3	56.3				
6-year	20.1	21.0	22.1	24.6	28.4	27.2	26.3	30.0	27.3	29.3	32.9	35.6	37.8	40.1	41.6	41.3	41.7	44.9	49.0	49.7	46.4					
7-year	16.3	16.7	17.6	20.5	23.7	23.7	22.4	24.7	23.7	25.4	29.5	32.1	33.3	35.8	37.1	36.7	37.5	42.0	44.2	45.9						
8-year	13.6	13.4	15.0	17.4	20.4	21.4	19.8	22.5	20.3	22.8	26.0	28.9	30.2	32.1	33.8	33.0	34.3	38.7	40.8							
9-year	11.5	11.3	11.9	15.4	18.2	19.5	17.1	19.8	18.5	20.6	23.4	26.0	27.4	30.1	31.3	31.0	31.0	35.8								
10-year	10.0	9.6	10.4	13.8	17.1	17.2	15.5	18.1	16.7	19.5	21.4	23.4	24.4	27.1	28.6	28.9	28.6									
11-year	8.8	8.7	9.2	12.8	15.6	15.6	14.3	17.0	14.4	18.6	20.1	20.5	22.1	25.1	25.5	27.6										
12-year	7.8	7.7	8.4	11.2	14.3	14.9	13.0	15.4	13.9	17.0	19.1	19.3	20.8	23.2	23.9											
13-year	6.6	6.6	7.6	10.4	13.6	13.3	11.9	14.4	13.0	16.0	17.1	18.4	19.1	22.2												
14-year	6.1	6.1	7.0	9.6	12.4	12.7	11.2	14.1	11.7	15.0	16.3	17.3	17.8													
15-year	5.4	5.7	6.1	9.3	12.0	11.6	11.0	12.5	11.4	14.2	15.2	16.1														
16-year	4.9	5.2	5.7	8.7	10.9	10.0	10.5	11.5	10.7	13.4	14.4															
17-year	4.6	4.7	5.5	8.2	10.1	9.0	9.9	10.8	10.4	12.7																
18-year	4.1	4.3	5.1	7.6	9.9	8.5	8.7	10.1	9.5																	
19-year	3.8	3.8	5.0	7.0	9.0	7.9	8.1	9.5																		
20-year	3.4	3.5	4.8	6.9	7.8	7.3	7.9																			

Figure A22: Myeloma (invasive). Relative survival (%) by year of diagnosis. All races, males and females.

Year of Diagnosis

Survival Time	1975-1979	1980-1984	1985-1989	1990-1994	1995	1996	1997	1998	1999	2000	2001	2002	2003	2004	2005	2006	2007	2008	2009	2010	2011	2012	2013	2014	2015	2016
1-year	70.2	73.0	71.2	70.4	70.4	73.2	75.5	76.6	77.5	77.2	78.6	80.9	80.3	81.0	81.6	80.9	81.7	84.1	83.8	84.3	83.5	84.1	84.4	83.5	84.6	85.0
2-year	60.2	63.7	62.8	62.2	61.9	65.8	68.4	70.1	71.1	71.4	73.3	76.1	75.6	76.0	77.3	76.9	77.6	78.9	79.5	79.6	79.8	80.0	80.1	79.8	80.8	
3-year	54.6	58.5	58.2	57.7	57.6	62.0	64.9	66.4	67.0	67.9	70.1	73.2	73.1	73.5	74.8	74.5	75.1	76.2	77.3	77.5	77.6	77.9	78.1	78.2		
4-year	50.6	54.6	54.4	54.4	54.8	58.8	62.1	63.6	64.6	65.5	68.1	71.2	71.6	71.5	73.1	72.6	73.3	74.5	75.6	75.3	75.7	76.9	76.6			
5-year	46.8	50.7	51.0	51.6	52.1	56.0	59.3	61.0	62.1	63.8	66.0	69.3	70.0	70.1	71.6	70.5	71.4	72.8	74.1	74.3	74.3	75.2				
6-year	43.6	47.5	48.3	49.3	49.9	54.1	57.1	58.3	59.6	62.1	64.6	68.1	68.3	68.7	70.3	68.9	70.3	71.8	72.2	72.4	72.3					
7-year	40.8	45.0	45.9	47.0	47.9	52.2	55.2	56.5	57.8	60.5	63.0	66.3	67.2	67.2	68.5	67.3	68.9	70.2	71.1	71.4						
8-year	38.5	42.9	43.4	45.1	46.5	50.5	53.0	55.2	56.3	58.9	61.3	65.1	65.7	66.6	67.1	66.2	67.4	69.1	70.2							
9-year	36.6	40.8	41.5	43.5	45.3	49.3	51.5	53.4	54.9	57.5	60.0	64.0	64.8	64.9	65.2	65.2	66.2	68.0								
10-year	35.0	38.9	39.8	42.1	44.4	48.1	50.1	52.2	53.7	56.2	58.9	63.2	63.3	63.8	63.5	63.8	65.3									
11-year	33.5	37.4	38.5	40.6	43.4	46.5	48.8	51.1	52.6	54.8	58.1	61.8	62.2	63.0	63.2	63.1										
12-year	32.4	36.0	36.9	39.6	42.3	45.7	48.2	50.2	51.3	53.6	56.9	60.1	61.6	61.7	62.3											
13-year	31.4	34.8	35.8	38.3	41.6	44.8	46.9	49.2	50.0	52.0	56.1	59.4	60.3	60.8												
14-year	30.2	33.5	34.7	37.2	40.7	44.3	45.7	48.7	48.9	51.3	54.9	58.3	59.6													
15-year	28.9	32.2	33.5	36.6	39.9	43.7	44.7	47.5	47.9	50.5	54.1	58.2														
16-year	28.0	31.2	32.8	35.7	39.4	43.6	43.4	47.3	47.2	49.2	53.6															
17-year	27.4	30.4	32.1	34.9	38.9	42.8	42.8	46.9	45.9	48.1																
18-year	26.8	29.6	31.4	34.1	37.9	42.0	42.0	46.3	45.4																	
19-year	26.1	28.8	30.6	33.3	37.1	41.3	41.3	45.8																		
20-year	25.2	28.0	30.1	32.9	35.9	40.9	40.9																			

Figure A23: Non-Hodgkin's lymphoma (invasive). Relative survival (%) by year of diagnosis. All races, males and females.

Year of Diagnosis

Survival Time	1975-1979	1980-1984	1985-1989	1990-1994	1995	1996	1997	1998	1999	2000	2001	2002	2003
1-year	79.3	79.1	80.4	81.0	82.7	81.7	83.2	80.7	83.7	84.3	82.4	83.3	84.0
2-year	65.9	65.9	67.5	69.4	71.6	69.6	71.1	70.2	72.3	73.5	72.9	72.8	75.0
3-year	59.8	59.5	61.1	63.2	65.4	63.5	66.2	64.3	65.6	68.7	67.1	67.5	70.4
4-year	55.9	55.5	57.1	59.0	61.6	60.0	62.1	60.2	62.2	64.9	62.6	64.1	67.1
5-year	52.8	52.3	54.0	56.1	58.5	57.6	58.5	57.1	58.9	62.0	60.0	61.7	65.2
6-year	50.5	49.6	51.5	53.6	56.0	55.2	56.0	55.7	56.3	59.9	57.9	60.0	63.9
7-year	48.4	47.5	49.0	51.6	53.5	53.3	54.3	54.0	53.9	57.9	56.1	58.5	62.4
8-year	46.2	45.9	47.0	49.4	51.3	51.1	53.2	52.9	52.1	55.6	53.9	56.2	60.8
9-year	44.3	43.9	45.1	47.1	49.4	49.8	51.7	52.2	50.3	54.0	51.6	54.5	58.4
10-year	42.3	42.2	43.3	45.4	47.8	47.6	49.9	51.1	49.0	52.3	49.5	52.6	57.5
11-year	40.4	40.7	42.1	43.6	46.0	46.3	47.8	49.0	48.0	50.6	48.4	50.9	55.5
12-year	39.1	39.3	40.5	42.4	44.1	45.3	46.1	47.9	46.7	48.9	46.6	50.3	53.1
13-year	37.7	37.6	39.0	40.8	42.5	43.1	44.6	46.2	46.0	47.9	46.1	49.9	51.7
14-year	36.7	36.5	37.7	39.6	40.0	41.4	43.8	44.4	44.7	46.6	45.4	48.4	50.6
15-year	35.3	35.0	36.6	38.6	38.6	40.0	42.5	43.8	43.1	45.8	44.3	47.9	
16-year	33.9	33.8	35.5	37.5	37.6	39.0	41.3	42.6	42.4	44.6	43.1		
17-year	32.9	32.8	34.5	36.4	36.6	38.1	40.4	41.0	40.9	43.4			
18-year	31.9	31.7	33.1	35.3	35.2	37.2	39.2	40.0	40.4				
19-year	31.1	30.2	32.0	34.3	34.6	35.7	37.6	39.1					
20-year	29.9	29.7	31.4	33.5	33.9	34.9	35.8						

Survival Time	2004	2005	2006	2007	2008	2009	2010	2011	2012	2013	2014	2015	2016
1-year	84.4	83.5	85.1	84.9	86.2	85.8	87.2	87.2	85.7	86.0	87.0	87.0	87.4
2-year	75.3	74.0	75.5	76.3	77.5	77.4	78.7	78.8	78.1	78.1	79.1	79.6	
3-year	70.7	68.8	70.6	71.3	71.9	73.1	74.4	74.5	73.1	72.7	74.9		
4-year	66.9	66.0	67.4	68.4	68.5	69.7	71.3	71.9	70.8	69.4			
5-year	64.9	64.0	65.1	66.5	66.5	67.3	69.0	69.9	67.6				
6-year	62.2	61.8	62.9	65.1	64.5	65.4	67.3	68.3					
7-year	60.5	60.1	61.5	62.9	62.7	64.0	65.6						
8-year	59.3	58.4	59.5	61.5	60.7	62.4							
9-year	57.2	56.8	57.9	59.4	59.9								
10-year	55.5	54.9	56.3	58.7									
11-year	54.0	54.4	53.7										
12-year	52.9	52.6											
13-year	51.4												

Figure A24: Cancer of the oral cavity and pharynx (invasive). Relative survival (%) by year of diagnosis. All races, males and females.

Year of Diagnosis

Survival Time	1975-1979	1980-1984	1985-1989	1990-1994	1995	1996	1997	1998	1999	2000	2001	2002	2003	2004	2005	2006	2007	2008	2009	2010	2011	2012	2013	2014	2015	2016
Cancer of the Tongue																										
1-year	74.2	73.4	76.7	79.4	79.3	79.2	81.6	79.9	80.9	84.0	81.0	80.9	84.3	86.8	84.8	84.3	86.0	86.6	87.2	86.5	88.0	87.8	87.2	87.7	87.1	88.0
2-year	54.8	57.5	60.5	64.7	65.2	66.7	71.0	65.8	68.2	71.0	70.4	69.5	71.4	75.4	72.2	73.9	76.5	76.9	79.0	76.7	78.6	80.8	78.4	79.8	79.3	
3-year	47.4	50.8	54.4	57.4	58.6	60.4	65.1	59.1	60.3	65.8	63.8	64.0	66.9	70.9	65.1	69.3	70.6	72.0	73.8	72.0	74.7	74.9	71.9	75.0		
4-year	43.5	47.4	49.9	53.6	56.4	57.4	62.1	55.5	57.9	63.1	61.0	61.7	64.2	66.5	61.7	66.9	68.9	69.7	71.1	69.5	72.1	73.4	68.8			
5-year	40.1	44.6	47.4	50.8	54.0	55.5	58.0	52.8	54.7	61.8	58.4	59.9	61.5	64.8	60.2	65.2	66.9	67.4	68.7	67.5	71.5	70.8				
6-year	38.4	41.8	45.1	48.5	50.4	52.9	56.4	52.3	52.8	59.8	56.3	57.8	61.0	61.6	59.2	63.8	65.6	66.0	67.0	66.5	70.1					
7-year	37.2	38.9	42.4	46.7	48.4	51.3	55.3	49.6	50.5	57.6	54.3	56.8	59.5	59.1	57.8	63.0	63.4	65.0	66.0	65.0						
8-year	35.7	37.4	40.5	44.3	46.5	49.9	53.8	48.0	50.4	55.5	50.6	53.6	58.9	56.9	56.1	60.3	61.8	63.1	64.5							
9-year	34.1	34.8	38.9	42.0	44.6	48.7	52.9	47.4	48.5	54.5	48.1	52.3	57.0	55.0	55.0	58.6	59.9	62.7								
10-year	32.5	33.3	37.5	40.0	43.3	46.7	52.3	45.9	47.7	53.2	46.2	50.0	56.2	53.9	52.4	56.8	58.5									
Cancer of the Lip																										
1-year	99.1	97.5	99.2	100.0	99.9	98.1	100.0	99.3	97.6	98.1	95.1	99.3	99.3	99.6	98.2	96.9	96.1	100.0	93.7	100.0	98.5	95.6	99.8	100.0	100.0	98.9
2-year	97.4	95.9	98.0	98.7	99.4	94.0	97.0	97.9	95.1	95.7	93.8	99.3	99.3	99.6	95.3	94.5	96.1	98.7	92.2	99.6	98.5	91.7	98.5	99.7	97.6	
3-year	96.9	93.5	97.2	98.4	98.7	92.8	95.8	95.1	93.6	95.1	93.8	97.1	98.2	98.7	95.2	93.2	95.7	98.2	91.7	94.8	98.5	89.7	95.6	99.1		
4-year	95.7	92.2	96.9	97.2	96.2	91.6	93.4	93.2	91.8	92.6	90.5	97.1	98.2	95.7	95.2	90.0	94.6	94.0	90.5	94.8	98.5	88.9	91.6			
5-year	94.1	90.0	95.9	95.8	95.0	88.1	91.2	90.8	90.5	90.7	90.5	93.8	98.2	94.2	95.2	89.0	94.6	94.0	87.8	93.7	98.5	87.0				
6-year	92.5	87.6	94.8	94.5	95.0	87.4	89.4	89.6	89.3	90.6	87.8	93.5	98.2	94.2	95.2	85.9	94.6	91.9	86.2	92.3	97.5					
7-year	91.1	86.4	93.4	93.3	95.0	86.6	88.2	89.6	88.8	90.4	84.7	93.5	97.3	93.5	95.2	85.9	94.6	89.8	83.5	90.3						
8-year	89.9	86.1	91.8	92.0	91.0	83.2	88.2	87.5	88.8	90.4	82.5	93.3	97.3	93.3	92.0	85.9	94.6	87.2	78.7							
9-year	87.4	85.0	91.2	89.8	86.0	82.5	88.0	86.8	88.5	89.1	79.0	90.6	96.1	93.3	89.5	85.9	92.0	84.4								
10-year	85.7	83.6	91.1	87.9	82.0	82.5	85.5	86.8	88.3	88.3	76.3	87.7	95.4	92.2	89.5	85.9	92.0									
Cancer of the Oropharynx and Tonsil																										
1-year	70.3	70.8	72.9	74.4	80.3	79.4	81.1	81.3	86.9	82.2	83.7	85.2	82.7	83.9	84.2	85.2	85.8	86.1	88.6	90.1	90.7	86.6	87.0	89.7	87.7	90.4
2-year	53.4	52.6	55.8	61.2	69.7	66.3	67.1	67.4	74.8	67.3	72.0	76.0	75.4	74.9	74.4	75.3	77.6	78.6	81.2	83.6	84.0	80.1	81.7	83.1	81.1	
3-year	44.5	44.4	45.7	53.7	62.7	60.6	61.8	60.2	67.2	62.3	66.7	71.4	71.4	71.2	70.7	70.4	72.2	73.3	76.5	80.0	79.7	75.6	77.5	78.8		
4-year	38.9	39.4	41.5	49.1	58.4	54.1	56.3	55.2	63.0	57.4	60.6	67.3	68.2	67.2	67.0	68.4	70.5	70.0	74.2	76.8	77.4	73.6	74.5			
5-year	35.4	35.3	37.8	44.9	53.9	52.6	51.3	53.2	59.7	53.6	58.7	64.4	65.2	64.6	65.5	65.6	68.0	67.4	72.6	74.9	75.4	70.1				
6-year	32.1	31.8	34.6	42.6	50.1	47.7	48.3	51.3	57.5	52.2	54.7	63.6	63.7	62.5	64.2	65.0	66.1	65.8	71.3	72.3	74.5					
7-year	30.2	29.7	31.9	41.3	46.6	43.7	47.3	50.1	55.7	48.8	52.8	62.7	62.8	61.7	61.7	62.9	63.5	62.5	70.2	70.8						
8-year	26.1	28.0	30.1	39.0	45.4	42.3	46.1	49.3	52.1	46.6	52.8	59.6	61.1	60.9	59.5	60.2	61.0	61.8	68.8							
9-year	23.7	25.8	28.3	36.6	44.1	41.3	44.3	48.8	51.9	45.0	49.6	59.0	60.7	57.6	57.8	58.8	58.9	60.2								
10-year	22.4	23.2	26.1	35.4	42.9	40.1	43.6	46.5	49.8	43.1	48.5	57.2	58.3	56.1	57.5	58.8	58.8									

Figure A25: Cancer of the oral cavity and pharynx for selected subsites (invasive). Relative survival (%) by year of diagnosis. All races, males and females.

Survival Time	Year of Diagnosis 1975-1979	1980-1984	1985-1989	1990-1994	1995	1996	1997	1998	1999	2000	2001	2002	2003	2004	2005	2006	2007	2008	2009	2010	2011	2012	2013	2014	2015	2016
1-year	64.0	69.4	69.4	71.9	73.1	73.7	76.6	75.0	73.9	76.8	76.1	75.3	76.0	75.3	75.1	75.3	75.4	75.8	76.2	76.9	78.4	78.4	78.1	81.1	78.2	79.4
2-year	48.9	53.2	54.1	58.7	60.2	61.4	65.2	64.2	61.6	63.6	65.2	63.6	66.0	64.8	64.6	64.1	64.2	64.9	67.3	67.1	69.4	67.2	68.0	69.6	69.0	
3-year	42.2	45.8	46.4	49.9	51.2	52.9	56.6	55.1	53.0	54.5	56.0	54.4	58.3	56.6	56.1	55.8	55.9	56.5	58.4	59.3	60.7	59.7	61.0	61.8		
4-year	38.6	41.5	41.5	44.5	46.2	46.2	48.7	50.3	47.6	47.7	50.4	48.0	50.0	49.9	50.1	49.8	49.3	50.7	52.8	52.5	54.2	53.3	55.6			
5-year	36.5	38.8	38.2	40.9	42.4	42.1	44.0	45.3	42.8	43.1	45.8	42.8	44.5	44.4	44.6	45.7	44.6	46.8	47.1	47.2	49.5	48.8				
6-year	35.2	37.0	36.0	38.3	39.0	38.8	40.4	42.4	40.0	40.1	42.9	39.5	41.2	41.5	41.3	42.6	42.0	44.1	44.1	43.9	46.6					
7-year	33.8	36.1	34.8	36.8	37.0	37.3	38.2	40.6	38.0	37.6	40.4	37.6	38.4	39.5	39.4	40.6	40.6	41.7	41.8	41.1						
8-year	33.0	34.6	33.8	35.7	35.2	36.2	36.7	38.6	36.4	36.0	38.1	35.8	37.1	37.7	38.2	38.9	38.8	39.4	39.9							
9-year	32.4	33.7	33.0	34.5	33.7	35.4	35.3	36.9	34.9	34.6	36.3	34.9	35.6	36.2	37.3	37.9	37.1	37.6								
10-year	31.9	32.9	32.4	33.7	32.8	34.4	34.2	35.9	34.1	33.3	35.2	34.1	34.3	35.2	36.2	37.4	36.0									
11-year	31.6	32.4	32.1	33.0	32.1	33.7	33.0	35.1	32.6	32.6	34.5	33.4	33.7	34.3	35.8	36.7										
12-year	30.9	31.8	31.4	32.4	32.0	33.2	32.2	34.6	31.8	32.0	33.3	33.0	33.0	33.5	34.7											
13-year	30.7	31.2	30.8	32.0	31.5	32.6	31.7	33.3	31.2	31.4	32.8	32.2	32.4	33.0												
14-year	30.2	30.8	30.4	31.8	31.5	31.8	31.4	32.6	30.2	31.1	32.6	31.5	32.3													
15-year	29.6	30.6	30.3	31.3	30.9	31.4	31.4	32.0	30.2	30.4	32.0	30.9														
16-year	29.3	30.4	30.3	31.1	30.7	30.8	30.6	31.4	29.3	30.4	32.0															
17-year	28.9	30.0	30.1	30.8	30.5	30.6	30.2	31.1	29.2	30.3																
18-year	28.8	29.8	29.9	30.5	30.1	30.6	30.0	30.7	28.5																	
19-year	28.5	29.5	29.9	30.1	29.6	30.4	29.9	30.5																		
20-year	28.3	29.2	29.6	29.9	29.0	29.9	29.8																			

Figure A26: Cancer of the ovary (invasive). Relative survival (%) by year of diagnosis. All races, females.

Year of Diagnosis

Survival Time	1975-1979	1980-1984	1985-1989	1990-1994	1995	1996	1997	1998	1999	2000	2001	2002	2003	2004	2005	2006	2007	2008	2009	2010	2011	2012	2013	2014	2015	2016
1-year	14.3	15.8	17.0	18.5	17.8	19.0	19.9	19.3	22.3	21.7	23.8	25.0	22.8	22.8	26.7	25.8	28.0	29.8	29.8	30.3	31.5	34.8	35.6	35.0	36.2	37.3
2-year	5.9	6.3	6.7	8.6	8.3	9.3	9.8	8.8	10.5	11.0	11.4	11.9	11.5	11.9	12.9	13.3	13.9	15.1	15.9	16.9	16.3	19.8	20.6	19.7	20.4	
3-year	3.6	4.1	4.7	5.8	5.3	6.2	7.1	5.6	6.6	7.3	7.4	8.5	7.6	8.3	9.1	9.5	9.8	10.5	11.9	12.0	11.7	14.8	15.6	15.2		
4-year	3.0	3.3	3.8	4.8	4.0	4.8	5.5	4.4	5.7	6.1	6.1	6.8	5.8	6.4	7.2	8.2	8.4	8.6	10.0	10.0	9.4	12.6	13.0			
5-year	2.5	2.8	3.3	4.2	3.6	4.2	4.9	3.8	5.1	5.2	5.1	6.2	5.1	5.5	6.4	7.6	7.7	7.8	8.7	8.4	8.2	11.5				
6-year	2.3	2.5	2.9	4.0	3.2	3.8	4.4	3.7	4.6	4.7	4.7	5.7	4.6	5.0	5.8	7.0	7.1	7.2	8.2	7.6	7.6					
7-year	2.2	2.3	2.7	3.6	3.1	3.5	4.3	3.4	4.3	4.3	4.2	5.0	4.3	4.7	5.5	6.2	6.5	6.6	7.5	7.1						
8-year	1.9	2.1	2.5	3.4	2.7	3.4	4.1	3.0	4.0	3.9	3.9	4.7	4.1	4.4	5.1	6.0	5.8	6.1	7.0							
9-year	1.8	2.0	2.2	3.2	2.7	3.3	4.0	2.7	3.8	3.8	3.8	4.6	3.8	4.2	4.8	5.9	5.3	6.0								
10-year	1.8	1.9	2.2	3.0	2.5	3.2	3.9	2.6	3.6	3.7	3.5	4.5	3.5	4.1	4.4	5.7	5.0									
11-year	1.7	1.7	2.1	2.8	2.5	2.9	3.8	2.5	3.3	3.6	3.3	4.1	3.3	4.1	4.0	5.5										
12-year	1.7	1.7	2.1	2.7	2.3	2.8	3.6	2.4	3.2	3.3	3.2	4.0	3.3	4.1	3.9											
13-year	1.6	1.6	2.0	2.6	2.2	2.7	3.3	2.2	3.1	3.1	3.2	3.8	3.1	3.9												
14-year	1.6	1.6	1.9	2.5	1.9	2.7	3.1	1.9	3.1	2.8	3.0	3.8	2.9													
15-year	1.5	1.5	1.8	2.4	1.8	2.7	3.0	1.9	3.0	2.7	2.7	3.5														
16-year	1.5	1.4	1.7	2.2	1.7	2.2	2.9	1.8	2.7	2.6	2.5															
17-year	1.4	1.4	1.7	2.2	1.5	1.9	2.7	1.5	2.7	2.5																
18-year	1.3	1.3	1.7	2.2	1.4	1.9	2.7	1.3	2.4																	
19-year	1.3	1.3	1.6	2.1	1.4	1.9	2.7	1.2																		
20-year	1.2	1.3	1.6	2.0	1.3	1.9	2.7																			

Figure A27: Cancer of the pancreas (invasive). Relative survival (%) by year of diagnosis. All races, males and females.

Year of Diagnosis

Survival Time	1975-1979	1980-1984	1985-1989	1990-1994	1995	1996	1997	1998	1999	2000	2001	2002	2003	2004	2005	2006	2007	2008	2009	2010	2011	2012	2013	2014	2015	2016
1-year	91.4	93.4	95.5	98.8	98.7	99.1	99.2	99.5	99.6	99.7	99.9	99.8	99.7	99.8	99.8	99.8	99.8	99.7	99.6	99.6	99.7	99.3	99.4	98.7	99.2	99.3
2-year	84.6	87.4	91.0	97.3	97.7	98.0	98.8	98.9	99.6	99.5	99.9	99.8	99.6	99.7	99.6	99.8	99.8	99.6	99.4	99.5	99.4	98.7	98.8	98.0	98.2	
3-year	78.4	81.5	86.9	96.0	97.0	97.4	98.4	98.5	99.6	99.1	99.9	99.7	99.4	99.7	99.5	99.7	99.8	99.3	99.3	99.5	99.3	98.4	98.3	97.5		
4-year	73.3	76.5	83.6	94.9	96.4	96.9	97.9	98.2	99.2	98.9	99.8	99.7	99.4	99.7	99.5	99.6	99.8	99.2	99.3	99.5	99.3	98.2	97.9			
5-year	68.8	71.9	80.4	93.7	95.7	96.4	97.4	98.1	99.2	98.8	99.8	99.7	99.1	99.7	99.1	99.6	99.8	99.2	99.3	99.5	99.2	98.1				
6-year	65.3	68.5	77.9	92.7	95.4	96.0	97.1	97.9	99.2	98.5	99.8	99.7	99.1	99.7	99.1	99.6	99.8	99.2	99.3	99.5	99.2					
7-year	61.8	65.0	75.4	91.6	95.0	95.4	97.1	97.7	99.1	98.5	99.8	99.7	99.1	99.7	99.1	99.6	99.8	99.2	99.3	99.5						
8-year	58.6	62.0	73.4	90.6	94.6	95.0	97.0	97.7	98.7	98.2	99.8	99.7	99.0	99.7	99.1	99.6	99.8	99.2	99.3							
9-year	55.7	59.7	71.4	89.4	94.3	94.4	96.6	97.0	98.5	98.2	99.8	99.7	98.5	99.6	99.1	99.6	99.8	99.2								
10-year	53.2	57.7	69.6	88.5	93.5	94.3	96.5	96.6	98.1	98.2	99.8	99.7	98.5	99.4	99.1	99.6	99.8									
11-year	50.7	55.5	67.9	87.5	92.8	93.9	96.4	96.2	97.9	97.3	99.8	99.7	98.3	99.2	99.1	99.6										
12-year	48.5	53.6	66.1	86.6	92.0	93.5	96.1	95.5	97.4	97.2	99.8	99.7	97.9	98.8	99.1											
13-year	46.5	51.6	64.3	85.9	91.0	93.0	95.3	95.3	97.2	96.6	99.8	99.7	97.9	98.8												
14-year	44.8	50.0	63.0	85.1	90.1	92.8	94.7	95.0	96.6	96.1	99.8	99.7	97.8													
15-year	43.6	48.7	61.5	84.2	89.4	92.1	94.7	94.9	96.0	95.4	99.8	99.7														
16-year	41.9	46.9	60.0	83.3	88.9	91.2	94.1	94.4	95.6	95.4	99.3															
17-year	40.1	45.8	59.4	82.4	88.0	90.6	93.6	94.1	95.6	95.4																
18-year	38.8	44.5	58.3	81.8	87.7	90.4	92.9	94.1	95.2																	
19-year	38.0	43.1	57.9	80.8	87.3	89.7	92.6	94.1																		
20-year	37.0	41.7	56.9	80.4	85.8	88.8	92.6																			
21-year	35.4	40.8	56.1	79.5	84.9	88.8																				
22-year	34.0	39.5	54.8	78.7	84.3																					
23-year	33.2	38.3	53.6	77.7																						
24-year	32.4	37.4	52.5																							
25-year	31.0	35.9	51.4																							
26-year	30.4	34.7	50.4																							
27-year	29.7	34.4	49.5																							
28-year	29.1	34.0	48.8																							
29-year	28.5	33.3																								
30-year	27.0	32.4																								

Figure A28: Cancer of the prostate (invasive). Relative survival (%) by year of diagnosis. All races, males.

Year of Diagnosis

Survival Time	1975-1979	1980-1984	1985-1989	1990-1994	1995	1996	1997	1998	1999	2000	2001	2002	2003	2004	2005	2006	2007	2008	2009	2010	2011	2012	2013	2014	2015	2016
1-year	38.6	41.5	43.8	44.8	45.9	45.4	45.3	47.6	46.7	47.0	48.6	49.6	50.8	53.0	49.9	52.2	52.1	55.6	55.6	54.4	58.2	57.5	59.4	60.5	58.5	60.0
2-year	24.9	26.7	29.1	31.3	31.5	31.9	32.4	33.6	34.0	32.6	33.3	35.3	38.4	38.4	36.4	40.0	39.5	41.8	40.6	40.4	43.5	45.2	44.8	45.9	44.3	
3-year	19.7	21.0	23.4	25.2	26.5	26.6	26.3	28.4	27.7	26.9	29.0	30.2	33.5	33.0	31.6	34.6	33.8	36.5	36.1	35.3	36.0	38.7	38.3	38.2		
4-year	17.3	18.3	20.6	22.5	23.3	23.8	23.4	25.2	24.8	24.7	27.5	27.1	30.5	30.1	28.0	33.3	31.2	34.0	33.7	32.1	32.9	35.9	34.4			
5-year	15.6	16.6	19.3	20.8	21.9	22.2	21.8	22.8	23.1	23.0	26.3	26.0	28.0	28.4	25.9	31.3	29.1	32.1	31.7	30.5	31.0	34.3				
6-year	14.6	15.8	18.3	19.3	20.7	20.5	20.7	21.9	21.1	22.2	25.3	24.5	26.6	26.7	24.8	29.9	27.2	31.0	30.6	29.6	29.6					
7-year	14.0	15.2	17.6	18.4	19.6	20.0	20.1	21.1	20.1	21.4	24.5	23.8	25.8	25.4	23.9	29.0	26.0	30.0	29.6	28.5						
8-year	13.4	14.5	17.0	17.5	19.4	19.7	19.4	20.6	19.0	20.7	22.8	22.5	24.6	24.5	23.4	28.4	25.1	29.3	29.0							
9-year	13.1	14.1	16.5	16.9	19.1	19.3	19.3	19.6	18.6	20.0	22.3	21.7	24.2	23.7	22.2	27.8	24.4	28.1								
10-year	12.7	13.9	15.8	16.1	18.2	18.8	18.3	18.7	18.0	19.5	21.6	21.0	23.6	22.5	21.5	26.4	23.9									
11-year	12.3	13.4	15.4	15.9	17.8	17.7	17.8	18.6	17.2	18.5	21.1	20.1	23.1	22.1	21.2	26.0										
12-year	12.0	12.9	15.0	15.6	17.2	17.1	17.7	17.8	17.0	18.0	20.5	19.4	22.5	21.1	20.3											
13-year	11.7	12.9	14.9	15.2	16.7	16.0	17.7	17.0	15.7	17.6	20.4	19.1	22.0	19.7												
14-year	11.2	12.8	14.6	14.9	16.4	15.7	17.3	17.0	15.6	17.5	20.0	18.1	21.5													
15-year	10.8	12.3	14.3	14.7	16.3	15.4	17.3	17.0	15.6	17.3	19.2	17.2														
16-year	10.5	11.9	14.0	14.1	16.3	14.7	16.6	16.6	14.9	17.2	18.4															
17-year	10.2	11.5	13.8	13.9	16.3	14.0	15.6	15.6	14.2	16.5																
18-year	9.8	11.5	13.6	13.3	16.1	13.3	15.2	15.4	13.8																	
19-year	9.6	11.3	13.2	13.1	15.7	13.0	15.2	14.2																		
20-year	9.2	10.8	13.0	12.8	14.8	12.7	15.2																			

Figure A29: Cancer of the stomach (invasive). Relative survival (%) by year of diagnosis. All races, males and females.

Year of Diagnosis

Survival Time	1975-1979	1980-1984	1985-1989	1990-1994	1995	1996	1997	1998	1999	2000	2001	2002	2003	2004	2005	2006	2007	2008	2009	2010	2011	2012	2013	2014	2015	2016
1-year	93.2	95.8	97.9	97.7	98.3	97.6	97.3	97.9	96.9	97.9	98.5	98.2	98.4	97.9	98.1	98.9	98.9	98.7	98.5	98.1	97.6	99.1	97.8	97.7	98.6	98.3
2-year	88.5	92.6	96.2	96.4	97.4	96.4	95.7	96.6	96.3	96.5	97.5	97.1	96.9	96.5	97.3	98.4	97.5	97.2	97.4	97.4	97.4	97.9	97.2	97.0	97.1	
3-year	86.5	92.0	95.5	95.8	96.4	95.8	95.1	96.2	95.7	95.9	97.1	96.9	96.6	95.7	97.1	98.2	96.8	97.2	97.2	97.4	96.8	97.6	96.8	97.0		
4-year	85.6	91.5	95.0	95.7	96.3	95.8	94.9	96.1	95.6	95.9	96.8	96.7	96.5	95.4	96.8	98.0	96.6	96.7	97.2	97.2	96.7	97.4	96.4			
5-year	85.0	91.1	94.7	95.7	96.3	95.8	94.9	95.8	95.6	95.9	96.6	96.3	96.1	95.2	96.8	98.0	95.6	96.7	96.9	97.0	96.4	97.4				
6-year	84.7	90.9	94.5	95.6	96.3	95.8	94.7	95.6	95.6	95.9	96.2	96.1	96.1	95.2	96.4	97.8	95.0	96.6	96.7	96.8	96.3					
7-year	84.6	90.7	94.3	95.4	96.3	95.8	94.7	95.4	95.5	95.9	96.2	96.1	96.1	95.2	96.3	97.4	95.0	96.6	96.7	96.8						
8-year	84.5	90.5	94.1	95.0	96.3	95.8	94.7	95.4	95.4	95.9	96.2	96.0	96.1	95.2	96.2	97.3	95.0	96.6	96.7							
9-year	84.5	90.1	94.0	94.9	96.2	95.8	94.7	95.4	95.2	95.9	96.2	95.7	96.1	95.2	96.1	97.3	95.0	96.6								
10-year	83.7	90.0	93.8	94.9	96.1	95.8	94.7	94.7	95.2	95.9	96.2	95.5	96.1	95.2	95.9	97.3	95.0									
11-year	83.3	89.8	93.7	94.9	96.1	95.8	94.7	94.2	95.1	95.9	96.1	95.2	96.1	95.2	95.8	97.2										
12-year	83.0	89.8	93.5	94.8	96.1	95.8	94.7	94.1	95.1	95.9	96.0	95.2	96.1	95.2	95.3											
13-year	82.6	89.4	93.2	94.8	95.9	95.8	94.7	94.1	95.1	95.9	95.8	95.2	96.1	94.7												
14-year	82.1	88.9	93.1	94.8	95.7	95.8	94.7	94.0	95.1	95.9	95.6	95.0	96.1													
15-year	81.8	88.6	93.1	94.8	95.7	95.8	94.7	94.0	95.1	95.9	95.4	94.9														
16-year	81.3	88.2	92.9	94.6	95.7	95.8	94.7	93.8	95.1	95.9	95.4															
17-year	80.8	87.9	92.7	94.4	95.7	95.8	94.7	93.4	95.1	95.9																
18-year	80.1	87.7	92.6	94.3	95.7	95.8	94.5	93.2	95.1																	
19-year	79.6	87.4	92.5	94.2	95.7	95.8	94.4	93.2																		
20-year	79.5	87.1	92.3	94.1	95.7	95.8	94.2																			

Figure A30: Cancer of the testis (invasive). Relative survival (%) by year of diagnosis. All races, males.

Year of Diagnosis

Survival Time	1975-1979	1980-1984	1985-1989	1990-1994	1995	1996	1997	1998	1999	2000	2001	2002	2003	2004	2005	2006	2007	2008	2009	2010	2011	2012	2013	2014	2015	2016
1-year	94.9	95.1	96.0	96.2	96.7	96.8	96.4	97.7	97.3	97.6	97.6	97.1	98.0	98.3	97.7	98.4	98.5	98.3	98.8	98.9	98.8	98.8	98.3	98.7	98.8	98.7
2-year	93.5	94.5	95.2	95.7	96.1	96.6	96.2	97.1	97.1	97.0	97.4	97.0	97.9	97.9	97.3	98.3	98.1	98.2	98.7	98.6	98.7	98.5	98.3	98.7	98.7	
3-year	92.9	93.7	94.8	95.3	96.0	96.0	96.2	96.9	97.1	96.9	97.0	97.0	97.7	97.9	97.3	98.2	98.1	98.2	98.7	98.6	98.6	98.4	98.2	98.7		
4-year	92.5	93.4	94.3	95.0	95.9	95.4	95.7	96.3	97.0	96.9	97.0	97.0	97.4	97.7	97.3	98.2	98.0	98.0	98.7	98.4	98.6	98.4	98.2			
5-year	92.0	93.3	94.1	94.6	95.8	95.0	95.3	96.2	96.8	96.3	96.2	97.0	97.3	97.4	97.3	98.2	97.9	97.8	98.7	98.1	98.6	98.4				
6-year	91.7	93.1	93.8	94.3	95.8	94.8	95.2	95.8	96.3	96.3	96.1	96.7	97.0	97.2	97.3	98.2	97.9	97.8	98.7	98.1	98.6					
7-year	91.5	92.7	93.5	94.2	95.6	94.5	95.2	95.8	96.2	96.1	96.1	96.5	96.8	96.9	97.3	98.1	97.9	97.7	98.7	98.1						
8-year	91.1	92.6	93.3	94.1	95.5	94.2	95.1	95.8	96.0	95.9	95.9	96.3	96.7	96.8	97.3	98.0	97.9	97.5	98.7							
9-year	90.7	92.3	93.1	93.9	95.5	93.8	95.0	95.8	95.7	95.9	95.4	96.3	96.7	96.8	97.3	98.0	97.9	97.5								
10-year	90.5	92.0	92.8	93.6	95.2	93.7	95.0	95.8	95.4	95.8	95.3	96.3	96.7	96.7	97.3	98.0	97.9									
11-year	90.3	91.7	92.7	93.3	95.1	93.3	95.0	95.8	95.0	95.8	95.3	96.1	96.7	96.5	97.2	98.0										
12-year	90.3	91.4	92.7	93.0	95.1	93.0	95.0	95.8	94.7	95.8	95.3	95.9	96.7	96.4	97.0											
13-year	90.0	91.3	92.5	93.0	95.1	92.8	95.0	95.1	94.6	95.8	95.3	95.9	96.7	96.4												
14-year	90.0	91.1	92.4	92.7	95.1	92.7	95.0	94.7	94.5	95.8	95.3	95.9	96.7													
15-year	89.7	90.7	92.4	92.4	94.3	92.4	95.0	94.2	94.4	95.8	95.2	95.9														
16-year	89.7	90.6	92.4	92.2	93.8	92.4	95.0	94.0	94.4	95.8	94.9															
17-year	89.6	90.6	92.3	92.2	93.8	92.4	95.0	93.8	94.4	95.7																
18-year	89.4	90.5	92.2	92.2	93.8	92.4	94.4	93.8	94.1																	
19-year	89.2	90.5	91.9	92.2	93.8	92.4	94.3	93.4																		
20-year	89.2	90.5	91.9	92.2	93.8	92.3	94.3																			

Figure A31: Cancer of the thyroid (invasive). Relative survival (%) by year of diagnosis. All races, males and females.

	Year of Diagnosis																									
Survival Time	1975-1979	1980-1984	1985-1989	1990-1994	1995	1996	1997	1998	1999	2000	2001	2002	2003	2004	2005	2006	2007	2008	2009	2010	2011	2012	2013	2014	2015	2016
1-year	86.9	88.6	89.9	90.0	90.7	89.5	91.2	90.9	90.0	90.9	90.9	89.7	90.4	90.7	90.6	91.1	90.2	89.2	89.5	89.1	90.5	89.0	89.2	89.9	89.9	88.3
2-year	80.8	83.5	85.1	85.4	86.6	85.4	86.2	86.6	85.0	86.5	86.4	85.7	85.7	86.0	85.5	86.6	85.1	84.8	83.8	85.1	86.1	84.3	84.9	85.5	85.8	
3-year	77.7	80.4	82.2	82.8	84.0	83.2	83.1	83.4	82.4	83.5	84.6	83.6	83.5	83.5	82.7	83.9	82.1	81.9	81.6	81.7	83.0	81.4	82.2	82.8		
4-year	74.9	78.0	80.0	81.2	81.8	81.1	80.7	81.5	80.4	81.6	82.4	81.2	81.5	81.6	81.2	82.4	80.5	79.5	79.5	79.7	80.2	79.6	80.1			
5-year	73.0	76.0	78.1	79.5	81.1	79.5	78.5	79.2	78.1	80.3	81.4	79.9	79.8	80.1	80.1	80.8	79.1	77.5	78.2	77.3	79.2	77.6				
6-year	71.1	74.5	76.4	78.0	79.3	78.3	76.8	77.6	76.7	79.3	80.1	78.5	78.3	78.2	78.6	79.8	77.1	75.4	75.6	75.8	77.2					
7-year	69.0	73.0	74.9	76.3	77.4	77.3	75.8	76.5	75.6	77.9	78.3	77.4	77.2	76.0	77.8	77.9	75.2	73.3	74.8	74.9						
8-year	67.5	71.8	73.4	75.0	76.2	75.9	74.8	74.8	74.3	76.4	77.2	76.2	75.9	74.4	75.8	76.3	74.0	72.5	73.3							
9-year	65.8	70.1	72.2	73.5	74.6	74.1	73.2	73.5	73.2	75.5	76.0	74.6	75.0	73.2	74.4	75.6	73.2	72.1								
10-year	63.8	68.9	70.8	72.1	73.7	72.7	72.1	71.8	72.6	73.6	74.5	73.7	74.0	72.1	73.6	74.5	72.2									
11-year	62.1	67.5	69.7	70.9	73.0	71.6	71.3	71.0	71.5	71.6	73.9	72.4	72.7	70.9	73.0	73.7										
12-year	60.7	66.2	68.2	69.9	71.6	70.0	70.2	68.9	69.7	70.6	72.7	71.2	72.0	69.9	71.5											
13-year	59.7	65.2	66.8	69.0	69.3	68.7	69.9	68.0	68.3	69.2	70.2	70.8	71.3	69.3												
14-year	58.5	63.6	65.4	68.0	67.6	67.2	68.7	67.2	67.0	67.8	68.6	70.1	70.1													
15-year	57.2	62.2	64.1	66.3	67.1	67.1	67.3	66.8	65.7	66.9	67.7	68.4														
16-year	56.2	60.7	62.8	65.1	65.7	66.8	65.6	66.0	65.4	66.1	66.3															
17-year	55.1	59.7	61.8	63.8	64.2	65.4	64.6	64.5	64.0	65.5																
18-year	54.3	58.6	60.8	62.7	63.4	64.0	63.3	64.2	62.7																	
19-year	53.7	57.8	60.2	61.6	62.1	63.0	62.9	63.3																		
20-year	52.8	57.3	59.6	60.6	60.0	61.3	61.4																			

Figure A32: Cancer of the urinary bladder (invasive). Relative survival (%) by year of diagnosis. All races, males and females.

APPENDIX B

CHEMOTHERAPY AGENTS

The below table lists categories of chemotherapy agents, which can be cross-referenced by color with subsequent tables that break down chemotherapy agents available for each type of cancer.

A	Alkylating Agents	
1	Alkyl sulfonates	Busulfan
2	Ethylenimines	Hexamethylmelamine and Thiotepa
3	Hydrazines and Triazines	Dacarbazine, Procarbazine and Temozolomide
4	Metal salts	Carboplatin, Cisplatin and Oxaliplatin
5	Nitrogen Mustards	Chlorambucil, Cyclophosphamide, Ifosfamide, Mechlorethamine and Melphalan
6	Nitrosoureas	Carmustine, Lomustine and Streptozocin
B	**Antimetabolites**	
1	Adenosine Deaminase Inhibitor	Cladribine, Fludarabine, Nelarabine and Pentostatin
2	Folate Antagonist	Methotrexate, Pemetrexed
3	Pyrimidine Antagonist	Capecitabine, Cytarabine, Decitabine, Fluorouracil, Floxuridine, and Gemcitabine
4	Purine Antagonist	Mercaptopurine and Thioguanine
C	**Anti-tumor antibiotics**	
1	Anthracyclines	Daunorubicin, Doxorubicin, Epirubicin, Idarubicin and Mitoxantrone
2	Chromomycins	Dactinomycin and Plicamycin
3	Miscellaneous	Bleomycin and Mitomycin
D	**Corticosteroids**	
1	Corticosteroids	Dexamethasone, Hydrocortisone, Methylprednisolone and Prednisone

E	Miscellaneous Antineoplastics	
1	Ribonucleotide Reductase Inhibitor	Hydroxyurea
2	Adrenocortical Steroid Inhibitor	Mitotane*
3	Enzymes	Asparaginase and Pegaspargase
4	Antimicrotubule Agent	Estramustine
5	Retinoids	Bexarotene†, Isotretinoin‡ and Tretinoin (ATRA) §
6	Monoclonal Antibody	Alemtuzumab, Bevacizumab, Cetuximab, Gemtuzumab, ozogamicin, Ipilimumab, Ofatumumab, Panitumumab, Pembrolizumab, Ramucirumab and Trastuzumab
7	Cytokine	Aldesleukin
8	mTOR Inhibitors	Everolimus, Ridaforolimus** Sirolimus and Temsirolimus
F	**Plant Alkaloids**	
1	Camptothecin Analogs	Irinotecan and Topotecan
2	Podophyllotoxins	Etoposide and Teniposide
3	Taxanes	Docetaxel and Paclitaxel
4	Vinca alkaloids	Vinblastine, Vincristine and Vinorelbine
G	**Topoisomerase Inhibitors**	
1	Topoisomerase I Inhibitors	Irinotecan and Topotecan
2	Topoisomerase II Inhibitors	Daunorubicin, doxorubicin, etoposide, mitoxantrone and teniposide

The following drugs are not cross-referenced due specific or rare applications:
*Used to treat adrenal cortical carcinoma (ACC).
†Used to treat cutaneous T-cell lymphoma.
‡Being investigated to treat various cancers.
§Used to treat acute promyelocytic leukemia (APL).
**Undergoing testing.

Individual Chemotherapy Agents

	Anal
	Gardasil (Recombinant Human Papillomavirus (HPV) Quadrivalent Vaccine)
	Gardasil 9 (Recombinant HPV Nonavalent Vaccine)

	Recombinant HPV Nonavalent Vaccine
	Recombinant HPV Quadrivalent Vaccine
Bladder	
	Atezolizumab
A4	Cisplatin
C1/G2	Doxorubicin Hydrochloride
A4	Platinol (Cisplatin)
A4	Platinol-AQ (Cisplatin)
	Tecentriq (Atezolizumab)
A2	Thiotepa
Bone	
B2	Abitrexate (Methotrexate)
C2	Cosmegen (Dactinomycin)
C2	Dactinomycin
	Denosumab
C1/G2	Doxorubicin Hydrochloride
B2	Folex (Methotrexate)
B2	Folex PFS (Methotrexate)
B2	Methotrexate
B2	Methotrexate LPF (Methotrexate)
B2	Mexate (Methotrexate)
B2	Mexate-AQ (Methotrexate)
	Xgeva (Denosumab)
Brain	
E8	Afinitor (Everolimus)
E8	Afinitor Disperz (Everolimus)
E6	Avastin (Bevacizumab)
A6	Becenum (Carmustine)
A6	BiCNU (Carmustine)
A6	Carmubris (Carmustine)
A6	Carmustine
A6	Carmustine Implant
	Everolimus
A6	Gliadel (Carmustine Implant)
A6	Gliadel wafer (Carmustine Implant)
A6	Lomustine
A3	Methazolastone (Temozolomide)

A3	Temodar (Temozolomide)
A3	Temozolomide
	Breast
B2	Abitrexate (Methotrexate)
F3	Abraxane (Paclitaxel Albumin-stabilized Nanoparticle Formulation)
E6	Ado-Trastuzumab Emtansine
E8	Afinitor (Everolimus)
	Anastrozole
	Aredia (Pamidronate Disodium)
	Arimidex (Anastrozole)
	Aromasin (Exemestane)
B3	Capecitabine
A5	Clafen (Cyclophosphamide)
A5	Cyclophosphamide
A5	Cytoxan (Cyclophosphamide)
F3	Docetaxel
C1/G2	Doxorubicin Hydrochloride
C1	Ellence (Epirubicin Hydrochloride)
C1	Epirubicin Hydrochloride
	Eribulin Mesylate
E8	Everolimus
	Exemestane
B3	5-FU (Fluorouracil Injection)
	Fareston (Toremifene)
	Faslodex (Fulvestrant)
	Femara (Letrozole)
B3	Fluorouracil Injection
B2	Folex (Methotrexate)
B2	Folex PFS (Methotrexate)
	Fulvestrant
B3	Gemcitabine Hydrochloride
B3	Gemzar (Gemcitabine Hydrochloride)
	Goserelin Acetate
	Halaven (Eribulin Mesylate)

E6	Herceptin (Trastuzumab)
	Ibrance (Palbociclib)
	Ixabepilone
	Ixempra (Ixabepilone)
E6	Kadcyla (Ado-Trastuzumab Emtansine)
	Lapatinib Ditosylate
	Letrozole
	Megestrol Acetate
B2	Methotrexate
B2	Methotrexate LPF (Methotrexate)
B2	Mexate (Methotrexate)
B2	Mexate-AQ (Methotrexate)
A5	Neosar (Cyclophosphamide)
	Nolvadex (Tamoxifen Citrate)
F3	Paclitaxel
F3	Paclitaxel Albumin-stabilized Nanoparticle Formulation
	Palbociclib
	Pamidronate Disodium
	Perjeta (Pertuzumab)
	Pertuzumab
	Tamoxifen Citrate
F3	Taxol (Paclitaxel)
F3	Taxotere (Docetaxel)
A2	Thiotepa
	Toremifene
E6	Trastuzumab
	Tykerb (Lapatinib Ditosylate)
F4	Velban (Vinblastine Sulfate)
F4	Velsar (Vinblastine Sulfate)
F4	Vinblastine Sulfate
B3	Xeloda (Capecitabine)
	Zoladex (Goserelin Acetate)
	Cervical
E6	Avastin (Bevacizumab)
C3	Blenoxane (Bleomycin)
C3	Bleomycin
F1	Hycamtin (Topotecan Hydrochloride)

F1	Topotecan Hydrochloride
Colon and Rectal	
E6	Avastin (Bevacizumab)
F1	Camptosar (Irinotecan Hydrochloride)
B3	Capecitabine
E6	Cetuximab
E6	Cyramza (Ramucirumab)
A4	Eloxatin (Oxaliplatin)
E6	Erbitux (Cetuximab)
B3	5-FU (Fluorouracil Injection)
B3	Floxuridine
B3	Fluorouracil Injection
F1	Irinotecan Hydrochloride
	Leucovorin Calcium
	Lonsurf (Trifluridine and Tipiracil Hydrochloride)
A4	Oxaliplatin
E6	Panitumumab
E6	Ramucirumab
	Regorafenib
	Stivarga (Regorafenib)
	Trifluridine and Tipiracil Hydrochloride
E6	Vectibix (Panitumumab)
	Wellcovorin (Leucovorin Calcium)
B3	Xeloda (Capecitabine)
	Zaltrap (Ziv-Aflibercept)
	Ziv-Aflibercept
Esophageal	
E6	Cyramza (Ramucirumab)
F3	Docetaxel
E6	Herceptin (Trastuzumab)
E6	Ramucirumab
F3	Taxotere (Docetaxel)
E6	Trastuzumab
Gastric (Stomach) Cancer	
E6	Cyramza (Ramucirumab)
F3	Docetaxel

C1/G2	Doxorubicin Hydrochloride
B3	Floxuridine
B3	5-FU (Fluorouracil Injection)
B3	Fluorouracil Injection
E6	Herceptin (Trastuzumab)
C3	Mitomycin C
C3	Mitozytrex (Mitomycin C)
C3	Mutamycin (Mitomycin C)
E6	Ramucirumab
F3	Taxotere (Docetaxel)
E6	Trastuzumab
	Gastroenteropancreatic Neuroendocrine Tumors
E8	Afinitor (Everolimus)
	Lanreotide Acetate
	Somatuline Depot (Lanreotide Acetate)
A6	Streptozocin
	Gastrointestinal Stromal Tumors
	Gleevec (Imatinib Mesylate)
	Imatinib Mesylate
	Regorafenib
	Stivarga (Regorafenib)
	Sunitinib Malate
	Sutent (Sunitinib Malate)
	Gestational Trophoblastic Disease
B2	Abitrexate (Methotrexate)
C2	Cosmegen (Dactinomycin)
C2	Dactinomycin
B2	Folex (Methotrexate)
B2	Folex PFS (Methotrexate)
B2	Methotrexate
B2	Methotrexate LPF (Methotrexate)
B2	Mexate (Methotrexate)
B2	Mexate-AQ (Methotrexate)
F4	Velban (Vinblastine Sulfate)
F4	Velsar (Vinblastine Sulfate)
F4	Vinblastine Sulfate
	Head and Neck
B2	Abitrexate (Methotrexate)

C3	Blenoxane (Bleomycin)
C3	Bleomycin
E6	Cetuximab
F3	Docetaxel
E6	Erbitux (Cetuximab)
B2	Folex (Methotrexate)
B2	Folex PFS (Methotrexate)
E1	Hydrea (Hydroxyurea)
E1	Hydroxyurea
E6	Keytruda (Pembrolizumab)
B2	Methotrexate
B2	Methotrexate LPF (Methotrexate)
B2	Mexate (Methotrexate)
B2	Mexate-AQ (Methotrexate)
E6	Pembrolizumab
F3	Taxotere (Docetaxel)
	Hodgkin's Lymphoma
	Adcetris (Brentuximab Vedotin)
A5	Ambochlorin (Chlorambucil)
A5	Amboclorin (Chlorambucil)
A6	Becenum (Carmustine)
A6	BiCNU (Carmustine)
C3	Blenoxane (Bleomycin)
C3	Bleomycin
	Brentuximab Vedotin
A6	Carmubris (Carmustine)
A6	Carmustine
A5	Chlorambucil
A5	Clafen (Cyclophosphamide)
A5	Cyclophosphamide
A5	Cytoxan (Cyclophosphamide)
A3	Dacarbazine
C1/G2	Doxorubicin Hydrochloride
A3	DTIC-Dome (Dacarbazine)
D1	Hydrocortisone
A5	Leukeran (Chlorambucil)

A5	Linfolizin (Chlorambucil)
A6	Lomustine
A3	Matulane (Procarbazine Hydrochloride)
A5	Mechlorethamine Hydrochloride
D1	Methylprednisolone
A5	Mustargen (Mechlorethamine Hydrochloride)
A5	Neosar (Cyclophosphamide)
	Nivolumab
	Opdivo (Nivolumab)
D1	Prednisone
A3	Procarbazine Hydrochloride
F4	Velban (Vinblastine Sulfate)
F4	Velsar (Vinblastine Sulfate)
F4	Vinblastine Sulfate
F4	Vincasar PFS (Vincristine Sulfate)
F4	Vincristine Sulfate
Kaposi Sarcoma	
C1/G2	Dox-SL (Doxorubicin Hydrochloride Liposome)
C1/G2	Doxil (Doxorubicin Hydrochloride Liposome)
C1/G2	Doxorubicin Hydrochloride Liposome
C1/G2	Evacet (Doxorubicin Hydrochloride Liposome)
	Intron A (Recombinant Interferon Alfa-2b)
C1/G2	LipoDox (Doxorubicin Hydrochloride Liposome)
F3	Paclitaxel
	Recombinant Interferon Alfa-2b
F3	Taxol (Paclitaxel)
F4	Velban (Vinblastine Sulfate)
F4	Velsar (Vinblastine Sulfate)
F4	Vinblastine Sulfate
Kidney (Renal Cell)	
E8	Afinitor (Everolimus)
E7	Aldesleukin
E6	Avastin (Bevacizumab)
	Axitinib
	Cabometyx (Cabozantinib-S-Malate)
	Cabozantinib-S-Malate

B3	Floxuridine
E8	Everolimus
E7	IL-2 (Aldesleukin)
	Inlyta (Axitinib)
E7	Interleukin-2 (Aldesleukin)
	Lenvatinib Mesylate
	Lenvima (Lenvatinib Mesylate)
	Nexavar (Sorafenib Tosylate)
	Nivolumab
	Opdivo (Nivolumab)
	Pazopanib Hydrochloride
E7	Proleukin (Aldesleukin)
	Sorafenib Tosylate
	Sunitinib Malate
	Sutent (Sunitinib Malate)
E8	Temsirolimus
E8	Torisel (Temsirolimus)
	Votrient (Pazopanib Hydrochloride)
Acute Lymphoblastic Leukemia (ALL)	
B2	Abitrexate (Methotrexate)
B1	Arranon (Nelarabine)
E3	Asparaginase Erwinia chrysanthemi
	Blinatumomab
	Blincyto (Blinatumomab)
C1/G2	Cerubidine (Daunorubicin Hydrochloride)
A5	Clafen (Cyclophosphamide)
	Clofarabine
	Clofarex (Clofarabine)
	Clolar (Clofarabine)
A5	Cyclophosphamide
B3	Cytarabine
B3	Cytosar-U (Cytarabine)
A5	Cytoxan (Cyclophosphamide)
	Dasatinib
C1/G2	Daunorubicin Hydrochloride
C1/G2	Doxorubicin Hydrochloride

E3	Erwinaze (Asparaginase Erwinia Chrysanthemi)
B2	Folex (Methotrexate)
B2	Folex PFS (Methotrexate)
D1	Hydrocortisone
	Gleevec (Imatinib Mesylate)
	Iclusig (Ponatinib Hydrochloride)
	Imatinib Mesylate
F4	Marqibo (Vincristine Sulfate Liposome)
B4	Mercaptopurine
B2	Methotrexate
B2	Methotrexate LPF (Methorexate)
B2	Mexate (Methotrexate)
B2	Mexate-AQ (Methotrexate)
B1	Nelarabine
A5	Neosar (Cyclophosphamide)
E3	Oncaspar (Pegaspargase)
E3	Pegaspargase
	Ponatinib Hydrochloride
D1	Prednisone
B4	Purinethol (Mercaptopurine)
B4	Purixan (Mercaptopurine)
C1/G2	Rubidomycin (Daunorubicin Hydrochloride)
	Sprycel (Dasatinib)
B3	Tarabine PFS (Cytarabine)
F2/G2	Teniposide
F4	Vincasar PFS (Vincristine Sulfate)
F4	Vincristine Sulfate
F4	Vincristine Sulfate Liposome
Acute Myeloid Leukemia (AML)	
	Arsenic Trioxide
C1/G2	Cerubidine (Daunorubicin Hydrochloride)
A1	Cladribine
A5	Clafen (Cyclophosphamide)
A5	Cyclophosphamide
B3	Cytarabine
B3	Cytosar-U (Cytarabine)

A5	Cytoxan (Cyclophosphamide)
C1/G2	Daunorubicin Hydrochloride
C1/G2	Doxorubicin Hydrochloride
E6	Gemtuzumab ozogamicin
D1	Hydrocortisone
C1	Idamycin (Idarubicin Hydrochloride)
C1	Idarubicin Hydrochloride
D1	Methylprednisolone
C1/G2	Mitoxantrone Hydrochloride
A5	Neosar (Cyclophosphamide)
C1/G2	Rubidomycin (Daunorubicin Hydrochloride)
B4	Tabloid (Thioguanine)
B3	Tarabine PFS (Cytarabine)
B4	Thioguanine
	Trisenox (Arsenic Trioxide)
F4	Vincasar PFS (Vincristine Sulfate)
F4	Vincristine Sulfate
	Chronic Lymphocytic Leukemia (CLL)
E6	Alemtuzumab
A5	Ambochlorin (Chlorambucil)
A5	Amboclorin (Chlorambucil)
E6	Arzerra (Ofatumumab)
	Bendamustine Hydrochloride
E6	Campath (Alemtuzumab)
A5	Chlorambucil
A1	Cladribine
A5	Clafen (Cyclophosphamide)
A5	Cyclophosphamide
A5	Cytoxan (Cyclophosphamide)
B1	Fludara (Fludarabine Phosphate)
B1	Fludarabine Phosphate
	Gazyva (Obinutuzumab)
D1	Hydrocortisone
	Ibrutinib
	Idelalisib
	Imbruvica (Ibrutinib)
A5	Leukeran (Chlorambucil)

A5	Linfolizin (Chlorambucil)
A5	Mechlorethamine Hydrochloride
D1	Methylprednisolone
A5	Mustargen (Mechlorethamine Hydrochloride)
A5	Neosar (Cyclophosphamide)
	Obinutuzumab
E6	Ofatumumab
D1	Prednisone
E6	Rituxan (Rituximab)
	Treanda (Bendamustine Hydrochloride)
	Venclexta (Venetoclax)
	Venetoclax
	Zydelig (Idelalisib)
Chronic Myelogenous Leukemia (CML)	
	Bosulif (Bosutinib)
	Bosutinib
A1	Busulfan
A1	Busulfex (Busulfan)
A5	Clafen (Cyclophosphamide)
A5	Cyclophosphamide
B3	Cytarabine
B3	Cytosar-U (Cytarabine)
A5	Cytoxan (Cyclophosphamide)
	Dasatinib
	Gleevec (Imatinib Mesylate)
D1	Hydrocortisone
E1	Hydrea (Hydroxyurea)
E1	Hydroxyurea
	Iclusig (Ponatinib Hydrochloride)
	Imatinib Mesylate
A5	Mechlorethamine Hydrochloride
D1	Methylprednisolone
A5	Mustargen (Mechlorethamine Hydrochloride)
A1	Myleran (Busulfan)
A5	Neosar (Cyclophosphamide)
	Nilotinib

	Omacetaxine Mepesuccinate
	Ponatinib Hydrochloride
	Sprycel (Dasatinib)
	Synribo (Omacetaxine Mepesuccinate)
B3	Tarabine PFS (Cytarabine)
	Tasigna (Nilotinib)
	Hairy Cell Leukemia
A1	Cladribine
	Intron A (Recombinant Interferon Alfa-2b)
	Recombinant Interferon Alfa-2b
A1	Pentostatin
	Meningeal Leukemia
B3	Cytarabine
B3	Cytosar-U (Cytarabine)
B3	Tarabine PFS (Cytarabine)
	Liver Cancer
	Nexavar (Sorafenib Tosylate)
	Sorafenib Tosylate
	Non-Small Cell Lung Cancer
B2	Abitrexate (Methotrexate)
F3	Abraxane (Paclitaxel Albumin-stabilized Nanoparticle Formulation)
	Afatinib Dimaleate
E8	Afinitor (Everolimus)
	Alecensa (Alectinib)
	Alectinib
B2	Alimta (Pemetrexed Disodium)
E6	Avastin (Bevacizumab)
A4	Carboplatin
A4	CARBOPLATIN-TAXOL
	Ceritinib
	Crizotinib
E6	Cyramza (Ramucirumab)
F3	Docetaxel
	Erlotinib Hydrochloride
E8	Everolimus
B2	Folex (Methotrexate)
B2	Folex PFS (Methotrexate)

	Gefitinib
	Gilotrif (Afatinib Dimaleate)
B3	Gemcitabine Hydrochloride
B3/A4	Gemcitabine-Cisplatin
B3	Gemzar (Gemcitabine Hydrochloride)
	Iressa (Gefitinib)
E6	Keytruda (Pembrolizumab)
A5	Mechlorethamine Hydrochloride
B2	Methotrexate
B2	Methotrexate LPF (Methotrexate)
B2	Mexate (Methotrexate)
B2	Mexate-AQ (Methotrexate)
A5	Mustargen (Mechlorethamine Hydrochloride)
F4	Navelbine (Vinorelbine Tartrate)
	Necitumumab
	Nivolumab
	Opdivo (Nivolumab)
	Osimertinib
F3	Paclitaxel
F3	Paclitaxel Albumin-stabilized Nanoparticle Formulation
A4	Paraplat (Carboplatin)
A4	Paraplatin (Carboplatin)
E6	Pembrolizumab
B2	Pemetrexed Disodium
	Portrazza (Necitumumab)
E6	Ramucirumab
	Tagrisso (Osimertinib)
	Tarceva (Erlotinib Hydrochloride)
F3	Taxol (Paclitaxel)
F3	Taxotere (Docetaxel)
F4	Vinorelbine Tartrate
	Xalkori (Crizotinib)
	Zykadia (Ceritinib)
Small-Cell Lung Cancer	
B2	Abitrexate (Methotrexate)
E8	Afinitor (Everolimus)

C1/G2	Doxorubicin Hydrochloride
F2/G2	Etopophos (Etoposide Phosphate)
F2/G2	Etoposide
F2/G2	Etoposide Phosphate
E8	Everolimus
B2	Folex (Methotrexate)
B2	Folex PFS (Methotrexate)
F1	Hycamtin (Topotecan Hydrochloride)
A5	Mechlorethamine Hydrochloride
B2	Methotrexate
B2	Methotrexate LPF (Methotrexate)
B2	Mexate (Methotrexate)
B2	Mexate-AQ (Methotrexate)
A5	Mustargen (Mechlorethamine Hydrochloride)
F1	Topotecan Hydrochloride
Malignant Mesothelioma	
B2	Alimta (Pemetrexed Disodium)
B2	Pemetrexed Disodium
Melanoma	
E7	Aldesleukin
	Cobimetinib
	Cotellic (Cobimetinib)
	Dabrafenib
A3	Dacarbazine
A3	DTIC-Dome (Dacarbazine)
E7	IL-2 (Aldesleukin)
	Imlygic (Talimogene Laherparepvec)
E7	Interleukin-2 (Aldesleukin)
	Intron A (Recombinant Interferon Alfa-2b)
E6	Ipilimumab
E6	Keytruda (Pembrolizumab)
	Mekinist (Trametinib)
	Nivolumab
	Opdivo (Nivolumab)
	Peginterferon Alfa-2b
	PEG-Intron (Peginterferon Alfa-2b)
E6	Pembrolizumab

E7	Proleukin (Aldesleukin)
	Recombinant Interferon Alfa-2b
	Sylatron (Peginterferon Alfa-2b)
	Tafinlar (Dabrafenib)
	Talimogene Laherparepvec
	Trametinib
	Vemurafenib
E6	Yervoy (Ipilimumab)
	Zelboraf (Vemurafenib)
Multicentric Castleman Disease	
	Siltuximab
	Sylvant (Siltuximab)
Multiple Myeloma and Other Plasma Cell Neoplasms	
A5	Alkeran for Injection (Melphalan Hydrochloride)
A5	Alkeran Tablets (Melphalan)
	Aredia (Pamidronate Disodium)
A6	Becenum (Carmustine)
A6	BiCNU (Carmustine)
	Bortezomib
	Carfilzomib
A6	Carmubris (Carmustine)
A6	Carmustine
A5	Clafen (Cyclophosphamide)
A5	Cyclophosphamide
A5	Cytoxan (Cyclophosphamide)
	Daratumumab
	Darzalex (Daratumumab)
C1/G2	Doxil (Doxorubicin Hydrochloride Liposome)
C1/G2	Doxorubicin Hydrochloride Liposome
C1/G2	Dox-SL (Doxorubicin Hydrochloride Liposome)
	Elotuzumab
	Empliciti (Elotuzumab)
C1/G2	Evacet (Doxorubicin Hydrochloride Liposome)
A5	Evomela (Melphalan Hydrochloride)
	Farydak (Panobinostat)
D1	Hydrocortisone

	Ixazomib Citrate
	Kyprolis (Carfilzomib)
	Lenalidomide
C1/G2	LipoDox (Doxorubicin Hydrochloride Liposome)
A5	Melphalan
A5	Melphalan Hydrochloride
D1	Methylprednisolone
	Mozobil (Plerixafor)
A5	Neosar (Cyclophosphamide)
	Ninlaro (Ixazomib Citrate)
	Pamidronate Disodium
	Panobinostat
	Plerixafor
	Pomalidomide
	Pomalyst (Pomalidomide)
	Revlimid (Lenalidomide)
	Synovir (Thalidomide)
	Thalidomide
	Thalomid (Thalidomide)
	Velcade (Bortezomib)
	Zoledronic Acid
	Zometa (Zoledronic Acid)
	Myeloproliferative Neoplasms
C1/G2	Adriamycin PFS (Doxorubicin Hydrochloride)
C1/G2	Adriamycin RDF (Doxorubicin Hydrochloride)
	Arsenic Trioxide
	Azacitidine
C1/G2	Cerubidine (Daunorubicin Hydrochloride)
A5	Clafen (Cyclophosphamide)
A5	Cyclophosphamide
B3	Cytarabine
B3	Cytosar-U (Cytarabine)
A5	Cytoxan (Cyclophosphamide)
B3	Dacogen *(Decitabine)*
	Dasatinib
C1/G2	Daunorubicin Hydrochloride
C1/G2	Doxorubicin Hydrochloride

	Gleevec (Imatinib Mesylate)
	Imatinib Mesylate
	Jakafi (Ruxolitinib Phosphate)
	Mylosar (Azacitidine)
A5	Neosar (Cyclophosphamide)
	Nilotinib
C1/G2	Rubidomycin (Daunorubicin Hydrochloride)
	Ruxolitinib Phosphate
	Sprycel (Dasatinib)
B3	Tarabine PFS (Cytarabine)
	Tasigna (Nilotinib)
	Trisenox (Arsenic Trioxide)
	Vidaza (Azacitidine)
F4	Vincasar PFS (Vincristine Sulfate)
F4	Vincristine Sulfate
Neuroblastoma	
A5	Clafen (Cyclophosphamide)
A5	Cyclophosphamide
A5	Cytoxan (Cyclophosphamide)
	Dinutuximab
C1/G2	Doxorubicin Hydrochloride
A5	Neosar (Cyclophosphamide)
	Unituxin (Dinutuximab)
F4	Vincasar PFS (Vincristine Sulfate)
F4	Vincristine Sulfate
Non-Hodgkin's Lymphoma	
B2	Abitrexate (Methotrexate)
	Adcetris (Brentuximab Vedotin)
A5	Ambochlorin (Chlorambucil)
A5	Amboclorin (Chlorambucil)
B1	Arranon (Nelarabine)
A6	Becenum (Carmustine)
	Beleodaq (Belinostat)
	Belinostat
	Bendamustine Hydrochloride
	Bexxar (Tositumomab and Iodine I-131 Tositumomab)

A6	BiCNU (Carmustine)
C3	Blenoxane (Bleomycin)
C3	Bleomycin
	Bortezomib
	Brentuximab Vedotin
A6	Carmubris (Carmustine)
A6	Carmustine
A5	Chlorambucil
A1	Cladribine
A5	Clafen (Cyclophosphamide)
A5	Cyclophosphamide
A5	Cytoxan (Cyclophosphamide)
B3	Cytarabine Liposome
	Denileukin Diftitox
B3	DepoCyt (Cytarabine Liposome)
D1	Dexamethasone
C1/G2	Doxorubicin Hydrochloride
B2	Folex (Methotrexate)
B2	Folex PFS (Methotrexate)
	Folotyn (Pralatrexate)
	Gazyva (Obinutuzumab)
D1	Hydrocortisone
	Ibritumomab Tiuxetan
	Ibrutinib
	Idelalisib
	Imbruvica (Ibrutinib)
	Intron A (Recombinant Interferon Alfa-2b)
	Istodax (Romidepsin)
	Lenalidomide
A5	Leukeran (Chlorambucil)
A5	Linfolizin (Chlorambucil)
A5	Mechlorethamine Hydrochloride
B2	Methotrexate
B2	Methotrexate LPF (Methotrexate)
D1	Methylprednisolone
B2	Mexate (Methotrexate)

B2	Mexate-AQ (Methotrexate)
	Mozobil (Plerixafor)
A5	Mustargen (Mechlorethamine Hydrochloride)
B1	Nelarabine
A5	Neosar (Cyclophosphamide)
	Obinutuzumab
	Ontak (Denileukin Diftitox)
	Plerixafor
	Pralatrexate
D1	Prednisone
	Recombinant Interferon Alfa-2b
	Revlimid (Lenalidomide)
E6	Rituxan (Rituximab)
	Romidepsin
	Tositumomab and Iodine I-131 Tositumomab
	Treanda (Bendamustine Hydrochloride)
F4	Velban (Vinblastine Sulfate)
	Velcade (Bortezomib)
F4	Velsar (Vinblastine Sulfate)
F4	Vinblastine Sulfate
F4	Vincasar PFS (Vincristine Sulfate)
F4	Vincristine Sulfate
	Vorinostat
	Zevalin (Ibritumomab Tiuxetan)
	Zolinza (Vorinostat)
	Zydelig (Idelalisib)
	Ovarian, Fallopian Tube or Primary Peritoneal Cancer
A5	Alkeran (Melphalan)
E6	Avastin (Bevacizumab)
A4	Carboplatin
A5	Clafen (Cyclophosphamide)
A4	Cisplatin
A5	Cyclophosphamide
A5	Cytoxan (Cyclophosphamide)
C1/G2	Doxorubicin Hydrochloride
C1/G2	Dox-SL (Doxorubicin Hydrochloride Liposome)

C1/G2	DOXIL (Doxorubicin Hydrochloride Liposome)
C1/G2	Doxorubicin Hydrochloride Liposome
C1/G2	Evacet (Doxorubicin Hydrochloride Liposome)
B3	Gemcitabine Hydrochloride
B3	Gemzar (Gemcitabine Hydrochloride)
F1	Hycamtin (Topotecan Hydrochloride)
C1/G2	LipoDox (Doxorubicin Hydrochloride Liposome)
	Lynparza (Olaparib)
A5	Melphalan
A5	Neosar (Cyclophosphamide)
	Olaparib
F3	Paclitaxel
A4	Paraplat (Carboplatin)
A4	Paraplatin (Carboplatin)
A4	Platinol (Cisplatin)
A4	Platinol-AQ (Cisplatin)
F3	Taxol (Paclitaxel)
A2	Thiotepa
F1	Topotecan Hydrochloride
Pancreatic Cancer	
F3	Abraxane (Paclitaxel Albumin-stabilized Nanoparticle Formulation)
E8	Afinitor (Everolimus)
	Erlotinib Hydrochloride
E8	Everolimus
B3	5-FU (Fluorouracil Injection)
B3	Fluorouracil Injection
B3	Gemcitabine Hydrochloride
B3	Gemzar (Gemcitabine Hydrochloride)
F1	Irinotecan Hydrochloride Liposome
C3	Mitomycin C
C3	Mitozytrex (Mitomycin C)
C3	Mutamycin (Mitomycin C)
F1	Onivyde (Irinotecan Hydrochloride Liposome)
F3	Paclitaxel Albumin-stabilized Nanoparticle Formulation
A6	Streptozocin
	Sunitinib Malate
	Sutent (Sunitinib Malate)

	Tarceva (Erlotinib Hydrochloride)
Gastroenteropancreatic Neuroendocrine Tumors	
	Lanreotide Acetate
	Somatuline Depot (Lanreotide Acetate)
Penile	
C3	Blenoxane (Bleomycin)
C3	Bleomycin
Prostate	
	Abiraterone Acetate
	Bicalutamide
	Cabazitaxel
	Casodex (Bicalutamide)
	Degarelix
F3	Docetaxel
	Enzalutamide
E4	Estramustine
	Flutamide
	Goserelin Acetate
	Jevtana (Cabazitaxel)
	Leuprolide Acetate
	Lupron (Leuprolide Acetate)
	Lupron Depot (Leuprolide Acetate)
	Lupron Depot-3 Month (Leuprolide Acetate)
	Lupron Depot-4 Month (Leuprolide Acetate)
	Lupron Depot-Ped (Leuprolide Acetate)
C1/G2	Mitoxantrone Hydrochloride
	Nilandron (Nilutamide)
	Nilutamide
	Provenge (Sipuleucel-T)
	Radium 223 Dichloride
	Sipuleucel-T
F3	Taxotere (Docetaxel)
	Viadur (Leuprolide Acetate)
	Xofigo (Radium 223 Dichloride)
	Xtandi (Enzalutamide)
	Zoladex (Goserelin Acetate)
	Zytiga (Abiraterone Acetate)

	Retinoblastoma
A5	Clafen (Cyclophosphamide)
A5	Cyclophosphamide
A5	Cytoxan (Cyclophosphamide)
A5	Neosar (Cyclophosphamide)
	Rhabdomyosarcoma
C2	Cosmegen (Dactinomycin)
C2	Dactinomycin
F4	Vincasar PFS (Vincristine Sulfate)
F4	Vincristine Sulfate
	Skin Cancer (Basal Cell Carcinoma)
	Aldara (Imiquimod)
B3	Efudex (Fluorouracil--Topical)
	Erivedge (Vismodegib)
B3	5-FU (Fluorouracil--Topical)
B3	Fluorouracil--Topical
	Imiquimod
	Odomzo (Sonidegib)
	Sonidegib
	Vismodegib
	Skin Cancer (Melanoma)
E7	Aldesleukin
	Cobimetinib
	Cotellic (Cobimetinib)
	Dabrafenib
A3	Dacarbazine
A3	DTIC-Dome (Dacarbazine)
E7	IL-2 (Aldesleukin)
	Imlygic (Talimogene Laherparepvec)
E7	Interleukin-2 (Aldesleukin)
	Intron A (Recombinant Interferon Alfa-2b)
E6	Ipilimumab
E6	Keytruda (Pembrolizumab)
	Mekinist (Trametinib)
	Nivolumab
	Opdivo (Nivolumab)
	Peginterferon Alfa-2b

E6	Pembrolizumab
E7	Proleukin (Aldesleukin)
	Recombinant Interferon Alfa-2b
	Sylatron (Peginterferon Alfa-2b)
	Tafinlar (Dabrafenib)
	Talimogene Laherparepvec
	Trametinib
	Vemurafenib
E6	Yervoy (Ipilimumab)
	Zelboraf (Vemurafenib)
Skin Cancer (Soft Tissue Sarcoma)	
C2	Cosmegen (Dactinomycin)
C2	Dactinomycin
C1/G2	Doxorubicin Hydrochloride
	Eribulin Mesylate
	Gleevec (Imatinib Mesylate)
	Halaven (Eribulin Mesylate)
	Imatinib Mesylate
	Pazopanib Hydrochloride
	Trabectedin
	Votrient (Pazopanib Hydrochloride)
	Yondelis (Trabectedin)
Testicular Cancer	
C3	Blenoxane (Bleomycin)
C3	Bleomycin
A4	Cisplatin
C2	Cosmegen (Dactinomycin)
A5	Cyfos (Ifosfamide)
C2	Dactinomycin
F2/G2	Etopophos (Etoposide Phosphate)
F2/G2	Etoposide
F2/G2	Etoposide Phosphate
A5	Ifex (Ifosfamide)
A5	Ifosfamide
A5	Ifosfamidum (Ifosfamide)
A4	Platinol (Cisplatin)
A4	Platinol-AQ (Cisplatin)

C2	Plicamycin
F4	Velban (Vinblastine Sulfate)
F4	Velsar (Vinblastine Sulfate)
F4	Vinblastine Sulfate
	Thyroid Cancer
	Cabozantinib-S-Malate
	Caprelsa (Vandetanib)
	Cometriq (Cabozantinib-S-Malate)
C1/G2	Doxorubicin Hydrochloride
	Lenvatinib Mesylate
	Lenvima (Lenvatinib Mesylate)
	Nexavar (Sorafenib Tosylate)
	Sorafenib Tosylate
	Vandetanib
	Vaginal
	N/A (preventative only)
	Vulvar
C3	Blenoxane (Bleomycin)
C3	Bleomycin
	Wilms Tumor and Other Childhood Kidney Cancers
C2	Cosmegen (Dactinomycin)
C2	Dactinomycin
C1/G2	Doxorubicin Hydrochloride
F4	Vincasar PFS (Vincristine Sulfate)
F4	Vincristine Sulfate

Combined Chemotherapy Treatments

	Bladder
B3/A4	Gemcitabine-Cisplatin
	Brain
	PCV
	Breast
	AC
	AC-T
	CAF
	CMF
	FEC
	TAC

	Cervical
B3/A4	Gemcitabine-Cisplatin
	Colon and Rectal
	CAPOX
	FOLFIRI
E6	FOLFIRI-BEVACIZUMAB
E6	FOLFIRI-CETUXIMAB
	FOLFOX
	FU-LV
	XELIRI
	XELOX
	Esophageal
	FU-LV
	XELIR
	Head and Neck
	TPF
	Hodgkin's Lymphoma
	ABVD
	ABVE
	ABVE-PC
	BEACOPP
	COPDAC
	COPP
	COPP-ABV
	ICE
	MOPP
	OEPA
	OPPA
	STANFORD V
	VAMP
	Acute Lymphoblastic Leukemia (ALL)
	Hyper-CVAD
	Acute Myeloid Leukemia (AML)
	ADE
	Chronic Lymphocytic Leukemia (CLL)
A5/D1	CHLORAMBUCIL-PREDNISONE
	CVP

	Chronic Myelogenous Leukemia (CML)
	Intron A (Recombinant Interferon Alfa-2b)
	Recombinant Interferon Alfa-2b
	Malignant Mesothelioma
B3/A4	Gemcitabine-Cisplatin
	Multiple Myeloma and Other Plasma Cell Neoplasms
	PAD
	Myeloproliferative Neoplasms
	ADE
	Neuroblastoma
	BuMel
	CEM
	Non-Hodgkin's Lymphoma
	CHOP
	COPP
	CVP
	EPOCH
	Hyper-CVAD
	ICE
	R-CHOP
	R-CVP
	R-EPOCH
	R-ICE
	Ovarian, Fallopian Tube or Primary Peritoneal Cancer
A4	CARBOPLATIN-TAXOL
B3/A4	Gemcitabine-Cisplatin
A2	Hexamethylmelamine
	JEB
	PEB
	VAC
	VeIP
	Pancreatic Cancer
B3/A4	Gemcitabine-Cisplatin
B3/A4	Gemcitabine –Oxaliplatin
	OFF
	Stomach (Gastric) Cancer
	FU-LV

	TPF
	XELIRI
Testicular Cancer	
	BEP
	JEB
	PEB
	VeIP
	VIP

Appendix C

Carcinogens

International Agency for Research on Cancer Group 1: Carcinogenic to Humans

- Acetaldehyde (from consuming alcoholic beverages)
- Acheson process, occupational exposure associated with
- Acid mists, strong inorganic
- Aflatoxins
- Alcoholic beverages
- Aluminum production
- 4-Aminobiphenyl
- Areca nut
- Aristolochic acid (and plants containing it)
- Arsenic and inorganic arsenic compounds
- Asbestos (all forms) and mineral substances (such as talc or vermiculite) that contain asbestos
- Auramine production
- Azathioprine
- Benzene
- Benzidine and dyes metabolized to benzidine
- Benzo[a]pyrene
- Beryllium and beryllium compounds
- Betel quid, with or without tobacco
- Bis (chloromethyl)ether and chloromethyl methyl ether (technical-grade)

- Busulfan
- 1,3-butadiene
- Cadmium and cadmium compounds
- Chlorambucil
- Chlornaphazine
- Chromium (VI) compounds
- Clonorchis sinensis (infection with), also known as the Chinese liver fluke
- Coal, indoor emissions from household combustion
- Coal gasification
- Coal-tar distillation
- Coal-tar pitch
- Coke production
- Cyclophosphamide
- Cyclosporine
- 1,2-dichloropropane
- Diethylstilbestrol
- Engine exhaust, diesel
- Epstein-Barr virus (infection with)
- Erionite
- Estrogen postmenopausal therapy
- Estrogen-progestogen postmenopausal therapy (combined)
- Estrogen-progestogen oral contraceptives (combined) (note: there is also convincing evidence in humans that these agents confer a protective effect against cancer in the endometrium and ovary)
- Ethanol in alcoholic beverages
- Ethylene oxide
- Etoposide
- Etoposide in combination with cisplatin and bleomycin
- Fission products, including strontium-90
- Fluoro-edenite fibrous amphibole

- Formaldehyde
- Haematite mining (underground)
- Helicobacter pylori (infection with)
- Hepatitis B virus (chronic infection with)
- Hepatitis C virus (chronic infection with)
- Human immunodeficiency virus type 1 (HIV-1) (infection with)
- Human papillomavirus (HPV) types 16, 18, 31, 33, 35, 39, 45, 51, 52, 56, 58, 59 (infection with) (note: the HPV types that have been classified as carcinogenic to humans can differ by an order of magnitude in risk for cervical cancer)
- Human T-cell lymphotropic virus type I (HTLV-1) (infection with)
- Ionizing radiation (all types)
- Iron and steel founding (workplace exposure)
- Isopropyl alcohol manufacture using strong acids
- Kaposi sarcoma herpesvirus (KSHV), also known as human herpesvirus 8 (HHV-8) (infection with)
- Leather dust
- Lindane
- Magenta production
- Melphalan
- Methoxsalen (8-methoxypsoralen) plus ultraviolet A radiation, also known as PUVA
- 4,4'-methylenebis(2-chloroaniline) (MOCA)
- Mineral oils, untreated or mildly treated
- MOPP and other combined chemotherapy including alkylating agents
- 2-naphthylamine
- Neutron radiation
- Nickel compounds
- N-nitrosonornicotine (NNN) and 4-(N-nitrosomethylamino)-1-(3-pyridyl)-1-butanone (NNK)

- Opisthorchis viverrini (infection with), also known as the Southeast Asian liver fluke
- Outdoor air pollution (and the particulate matter in it)
- Painter (workplace exposure as a)
- 3,4,5,3',4'-pentachlorobiphenyl (PCB-126)
- 2,3,4,7,8-pentachlorodibenzofuran
- Phenacetin (and mixtures containing it)
- Phosphorus-32, as phosphate
- Plutonium
- Polychlorinated biphenyls (PCBs), dioxin-like, with a toxicity equivalency factor according to WHO (PCBs 77, 81, 105, 114, 118, 123, 126, 156, 157, 167, 169, 189)
- Processed meat (consumption of)
- Radioiodines, including iodine-131
- Radionuclides, alpha-particle-emitting, internally deposited (note: specific radionuclides for which there is sufficient evidence for carcinogenicity to humans are also listed individually as Group 1 agents)
- Radionuclides, beta-particle-emitting, internally deposited (note: specific radionuclides for which there is sufficient evidence for carcinogenicity to humans are also listed individually as Group 1 agents)
- Radium-224 and its decay products
- Radium-226 and its decay products
- Radium-228 and its decay products
- Radon-222 and its decay products
- Rubber manufacturing industry
- Salted fish (Chinese-style)
- Schistosoma haematobium (infection with)
- Semustine (methyl-CCNU)
- Shale oils
- Silica dust, crystalline, in the form of quartz or cristobalite

- Solar radiation
- Soot (as found in workplace exposure of chimney sweeps)
- Sulfur mustard
- Tamoxifen (note: there is also conclusive evidence that tamoxifen reduces the risk of contralateral breast cancer in breast cancer patients)
- 2,3,7,8-tetrachlorodibenzo-para-dioxin
- Thiotepa
- Thorium-232 and its decay products
- Tobacco, smokeless
- Tobacco smoke, secondhand
- Tobacco smoking
- Ortho-toluidine
- Treosulfan
- Trichloroethylene
- Ultraviolet (UV) radiation, including UVA, UVB, and UVC rays
- Ultraviolet-emitting tanning devices
- Vinyl chloride
- Wood dust
- X- and Gamma-radiation

National Toxicology Program 14th Report on Carcinogens Group 1: Known to be Human Carcinogens

- Aflatoxins
- Alcoholic beverage consumption
- 4-aminobiphenyl
- Analgesic mixtures containing phenacetin
- Aristolochic acids
- Arsenic and inorganic arsenic compounds
- Asbestos
- Azathioprine

- Benzene
- Benzidine
- Beryllium and beryllium compounds
- Bis(chloromethyl) ether and technical-grade chloromethyl methyl ether
- 1,3-butadiene
- 1,4-butanediol dimethyl sulfonate (also known as busulfan)
- Cadmium and cadmium compounds
- Chlorambucil
- 1-(2-chloroethyl)-3-(4-methylcyclohexyl)-1-nitrosourea (MeCCNU)
- Chromium hexavalent compounds
- Coal tar pitches
- Coal tars
- Coke oven emissions
- Cyclophosphamide
- Cyclosporin A
- Diethylstilbestrol (DES)
- Dyes metabolized to benzidine
- Epstein-Barr virus (EBV)
- Erionite
- Estrogens, steroidal
- Ethylene oxide
- Formaldehyde
- Hepatitis B virus
- Hepatitis C virus
- Human immunodeficiency virus type 1 (HIV-1)
- Human papillomaviruses: some genital-mucosal types
- Human T-cell lymphotropic virus type 1 (HTLV-1)

- Kaposi sarcoma-associated herpesvirus (KSHV) (also known as human herpesvirus 8, or HHV-8)
- Melphalan
- Merkel cell polyomavirus (MCV)
- Methoxsalen with ultraviolet A therapy (PUVA)
- Mineral oils (untreated and mildly treated)
- Mustard gas
- 2-naphthylamine
- Neutrons
- Nickel compounds
- Oral tobacco products
- Radon
- Silica, crystalline (respirable size)
- Solar radiation
- Soot
- Strong inorganic acid mists containing sulfuric acid
- Sunlamps or sunbeds, exposure to
- Tamoxifen
- 2,3,7,8-tetrachlorodibenzo-p-dioxin (TCDD); "dioxin"
- Thiotepa
- Thorium dioxide
- Tobacco smoke, environmental
- Tobacco, smokeless
- Tobacco smoking
- o-Toluidine
- Trichloroethylene (TCE)
- Vinyl chloride
- Ultraviolet (UV) radiation, broad spectrum
- Wood dust
- X-radiation and gamma radiation

International Agency for Research on Cancer Group 2A: Probably Carcinogenic to Humans

- Acrylamide
- Adriamycin (doxorubicin)
- Androgenic (anabolic) steroids
- Art glass, glass containers, and press ware (manufacture of)
- Azacitidine
- Biomass fuel (primarily wood), emissions from household combustion
- Bischloroethyl nitrosourea (BCNU), also known as carmustine
- Captafol
- Carbon electrode manufacture
- Chloral
- Chloral hydrate
- Chloramphenicol
- alpha-chlorinated toluenes (benzal chloride, benzotrichloride, benzyl chloride) and benzoyl chloride (combined exposures)
- 1-(2-chloroethyl)-3-cyclohexyl-1-nitrosourea (CCNU)
- 4-chloro-ortho-toluidine
- Chlorozotocin
- Cisplatin
- Cobalt metal with tungsten carbide
- Creosotes
- Cyclopenta[cd]pyrene
- DDT (4,4'-dichlorodiphenyltrichloroethane)
- Diazinon
- Dibenz[a,j]acridine
- Dibenz[a,h]anthracene
- Dibenzo[a,l]pyrene
- Dichloromethane (methylene chloride)

- Diethyl sulfate
- Dimethylcarbamoyl chloride
- 1,2-dimethylhydrazine
- Dimethyl sulfate
- Epichlorohydrin
- Ethyl carbamate (urethane)
- Ethylene dibromide
- N-ethyl-N-nitrosourea
- Frying, emissions from high-temperature
- Glycidol
- Glyphosate
- Hairdresser or barber (workplace exposure as)
- Human papillomavirus (HPV) type 68 (infection with)
- Indium phosphide
- IQ (2-Amino-3-methylimidazo[4,5-f]quinoline)
- Lead compounds, inorganic
- Malaria (caused by infection with Plasmodium falciparum)
- Malathion
- Merkel cell polyomavirus (MCV)
- 5-Methoxypsoralen
- Methyl methanesulfonate
- N-methyl-N'-nitro-N-nitrosoguanidine (MNNG)
- N-methyl-N-nitrosourea
- Nitrate or nitrite (ingested) under conditions that result in endogenous nitrosation
- 6-Nitrochrysene
- Nitrogen mustard
- 1-Nitropyrene
- N-nitrosodiethylamine
- N-nitrosodimethylamine

- 2-nitrotoluene
- Non-arsenical insecticides (workplace exposures in spraying and application of)
- Petroleum refining (workplace exposures in)
- Pioglitazone
- Polybrominated biphenyls (PBBs)
- Procarbazine hydrochloride
- 1,3-Propane sultone
- Red meat (consumption of)
- Shiftwork that involves circadian disruption
- Silicon carbide whiskers
- Styrene-7,8-oxide
- Teniposide
- Tetrachloroethylene (perchloroethylene)
- Tetrafluoroethylene
- Trichloroethylene
- 1,2,3-trichloropropane
- Tris(2,3-dibromopropyl) phosphate
- Very hot beverages (above 65°C)
- Vinyl bromide (note: for practical purposes, vinyl bromide should be considered to act similarly to the human carcinogen vinyl chloride.)
- Vinyl fluoride (note: for practical purposes, vinyl fluoride should be considered to act similarly to the human carcinogen vinyl chloride.)

National Toxicology Program 14th Report on Carcinogens Group 2: Reasonably Anticipated to be Human Carcinogens

- Acetaldehyde
- 2-Acetylaminofluorene
- Acrylamide
- Acrylonitrile

- Adriamycin (doxorubicin hydrochloride)
- 2-Aminoanthraquinone
- o-aminoazotoluene
- 1-Amino-2,4-dibromo anthraquinone
- 1-Amino-2-methylanthraquinone
- 2-Amino-3,4-dimethylimidazo[4,5-f]quinoline (MeIQ)
- 2-Amino-3,8-dimethylimidazo[4,5-f]quinoxaline (MeIQx)
- 2-Amino-3-methylimidazo[4,5-f]quinoline (IQ)
- 2-Amino-1-methyl-6-phenylimidazo[4,5-b]pyridine (PhIP)
- Amitrole
- o-Anisidine and its hydrochloride
- Azacitidine (5-azacytidine, 5-AzaC)
- Basic Red 9 monohydrochloride
- Benz[a]anthracene
- Benzo[b]fluoranthene
- Benzo[j]fluoranthene
- Benzo[k]fluoranthene
- Benzo[a]pyrene
- Benzotrichloride
- 2, 2-bis-(bromomethyl)-1,3-propanediol (technical grade)
- Bromodichloromethane
- 1-Bromopropane
- Butylated hydroxyanisole (BHA)
- Captafol
- Carbon tetrachloride
- Ceramic fibers (respirable size)
- Chloramphenicol
- Chlorendic acid
- Chlorinated paraffins (C_{12}, 60% chlorine)
- Chloroform
- 1-(2-chloroethyl)-3-cyclohexyl-1-nitrosourea

- Bis(chloroethyl) nitrosourea
- 3-chloro-2-methylpropene
- 4-chloro-o-phenylenediamine
- Chloroprene
- p-chloro-o-toluidine and p-chloro-o-toluidine hydrochloride
- Chlorozotocin
- Cisplatin
- Cobalt and cobalt compounds that release cobalt ions in vivo
- Cobalt-tungsten carbide: powders and hard metals
- p-cresidine
- Cumene
- Cupferron
- Dacarbazine
- Danthron (1,8-dihydroxyanthraquinone)
- 2,4-diaminoanisole sulfate
- 2,4-diaminotoluene
- Diazoaminobenzene
- Dibenz[a,h]acridine
- Dibenz[a,j]acridine
- Dibenz[a,h]anthracene
- 7H-Dibenzo[c,g]carbazole
- Dibenzo[a,e]pyrene
- Dibenzo[a,h]pyrene
- Dibenzo[a,i]pyrene
- Dibenzo[a,l]pyrene
- 1,2-dibromo-3-chloropropane
- 1,2-dibromoethane (ethylene dibromide)
- 2,3-dibromo-1-propanol
- Tris (2,3-dibromopropyl) phosphate
- 1,4-dichlorobenzene
- 3,3'-dichlorobenzidine and 3,3'-dichlorobenzidine dihydrochloride

- Dichlorodiphenyltrichloroethane (DDT)
- 1,2-dichloroethane (ethylene dichloride)
- Dichloromethane (methylene chloride)
- 1,3-dichloropropene (technical grade)
- Diepoxybutane
- Diesel exhaust particulates
- Diethyl sulfate
- Diglycidyl resorcinol ether
- 3,3'-dimethoxybenzidine
- 4-dimethylaminoazobenzene
- 3,3'-dimethylbenzidine
- Dimethylcarbamoyl chloride
- 1,1-dimethylhydrazine
- Dimethyl sulfate
- Dimethylvinyl chloride
- 1,6-dinitropyrene
- 1,8-dinitropyrene
- 1,4-dioxane
- Disperse blue 1
- Dyes metabolized to 3,3'-dimethoxybenzidine
- Dyes metabolized to 3,3'-dimethylbenzidine
- Epichlorohydrin
- Ethylene thiourea
- Ethyl methanesulfonate
- Furan
- Glass wool fibers (inhalable)
- Glycidol
- Hexachlorobenzene
- Hexachlorocyclohexane isomers
- Hexachloroethane
- Hexamethylphosphoramide

- Hydrazine and hydrazine sulfate
- Hydrazobenzene
- Indeno[1,2,3-cd]pyrene
- Iron dextran complex
- Isoprene
- Kepone (chlordecone)
- Lead and lead compounds
- Lindane, hexachlorocyclohexane
- 2-methylaziridine (propylencimine)
- 5-methylchrysene
- 4,4'-methylenebis(2-chloroaniline)
- 4-4'-methylenebis(N,N-dimethyl)benzenamine
- 4,4'-methylenedianiline and its dihydrochloride salt
- Methyleugenol
- Methyl methanesulfonate
- N-methyl-N'-nitro-N-nitrosoguanidine
- Metronidazole
- Michler's ketone [4,4'-(dimethylamino) benzophenone]
- Mirex
- Naphthalene
- Nickel, metallic
- Nitrilotriacetic acid
- o-nitroanisole
- Nitrobenzene
- 6-nitrochrysene
- Nitrofen (2,4-dichlorophenol-p-nitrophenyl ether)
- Nitrogen mustard hydrochloride
- Nitromethane
- 2-nitropropane
- 1-nitropyrene
- 4-nitropyrene

- N-nitrosodi-n-butylamine
- N-nitrosodiethanolamine
- N-nitrosodiethylamine
- N-nitrosodimethylamine
- N-nitrosodi-n-propylamine
- N-nitroso-N-ethylurea
- 4-(N-nitrosomethylamino)-1-(3-pyridyl)-1-butanone
- N-nitroso-N-methylurea
- N-nitrosomethylvinylamine
- N-nitrosomorpholine
- N-nitrosonornicotine
- N-nitrosopiperidine
- N-nitrosopyrrolidine
- N-nitrososarcosine
- o-nitrotoluene
- Norethisterone
- Ochratoxin A
- 4,4'-oxydianiline
- Oxymetholone
- Pentachlorophenol and by-products of its synthesis
- Phenacetin
- Phenazopyridine hydrochloride
- Phenolphthalein
- Phenoxybenzamine hydrochloride
- Phenytoin and phenytoin sodium
- Polybrominated biphenyls (PBBs)
- Polychlorinated biphenyls (PCBs)
- Polycyclic aromatic hydrocarbons (PAHs)
- Procarbazine and its hydrochloride
- Progesterone
- 1,3-propane sultone

- beta-propiolactone
- Propylene oxide
- Propylthiouracil
- Reserpine
- Riddelliine
- Safrole
- Selenium sulfide
- Streptozotocin
- Styrene
- Styrene-7,8-oxide
- Sulfallate
- Tetrachloroethylene (perchloroethylene)
- Tetrafluoroethylene
- Tetranitromethane
- Thioacetamide
- 4,4'-thiodianaline
- Thiourea
- Toluene diisocyanates
- Toxaphene
- 2,4,6-trichlorophenol
- 1,2,3-trichloropropane
- Tris(2,3-dibromopropyl) phosphate
- Ultraviolet A radiation
- Ultraviolet B radiation
- Ultraviolet C radiation
- Urethane
- Vinyl bromide
- 4-Vinyl-1-cyclohexene diepoxide
- Vinyl fluoride
- Progesterone
- 1,3-propane sultone

- beta-propiolactone
- Propylene oxide
- Propylthiouracil
- Reserpine
- Riddelliine
- Safrole
- Selenium sulfide
- Streptozotocin
- Styrene
- Styrene-7,8-oxide
- Sulfallate
- Tetrachloroethylene (perchloroethylene)
- Tetrafluoroethylene
- Tetranitromethane
- Thioacetamide
- 4,4'-thiodianaline
- Thiourea
- Toluene diisocyanates
- Toxaphene
- 2,4,6-trichlorophenol
- 1,2,3-trichloropropane
- Tris(2,3-dibromopropyl) phosphate
- Ultraviolet A radiation
- Ultraviolet B radiation
- Ultraviolet C radiation
- Urethane
- Vinyl bromide
- 4-vinyl-1-cyclohexene diepoxide
- Vinyl fluoride

Bibliography

Adler, Elizabeth M. *Living with Lymphoma*. Baltimore: John Hopkins University Press, 2015.

Anderson, Greg. *The Cancer Conqueror: An Incredible Journey to Wellness.* Kansas City: Andrews and McMeel, 1988.

Anderson, Greg. *Cancer: 50 Essential Things to Do*, 3rd ed. London: Penguin Books Ltd., 2009.

Besser, Jeanne, Kristina Ratley, Sheri Knecht, and Michele Szafranski. *What to Eat During Cancer Treatment*. Atlanta: American Cancer Society, 2009.

Burns, David D. *Feeling Good: The New Mood Therapy.* New York: Avon Books, 1992.

Canfield, Jack, Mark Victor Hansen, Patty Aubery, Nancy Mitchell, and Beverly Kirkhart. *Chicken Soup for the Cancer Survivor's Soul: 101 Healing Stories About Those Who Have Survived Cancer.* Deerfield Beach: Health Communications, Inc., 1996.

Columbu, Franco. *The Bodybuilder's Nutrition Book.* New York: McGraw-Hill Education, 1985.

Dhammananda, K. Sri. *How to Live without Fear & Worry.* Taipei: The Corporate Body of the Buddha Educational Foundation, 1989.

Ilardi, Stephen S. *The Depression Cure: The 6-Step Program to Beat Depression Without Drugs.* Philadelphia: Da Capo Press, 2009.

Jacobs, Gregg D. *Say Good Night to Insomnia.* New York: Henry Holt and Company, LLC, 1998.

Jochems, Ruth. *Dr. Moerman's Anti-Cancer Diet: Holland's Revolutionary Nutritional Program for Combating Cancer.* New York: Avery Publishing Group, Inc., 1990.

Lejeune, Chad. *The Worry Trap: How to Free Yourself from Worry & Anxiety Using Acceptance & Commitment Therapy.* Oakland: New Harbinger Publications, Inc., 2007.

Lichtblau, Leonard. *Psychopharmacology Demystified*. Clifton Park: Delmar, Cengage Learning, 2011.

Mallatt, Jon, Elaine Marieb, and Brady Wilhelm. *Human Anatomy*, 7th Edition. Glenview: Pearson Education, Inc., 2014.

O'Connor, Richard. *Undoing Depression: What Therapy Doesn't Teach You and Medication Can't Give You,* 2nd ed. New York: Little, Brown and Company, 2010.

Schwarzenegger, Arnold. *The New Encyclopedia of Modern Bodybuilding.* New York: Simon & Schuster, 2012.

Servan-Schreiber, David. *Anti-Cancer: A New Way of Life.* London: Penguin Books Ltd., 2008.

Silver, Julie K. *What Helped Get Me Through: Cancer Survivors Share Wisdom and Hope.* Atlanta: American Cancer Society, 2009.

Specter, Arlen. *Never Give In: Battling Cancer in the Senate.* New York: St. Martin's Press, 2008.

Williams, Jason R. *The Immunotherapy Revolution: The Best New Hope For Saving Cancer Patients' Lives.* Atlanta: Williams Cancer Institute, 2019.

Williams, Mark, John Teasdale, Zindel Segal, and Jon Kabat-Zinn. *The Mindful Way Through Depression: Freeing Yourself from Chronic Unhappiness.* New York: The Guilford Press, 2007.

Zaret, Barry L., Peter I. Jatlow, and Lee D. Katz. *The Yale University School of Medicine Patient's Guide to Medical Tests.* New York: Houghton Mifflin Company, 1997.

"Types of Cancer Treatment." *National Cancer Institute.* https://www.cancer.gov/about-cancer/treatment/types.

"University of Maryland Medical Center." *University of Maryland Medical System.* https://www.umms.org/ummc/locations/university-maryland-medical-center.

"Diet, Nutrition, Physical Activity and Cancer: a Global Perspective." *World Cancer Research Fund and the American Institute for Cancer.* https://www.wcrf.org/dietandcancer.

Index

C

D

E

F

I

J

K

L

M

N

O

P

R

S

T

U

V

W

X

Y

Made in the USA
Las Vegas, NV
20 May 2021